PERGAMON INTERNATIONAL LIBRARY
of Science, Technology, Engineering and Social Studies
The 1000-volume original paperback library in aid of education, industrial training and the enjoyment of leisure
Publisher: Robert Maxwell, M.C.

Fundamentals of Metallurgical Processes

Second Edition

International Series on
MATERIALS SCIENCE AND TECHNOLOGY

Volume 27 — Editor: D. W. HOPKINS

Other Titles in the Series

ASHBY & JONES
Engineering Materials

BARRETT & MASSALSKI
Structure of Metals, 3rd Edition

BISWAS & DAVENPORT
Extractive Metallurgy of Copper, 2nd Edition

CHRISTIAN
The Theory of Transformations in Metals and Alloys, Part 1, 2nd Edition

GABE
Principles of Metal Surface Treatment and Protection, 2nd Edition

GILCHRIST
Extraction Metallurgy, 2nd Edition

HARRIS
Mechanical Working of Metals

HEARN
Mechanics of Materials, 2nd Edition

HULL & BACON
Introduction to Dislocations, 3rd Edition

IIW
Physics of Welding

KUBASCHEWSKI & ALCOCK
Metallurgical Thermochemistry, 5th Edition

MASUBUCHI
Analysis of Welded Structures

PARKER
An Introduction to Chemical Metallurgy, 2nd Edition

RAYNOR
The Physical Metallurgy of Magnesium and its Alloys

SARKAR
Wear of Metals

SCULLY
Fundamentals of Corrosion, 2nd Edition

WILLS
Mineral Processing Technology, 3rd Edition

Fundamentals of Metallurgical Processes

Second Edition

by

LUCIEN COUDURIER M.Sc., Ph.D.
Grenoble University, France

DONALD W. HOPKINS, M.Sc., F.I.M., C.Eng., M.I.M.M.
University College of Swansea, Wales

IGOR WILKOMIRSKY, B.Sc., M.Sc., Ph.D.
University of British Columbia, Canada

PERGAMON PRESS

OXFORD · NEW YORK · TORONTO · SYDNEY · FRANKFURT

U.K. — Pergamon Press Ltd., Headington Hill Hall, Oxford OX3 0BW, England

U.S.A. — Pergamon Press Inc., Maxwell House, Fairview Park, Elmsford, New York 10523, U.S.A.

CANADA — Pergamon Press Canada Ltd., Suite 104, 150 Consumers Road, Willowdale, Ontario M2J 1P9, Canada

AUSTRALIA — Pergamon Press (Aust.) Pty. Ltd., P.O. Box 544, Potts Point, N.S.W. 2011, Australia

FEDERAL REPUBLIC OF GERMANY — Pergamon Press GmbH, Hammerweg 6, D-6242 Kronberg-Taunus, Federal Republic of Germany

First edition 1978
Second edition 1985

Library of Congress Cataloging in Publication Data
Coudurier, Lucien.
Fundamentals of metallurgical processes.
(International series on materials science and technology; v. 27) (Pergamon international library of science, technology, engineering, and social studies)
Includes bibliographical references and index.
1. Metallurgy. I. Hopkins, Donald W. (Donald Walter), 1913– . II. Wilkomirsky, Igor. III. Title. IV. Series. V. Series: Pergamon international library of science, technology, engineering and social studies.
TN665.C7 1985 669 84–20614

British Library Cataloguing in Publication Data
Coudurier, Lucien
Fundamentals of metallurgical processes.—2nd ed.
—(Pergamon international library)—
(International series on materials science and technology; v. 27)
1. Metallurgy
I. Title II. Hopkins, Donald W.
III. Wilkomirsky, Igor IV. Series V. Series
669 TN665

ISBN 0-08-032536-X Hardcover
ISBN 0-08-032537-8 Flexicover

Printed in Great Britain by A. Wheaton & Co. Ltd., Exeter

ACKNOWLEDGEMENTS

We acknowledge with gratitude the assistance received from the many colleagues, professional acquaintances and authors whose publications and advice have contributed to this work. It is not possible to thank them individually, but we wish to make special mention of Professor E. Bonnier of the Institut National Polytechnique de Grenoble, Professor A. Rist of the École Centrale des Arts et Manufactures de Paris and Sr. A. Sutulov, Director of Centro de Investigacion Minera y Metalurgica, Santiago, Chile. Our thanks are also due to the Metals Society of London for permission to use illustrations from *Refractories for Iron and Steelmaking* by J.H. Chesters for Figures 39 on p. 80 and 60 on p. 113 and from I.S.I. Special Report No. 102 for Figure 67 on p. 161.

Substantial use has been made of material which appeared originally in *Fundamentos de los Procesos Metalurgicos* by L. Coudurier and I. Wilkomirsky published by the University of Concepcion, Chile, in 1971.

The important contribution made by Mrs. E. Phillips in deciphering the bilingual amendments to the manuscript and typing this document has earned our deepest gratitude.

CONTENTS

INTRODUCTION TO THE SECOND EDITION

The preparation of this edition has provided the opportunity for inclusion of a section based on the work of F.D. Richardson, and the permission of the Academic Press publishers to use a number of diagrams is gratefully acknowledged. Brief sections have also been added on non-ferrous process metallurgy, in order to present a more complete statement of the conditions necessary for the production of crude and pure metal. The iron and steel section has also been expanded. A number of errors in the first edition have been corrected and the assistance of readers in pointing them out is gratefully acknowledged. An extra problem has been inserted as No. 17 and the former No. 17 has become 18a.

INTRODUCTION TO THE FIRST EDITION

Rapid developments have taken place in recent years in the design, control, and efficiency of metal extraction and refining processes. These have been substantially due to the application of fundamental knowledge of thermodynamics and kinetics in place of the empirical data previously used. This application has been greatly facilitated by the increased sensitivity and accuracy of the newer methods of process control and analysis of reacting materials and products.

The application of thermodynamics to a reacting system makes it possible to calculate the composition at equilibrium and the direction and extent of change which can take place under specified conditions. A thermodynamic system is defined as that part of the universe within which the reactions being studied are taking place. When there is no interchange of matter or energy with the environment, the system is isolated. A closed system is defined as one where energy, but not material, can be exchanged with the environment. Exchange of both takes place freely when the system is open. For example, a fluidized bed reactor is an open system; the pressure container of a calorimeter is a closed system, and the whole calorimeter can be considered as an isolated system.

Equilibrium is a state in which there is no nett reaction and zero change of energy. It is the state of minimum available energy, and is that to which every system approaches, or returns, after a temporary displacement. Attainment of the equilibrium condition will involve one or more of the following changes, either within the system or between two systems:

(a) a change in volume;
(b) the generation or absorption of energy;
(c) an exchange of matter.

For example, a mixture of CO and CO_2 in contact with C will tend toward the equilibrium gas composition resulting from the reaction:

$$\langle C \rangle + |CO_2| = 2|CO|$$

at the ruling temperature and pressure. This will result in the formation of either CO or CO_2 with the consequential changes in volume and available energy.

Thermodynamics cannot be used to calculate the time required to reach equilibrium or to determine the method by which it is achieved. These are governed by kinetics; the study of reaction rates and mechanisms. Since reaction in metallurgy take place most frequently between two or more phases, knowledge of the relevant aspects of heterogeneous kinetics is most important.

The general arrangement of this book is such that the fundamentals of thermodynamics (functions, relationships and behaviour of solutions) are discussed in the first two chapters and methods of obtaining data from tables and graphs and by experiment are described in Chapter 3. The kinetics of heterogeneous reactions in metallurgy are examined in Chapter 4, with particular reference to reaction between a particle and a fluid. The argument is developed to include the study of reaction between a moving fluid passing through a bed of particles. The relationships and functions examined and developed in Chapters 1 and 2 are applied to the extraction of metals from oxides, sulphides, halides and carbonyls in Chapter 5, and to the production of iron and steel in Chapters 6 and 7. Chapter 6 concentrates upon the topic of slag structure and the behaviour of constituents, while Chapter 7 deals with slag/metal reactions and equilibria in iron and steel production. Chapter 8 is a combination of development and revision in that it consists entirely of solved problems. A conscious decision was made to exclude electrochemistry when it became apparent that some choice was necessary in order to keep the cost and size of the book within reasonable dimensions.

Chapter 1

THERMODYNAMIC FUNCTIONS

INTRODUCTION

Thermodynamics describes systems by means of functions or variables of state. These are quantities associated with the state of a body and the values are independent of the means by which the body has been brought to that state.

The variables or functions of state may be extensive or intensive:

(a) Extensive variables are those which depend on the mass of the system, the volume or the number of moles, n_1, n_2, n_3, ..., n_C, of the constituents 1, 2, 3, ..., C.

(b) Intensive variables are those which are independent of the mass or the number of moles in the system; for example: temperature T, pressure P, and mole fraction $n_i/\Sigma n$ (Σn = total number of moles).

When the intensive variables have uniform values throughout a system, its state is defined by $(C + 2)$ variables (n_1, n_2, n_3, ..., n_C, T, P, or V) where C is the number of components as required in the definition of the Phase Rule. Any function X depends on these $(C + 2)$ variables:

$$X = f(P, T, n_1, n_2, \ldots, n_C) \tag{1.1}$$

or

$$X = g(V, T, n_1, n_2, \ldots, n_C) \tag{1.2}$$

and

$$dX = \left\{\frac{\partial X}{\partial P}\right\}_{T, V, n_i} dP + \left\{\frac{\partial X}{\partial T}\right\}_{P, V, n_i} dT + \sum_{1}^{n_C} \left\{\frac{\partial X}{\partial n_i}\right\}_{T, P, n_j} dn_i \tag{1.3}$$

or
$$dX = \left\{\frac{\partial X}{\partial V}\right\}_{T,\,P,\,n_i} dV + \left\{\frac{\partial X}{\partial T}\right\}_{P,\,V,\,n_i} dT + \sum_{1}^{n_C} \left\{\frac{\partial X}{\partial n_i}\right\}_{T,\,V,\,n_j} dn_i \tag{1.4}$$

This chapter deals with the thermodynamic functions and their variation with T and P (or V), while Chapter 2 will be concerned with the change of these functions with the mole fraction of the constituents in solution.

All thermodynamic functions are derived from two fundamental principles; the first is that of conservation of energy and of its equivalence in the various forms. The second is that of the degradation of energy.

1.1. CONSERVATION OF ENERGY

The total energy of an isolated system, which is the sum of all external (kinetic) and internal energy, is constant. The kinetic energy is of little interest in metallurgical processes, and will be assumed constant. The first law of thermodynamics states that *"energy cannot be created or destroyed"*. To translate this law into mathematical terms, two functions are defined: the internal energy U and the enthalpy H. It must be emphasized that the energy is defined taking an arbitrary state as reference, i.e. it is expressed as a difference between two states and not as an intrinsic quantity.

1.1.1. THE INTERNAL ENERGY FUNCTION

The internal energy of a system is the sum of various types of energy: thermal, electrical, mechanical, etc., which are not kinetic. If Q is the thermal energy and W the work which a system performs on passing from one state to another, the change in internal energy may be expressed by:

$$\Delta U = Q + W \tag{1.5a}$$

and, in the case of an infinitesimal change:

$$dU = \delta Q + \delta W \tag{1.5b}$$

The changes δQ and δW depend on the path followed, while the sum dU depends only on the initial and final states. Therefore, the internal energy is a state function, and its change in a cyclic process is zero ($\oint \phi\, dU = 0$).

The work W may be mechanical or of some other type (electrical, chemical, etc.) The first of these is due to the pressure, and is equal to -PdV. If W' is the expression for non-mechanical work (1.5b) may be expressed as:

$$dU = \delta Q - PdV + \delta W' \tag{1.6}$$

In many instances, the work is considered to be completely mechanical, from which:

$$dU = dQ - PdV \tag{1.7}$$

or

$$\Delta U = Q - P\Delta V \tag{1.8}$$

By convention, the heat transferred from a system to the environment is assumed negative, and the energy received by the system is positive. For example, if the volume increases as the result of a reaction ($\Delta V > 0$), the system is able to produce work, and therefore the mechanical energy W is negative:

$$W = -P\Delta V \quad \text{and} \quad W < 0 \quad \text{when} \quad \Delta V > 0 \tag{1.9}$$

The change of internal energy ΔU may be measured by calorimetry. When a reaction takes place in a calorimeter at constant volume, the work due to pressure forces is nil, from which:

$$\Delta U = Q_V \tag{1.10}$$

The internal energy change due to this reaction is then equal to the heat given off or absorbed at constant volume.

For example, the combustion of methane with oxygen at 298 K:

$$|CH_4| + 2|O_2| \rightleftarrows |CO_2| + 2|H_2O|$$

may occur in various ways. In a calorimeter (at constant volume) combustion produces only heat and $Q_V = \Delta U$. If combustion takes place in an open vessel, the volume will decrease; three moles of gas are reduced to one, since water condenses at 298 K. The thermal energy produced at constant pressure Q_P is different from Q_V:

$$\Delta U = Q_P - P\Delta V = Q_V \tag{1.11}$$

Since the work done ($W = -P\Delta V$) is positive in this example, Q_P is more negative then Q_V, and the reaction produces more thermal energy at constant pressure than at constant volume. ΔV is an extensive function, while P is intensive: thus W is an extensive function, in addition to ΔU and Q_V.

By definition, the specific heat of a substance at constant volume is:

$$C_V = \frac{\delta Q_V}{\delta T} \tag{1.12}$$

Since $dQ_V = dU$, then

$$C_V = \left\{\frac{\delta U}{\delta T}\right\}_V \tag{1.13}$$

1.1.2. THE ENTHALPY FUNCTION

When reactions take place at constant pressure, which is the most common case in metallurgy, (1.11) may be expressed as follows:

$$Q_P = \Delta U + P\Delta V \tag{1.14}$$

or

$$dQ_P = dU + PdV = d(U + PV)_P \tag{1.15}$$

The quantity $(U + PV)$ appears very often in thermodynamics. It is the "enthalpy" or heat content, and is expressed by the letter H. Enthalpy represents the total energy of the system, minus the work done by pressure:

$$H = U + PV \tag{1.16}$$

Like U and PV, H is an extensive state function. As with internal energy, it is impossible to state the intrinsic value of H; only its change from a state taken as reference. For an element, the reference state is that in which it is stable at 25°C (298 K) and a pressure of one atmosphere.

Example: For an element in the solid state at 298 K, $H^o_{298} = 0$
For a gaseous element at 298K (at 1 atm), $H^o_{298} = 0$

By differentiation of (1.16) we obtain:

$$dH = dU + PdV + VdP = dQ_P + VdP \tag{1.17}$$

It may be seen that dH is equal to dQ_P at constant pressure, and $\Delta H = Q_P$.

SPECIFIC HEAT AT CONSTANT PRESSURE

The specific heat of a pure substance at constant pressure is defined as follows:

$$C_P = \frac{dQ_P}{dT} = \left\{\frac{\partial H}{\partial T}\right\}_P \tag{1.18}$$

The dependence of the specific heat at constant pressure on temperature may be expressed by:

$$C_P = a + bT + cT^{-2} + dT^{\frac{1}{2}} + eT^{-2} \tag{1.19}$$

where a, b, c, ... are constants which can be found in reference tables for most pure substances. Usually, only the first three terms are included, since the contribution of later terms is extremely small.

HEAT REQUIRED FOR CHANGE OF TEMPERATURE

To heat a pure substance at constant pressure from the initial temperature (for example, ambient temperature) to a final temperature T, it is necessary to supply a certain amount of heat. The quantity required is obtained by integration of (1.18):

$$\Delta H = H_T - H_{298} = \int_{298}^{T} C_P dT = \int_{298}^{T} (a + bT + cT^{-2})dT \tag{1.20}$$

In this equation, H_{298} and H_T represent the enthalpies at 298 K and T, respectively, and at the same pressure.

HEAT REQUIRED FOR CHANGE OF STATE

When a substance undergoes a state transformation: fusion, crystallisation, vaporisation, condensation, or allotropic change, it absorbs or emits a certain amount of heat. For a pure substance at a given pressure, the transformation takes place at a fixed temperature, T_t. At this temperature, the enthalpy of transformation,

ΔH_t, is constant. When the transformation occurs at a temperature T, different from the normal temperature of transformation T_t, as for example the solidification of a violently shaken liquid, the enthalpy of transformation ΔH_t^T is slightly different from $\Delta H_t^{T_t}$ at the normal temperature. Consider a liquid A in equilibrium with a small crystal of solid A at the melting temperature T_f:

$$\langle A \rangle \leftrightarrows (A)$$

To reach this equilibrium, starting from solid A at 298 K, it is possible to follow two routes. The normal method involves heating solid A to T_f and then applying sufficient heat to melt it. The total heat supplied is then equal to:

$$\Delta H = \int_{298}^{T_f} C_{P(s)} dT + \Delta H_t^{T_f} \tag{1.21}$$

Assume now that, by some means, it is possible to melt solid A at a temperature T which is below T_f, and then to heat the liquid to T_f. The heat supplied will be:

$$\Delta H = \int_{298}^{T} C_{P(s)} dT + \Delta H_f^T + \int_{T}^{T_f} C_{P(l)} dT \tag{1.22}$$

Since the initial state (solid A at 298 K) and final state (liquid A at temperature T_f) are the same, the heat supplied must be the same in both cases. On equating (1.21) and (1.22), we find:

$$\Delta H_t^T = \Delta H_f^{T_f} + \int_{T}^{T_f} |C_{P(s)} - C_{P(l)}| dT \tag{1.23}$$

In general, the differences in specific heats of the liquid and solid are very small, and the enthalpy of fusion is considered to be independent of temperature. In the same way, the enthalpies of vaporisation or of change of crystalline state are considered independent of T, particularly when T is very near to the standard transformation temperature.

ENTHALPY OF A SUBSTANCE AT TEMPERATURE T

If T is above the boiling point, this may be calculated by means of the expression:

$$H_T - H_{298} = \int_{298}^{T_c} C_{P(s)} dT + \Delta H_c + \int_{T_c}^{T_f} C_{P(s)} dT + \Delta H_f$$

$$+ \int_{T_f}^{T_b} C_{P(l)} dT + \Delta H_b + \int_{T_b}^{T} C_{P(g)} dT \tag{1.24}$$

where ΔH_c, ΔH_f, and ΔH_b are the heat changes due to recrystallisation, fusion, and boiling.

REACTION ENTHALPY AND HEAT OF FORMATION OF A SUBSTANCE

The change of enthalpy in a reaction at constant temperature and pressure may be calculated from the differences in enthalpies of the products (final state) and the reactants (initial state):

$$\Delta H_T = \Sigma H_{prod.} - \Sigma H_{react.} \tag{1.25}$$

Therefore,

$$\Delta H_T = \Delta H_{298} + \int_{298}^{T} \Delta C_P dT \tag{1.26}$$

from which:

$$\Delta C_P = \Sigma C_{P_{prod.}} - \Sigma C_{P_{react.}} \tag{1.27}$$

The simplest reactions which can be considered are the formation of compounds from elements. Since, in general, the enthalpies of the elements at 298 K serve as zero reference states, the enthalpies of these reactions at 298 K are the heats of formation of the compounds themselves.

HESS'S LAW

The heat of formation of a substance is independent of the process and the number of steps involved in the formation. This principle is known as the Hess's Law. For example, to produce copper sulphate at 298 K, one may consider:

(a) $<CuO> + |SO_2| + \frac{1}{2}|O_2| \rightleftarrows <CuSO_4> \quad \Delta H_{(a)} = H_{CuSO_4} - \left\{H_{CuO} + H_{SO_2}\right\}$

(b) $<CuS> + 2|O_2| \rightleftarrows <CuSO_4> \quad \Delta H_{(b)} = H_{CuSO_4} - H_{CuS}$

Because the heat of formation of copper sulphate is independent of the manner in which it is obtained,

$$H_{CuSO_4} = \Delta H_{(b)} + H_{CuS} = \Delta H_{(a)} + H_{CuO} + H_{SO_2}$$

1.2. ENERGY DEGRADATION

The second law of thermodynamics is based on the simple fact that when two sources of energy at different temperatures are put in contact, the heat energy is spontaneously and irreversibly transferred from the source at higher temperature to that at lower temperature, until both are at the same temperature. More generally, the second law indicates that any process capable of doing work without absorbing external energy is spontaneous and irreversible.

1.2.1. THE ENTROPY FUNCTION

Entropy is an abstract concept which may be related to the notion of order and disorder of a system. In statistical thermodynamics, it has been shown that the increase of entropy of a body corresponds to an increase in the disorder of the atoms which form the body.

The entropy of a system is designated by the symbol S, and the change of entropy by dS and ΔS. An increase of entropy involves the absorption of energy, and vice versa.

A quantity of energy such as EI, $P\Delta V$, $T\Delta S$, etc., is the product of two terms; one intensive (E, P, T, etc.) and the other extensive (I, ΔV, ΔS, ...). Entropy is the extensive function of heat and energy and the corresponding intensive variable is the temperature T. On the other hand, mechanical and electrical work have losses associated with them (friction, Joule effect, etc.) which reduce the energy efficiency. In a chemical process in which ΔS is positive, the term $T\Delta S$ represents the energy loss.

Quantitatively, the concept of entropy is derived from the Carnot cycle. This consists of a reversible cyclic transformation between two temperatures T_1 and T_2, where there are two isothermal stages (A - B and C - D), and two adiabatic (B - C and D - A) (Fig. 1). The maximum efficiency of a thermal machine in perfectly reversible conditions can be expressed in the form:

$$\eta = \frac{Q_2 + Q_1}{Q_2} = \frac{T_2 - T_1}{T_2} \quad (1.28)$$

where Q_2 and Q_1 are the quantities of heat liberated by the hot source at temperature T_2 and absorbed by the cold source at temperature T_1, respectively. Equation (1.28) may be written as:

$$\frac{Q_2}{T_2} + \frac{Q_1}{T_1} = 0 \quad (1.29)$$

or:

$$\Sigma \frac{Q}{T} = 0 \quad (1.30)$$

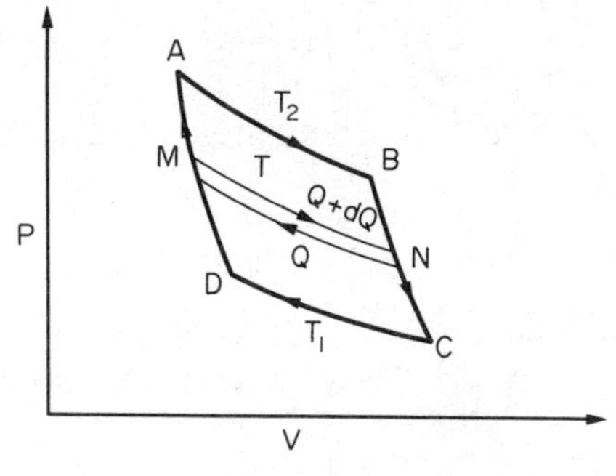

Fig. 1. Graphical representation of the Carnot cycle

The Carnot cycle can be broken up into a number of infinitely small reversible cycles, where Q_2 and Q_1 have values of $(Q + dQ)$ and $-Q$, and where T_2 and T_1 are equal to T (Fig. 1). For a reversible cycle of such a type, (1.30) may be written as:

$$\oint \frac{dQ}{T} = 0 \tag{1.31}$$

This cycle can be divided into two half-cycles of M to N, and of N to M, such that:

$$\int_M^N \frac{dQ}{T} + \int_N^M \frac{dQ}{T} = 0 \tag{1.32}$$

This equation shows that, to pass from one state to another in a reversible change, the integral of dQ/T does not depend on the path followed. The quantity dQ/T is then, by definition, a state function which is called entropy:

$$dS = \left(\frac{dQ}{T}\right)_{rev.} \tag{1.33}$$

Consequently, the change in entropy between M and N is:

$$\Delta S = S_N - S_M = \int_M^N \left(\frac{dQ}{T}\right)_{rev.} \tag{1.34}$$

The terms $dQ_{rev.}$ and TdS of (1.33) represent that part of the internal energy, or enthalpy, which cannot be transformed into useful energy in a reversible change of state, that is, in a change where there are no external heat losses. If part of the energy is dissipated to the outside, the change of state will be irreversible, and $dQ_{irrev.}$ will be smaller in absolute value than $dQ_{rev.}$

$$dQ_{rev.} > dQ_{irrev.} \tag{1.35}$$

or:

$$dS > \left(\frac{dQ}{T}\right)_{irrev.} \tag{1.36}$$

Entropy is expressed in cal $mole^{-1}\ {}^oC^{-1}$ (called Entropy Unit EU), or in joules $mole^{-1}\ K^{-1}$.

CHANGE OF ENTROPY WITH TEMPERATURE

The third law of thermodynamics, or Nernst's Law, states that the entropy of all crystalline substances at 0 K is zero. Therefore, by integrating (1.33) it is possible to obtain the value of the entropy at temperature T:

$$S_T = 0 + \int_0^T \frac{dQ}{T} = \int_0^T \frac{C_p dT}{T} \tag{1.37}$$

Generally, it is more practical to take the entropy at 298 K as the reference state. The entropy of a substance at a temperature above 298 K will be:

$$S_T = S_{298} + \int_{298}^{T} \frac{C_p dT}{T} \tag{1.38}$$

ENTROPY OF CHANGE OF STATE

A change of entropy, sometimes of substantial value, accompanies every change of state. The entropy of a substance which undergoes a series of transformations (allotropic changes c, melting f, vaporisation v) when the temperature is raised from 298 K to T, will be equal to:

$$S_T = S_{298} + \int_{298}^{T_c} \frac{C_{P(s)}}{T} dT + \Delta S_c + \int_{T_c}^{T_f} \frac{C_{P(s)}}{T} dT$$

$$+ \Delta S_f + \int_{T_f}^{T_v} \frac{C_{P(l)}}{T} dT + \Delta S_v + \int_{T_v}^{T} \frac{C_{P(v)}}{T} dT \tag{1.39}$$

Graphically, this may be represented by Fig. 2.

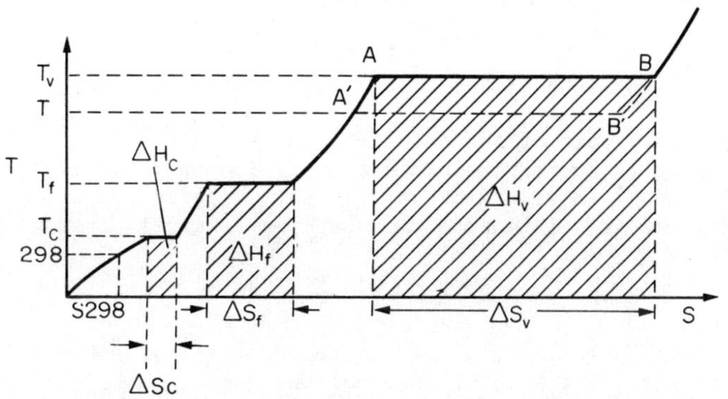

Fig. 2. Change of entropy with temperature

The entropy of change of state of a substance is practically independent of the temperature, as may be seen in Fig. 2 for the entropy of vaporisation; at temperature T_v the vaporisation entropy is represented by the segment AB, and at the temperature T by A'B'. If the sides AA' and BB' of the rectangle AA'BB' are parallel, which is the case when the slope of AA': $(dS/dT = C_p(\text{liq.})/T)$, and that of BB': $(dS/dT = C_p(\text{gas})/T)$ are equal, the entropy of change of state does not change with temperature.

It will be shown later that the entropy of transformation of a substance is related to the transformation enthalpy by:

$$\Delta H_t = T_t \cdot \Delta S_t \tag{1.40}$$

The shaded areas in Fig. 2 represent, therefore, the values of the different enthalpies of transformation.

Trouton's empirical "law" gives approximate values of the entropies of melting and vaporisation for pure metals.

$$\Delta S_f \simeq 2.3 \text{ EU} \qquad \Delta S_v \simeq 23 \text{ EU}$$

It may be noted that the entropy of vaporisation is greater than that of fusion. From this, one deduces that the entropy of a vapour or a gas is much greater than that of the condensed body. This property is important for its effect on the slopes of the ΔG^o_T curves (Fig. 23).

REACTION ENTROPY

In the same way as for enthalpy, the change in entropy of a reaction may be calculated from the entropy of the products and reactants, at the reaction temperature:

$$\Delta S_T = \Sigma S^T_{prod.} - \Sigma S^T_{react.} \tag{1.41}$$

and

$$\Delta S_T = \Delta S_{298} + \int_{298}^{T} \frac{\Delta C_P}{T} dT \tag{1.42}$$

1.2.2. AVAILABLE ENERGY

As has been seen, only part of the integral energy and of the enthalpy may be used to do work; the rest is converted into heat. It is now possible to define two thermodynamic functions in reversible processes, which correspond to useful or available energy; one at constant volume, and the other at constant pressure.

1.2.3. FREE ENERGY OR HELMHOLTZ FUNCTION

In a reversible transformation at constant temperature, if one substitutes TdS for dQ, then for a small change of internal energy:

$$dU = TdS + dW \tag{1.43}$$

or

$$dW = dU - TdS = d(U - TS) \tag{1.44}$$

The expression $(U - TS)$ is by definition the "Helmholtz Free Energy" and is generally denoted by the letter F, although the letter A is also used (Arbeit = work (in German)). As U and S are extensive state functions, so is $F = U - TS$.

The differential of the free energy function is:

$$dF = dU - TdS - SdT = - SdT + dW \tag{1.45}$$

If the only work considered is due to pressure forces $(dW = - PdV)$, then:

$$dF = - SdT - PdV \tag{1.46}$$

Equation (1.46) shows that F is a function of volume and temperature. In an isothermal process, the maximum energy which may be used is -PdV, whereas in an adiabatic process, at constant volume, this is equal to -SdT. By differentiating the free energy function F with respect to V and T, we obtain:

$$dF = \left(\frac{\partial F}{\partial T}\right)_V dT + \left(\frac{\partial F}{\partial V}\right)_T dV \qquad (1.47)$$

and comparing (1.46) and (1.47), it can be seen that:

$$\left(\frac{\partial F}{\partial T}\right)_V = -S \qquad (1.48)$$

and

$$\left(\frac{\partial F}{\partial V}\right)_T = -P \qquad (1.49)$$

1.2.4. FREE ENTHALPY FUNCTION OR GIBBS FREE ENERGY

The Helmholtz free energy is the total work which a system can produce or absorb, and is a function of temperature and volume. In metallurgy, processes are often carried out at constant pressure, and mechanical work -PdV due to the pressure forces, is not used. The non-mechanical work dW' is given by:

$$dW' = dW - (-PdV) = dF + d(PV)_{T,P} \qquad (1.50)$$

or

$$dW' = d(F + PV)_{P,T} \qquad (1.51)$$

The function (F + PV) is indicated by the letter G; thus:

$$G = F + PV = U - TS + PV \qquad (1.52)$$

or, since H = U + PV,

$$G = H - TS \qquad (1.53)$$

Like F and PV, G is an extensive state function. In the same way as the free energy function F corresponds to the available portion of the internal energy U(F = U - TS), G is that part of the enthalpy H which can be used and for this reason it is called "free enthalpy". G is also called "Gibbs Free Energy"; the Gibbs free energy represents the maximum work which a system can give, excluding the mechanical work due to pressure. In future references, the term "free energy" will be used to describe this property.

According to the second law of thermodynamics in an open system, a reversible transformation $A \rightleftarrows B$ is spontaneous in the direction $A \rightarrow B$ when this system is able to do work; that is, when dW' or dG is negative. If dW' or dG is zero, the system will be in equilibrium, and if dW' or dG is positive, it will be necessary to supply energy to the system for the transformation to take place. The free energy balance of a system indicates, therefore, the possibility of a transformation taking place spontaneously.

CHANGE OF FREE ENERGY WITH PRESSURE AND TEMPERATURE

When (1.52) is differentiated, then:

$$dG = dU + PdV + VdP - TdS - SdT \tag{1.54}$$

$$= VdP - SdT$$

The free energy G is a function of T and P. In an isothermal process, the available energy is equal to VdP, while in an adiabatic process at constant pressure, it is equal to $-SdT$. Using the partial derivatives of G with respect to T and P, one obtains:

$$dG = \left(\frac{\partial G}{\partial T}\right)_P dT + \left(\frac{\partial G}{\partial P}\right)_T dP \tag{1.55}$$

If (1.54) and (1.55) are compared, we find:

$$\left(\frac{\partial G}{\partial T}\right)_P = -S \tag{1.56}$$

and

$$\left(\frac{\partial G}{\partial P}\right)_T = V \tag{1.57}$$

CHANGE OF FREE ENERGY WITH PRESSURE AT CONSTANT TEMPERATURE

Equation (1.57) relates the change of free energy of a substance with pressure. At constant temperature, $dG = VdP$. If the substance is liquid or solid, the change of volume with pressure is small, and V may be considered constant. Integrating (1.57) between P_1 and P_2, one obtains:

$$\Delta G = G_{P_2} - G_{P_1} = V(P_2 - P_1) = V\Delta P \tag{1.58}$$

Variation of the free energy of a condensed substance with pressure is generally very small. For example, a change in pressure of 10 atm produces a change in the free energy of 1 mole of iron ($V_{Fe} = 7.2$ cm^3 mole^{-1}) of:

$$\Delta G = 7.2 \times 10^{-6} \times 10 \times 1.013 \times 10^5 = 7.3 \text{ joules mole}^{-1}$$

$$= 1.7 \text{ cal mole}^{-1}$$

For an ideal gas, V is equal to RT/P; if this is substituted in (1.57), then:

$$dG = RT\,\frac{dP}{P} \tag{1.59}$$

Integrating between P_1 and P_2 gives:

$$G = G_{P_2} - G_{P_1} = RT \ln \frac{P_2}{P_1}^{*} \quad (1.60)$$

In metallurgy, the approximation that the gas is ideal is sufficient in most cases because reaction is generally at high temperature and at atmospheric or relatively slightly elevated pressure.

FREE ENERGY AND STANDARD FREE ENERGY OF REACTION

When the reactants and products of a reaction are in their standard states (i.e. pure liquid or solid and the gases at 1 atm), the free energy of the reaction is the standard free energy, at the temperature considered. For example, in the oxidation of a metal at temperature T:

$$2\langle M\rangle \quad |O_2| \rightleftarrows 2\ \langle MO\rangle \quad (1.61)$$

if M, MO and O_2 form separate phases and O_2 is at 1 atm, the free energy of the reaction has the standard value:

$$\Delta G^o_T = \Delta H^o_T - T\Delta S^o_T \quad (1.62)$$

ΔH^o_T and ΔS^o_T are the standard enthalpy and standard entropy respectively, at the temperature considered, and are given by (1.26) and (1.42).

If the oxygen in (1.61) is at a pressure different from 1 atm, the only variation in the free energy will result from the difference between the actual pressure P and 1 atm, i.e. from the reaction:

$$|O_2|_{(1\ atm)} \rightleftarrows |O_2|_{(P)} \quad (1.63)$$

The free energy of this reaction according to (1.60) is $RT \ln (p/p_0)$ or $RT \ln p$, since $p_0 = 1$ atm. In (1.61), where $p_{O_2} \neq 1$ atm, ΔG will have the value:

$$\Delta G = \Delta G^o + RT \ln \frac{1}{p}$$

In the following chapter, this result will be extended to condensed substances (liquids and solids).

CHANGE OF FREE ENERGY WITH TEMPERATURE: CLAUSIUS-CLAPEYRON RELATIONSHIP

(1.56) relates the changes of free energy of a substance to the temperature. It is possible to express the value of dG as a function of H and T by substituting for S, $\left(\frac{-(G - H)}{T}\right)$:

*The value of R is equal to 1.987 cal $mole^{-1}$ K^{-1} or 8.32 joules $mole^{-1}$ K^{-1} when the natural logarithm is used, to 4.575 cal $mole^{-1}$ K^{-1} or 19.16 joules $mole^{-1}$ K^{-1} with the decimal logarithm.

$$dG = \left(\frac{G - H}{T}\right) dT \tag{1.64}$$

or

$$TdG - GdT = - HdT \tag{1.65}$$

This expression can be changed to the form:

$$\frac{dG}{T} - G\left(\frac{dT}{T^2}\right) = - H\left(\frac{dT}{T^2}\right) \tag{1.66}$$

The left-hand side is the differential $d(G/T)$, and the right is $Hd(1/T)$. The Clausius-Clapeyron relationship between H, G, and T can be deduced from the above:

$$H = \left|\frac{\partial(G/T)}{\partial(1/T)}\right|_P \tag{1.67}$$

If, instead of considering the thermodynamic functions of a pure substance, one considers the enthalpy, ΔH, and free energy ΔG, of a reaction or a change of state, the Clausius-Clapeyron relationship will still be valid on replacing H and G by ΔH and ΔG.

FREE ENERGY OF CHANGE OF STATE

Consider a change of state, for example the melting of A at the melting temperature T_f:

$$\langle A \rangle \rightleftarrows (A) \tag{1.68}$$

Solid and liquid are in equilibrium, and therefore $G_{(l)} = G_{\langle s \rangle}$, i.e.:

$$\Delta G_f^{T_f} = G_{(l)} - G_{\langle s \rangle} = 0$$

At a temperature T, different from T_f, the Clausius-Clapeyron equation can be expressed in the form:

$$\frac{\partial(\Delta G_f/T)}{\partial(1/T)} = \Delta H_f \tag{1.69}$$

and therefore

$$\Delta G_f^T = \frac{\Delta H_f(T_f - T)}{T_f} \tag{1.70}$$

The same result may be reached if one considers the free energy of (1.68) expressed as a function of enthalpy and entropy:

$$\Delta G_f = \Delta H_f - T\Delta S_f \tag{1.71}$$

Since the free energy of fusion is zero at the melting temperature T_f:

$$\Delta H_f^{T_f}/T_f = \Delta S_f^{T_f} \tag{1.71a}$$

If ΔH_f and ΔS_f are considered constant with temperature, (1.71) may be written as (1.70), or as:

$$\Delta G_f^T = \Delta S_f(T_f - T) \tag{1.72}$$

This expression applies to any transformation or change of state. In Fig. 2, the free energy of vaporisation at $T < T_v$ is represented by the area AA'BB'.

ANOTHER FORM OF THE CLAUSIUS-CLAPEYRON RELATIONSHIP

At the equilibrium between two states A_1 and A_2 of a substance A:

$$A_1 \rightleftarrows A_2 \tag{1.73}$$

the free energy is zero, and (1.54) may be written:

$$V_1\ dP - S_1\ dT = V_2\ dP - S_2\ dT$$

or

$$(V_2 - V_1)\ dP = (S_2 - S_1)\ dT \tag{1.74}$$

and since $\Delta G = \Delta H - T\Delta S = 0$, $(S_2 - S_1)$ can be exchanged for $(H_2 - H_1)/T = \Delta H/T$, giving:

$$\frac{H_2 - H_1}{T(V_2 - V_1)} = \frac{\Delta H}{T\Delta V} = \frac{dP}{dT} \tag{1.75}$$

1.2.5. APPLICATION TO THE SOLID-LIQUID-VAPOUR EQUILIBRIUM

In the case of solid-liquid equilibrium:

$$\langle A \rangle \rightleftarrows (A) \tag{1.76}$$

ΔV and ΔH in (1.75) represent the variation of volume and enthalpy of fusion. Consequently:

$$\frac{dP}{dT} = \frac{\Delta H_f}{T\Delta V_f} \tag{1.77}$$

dP/dT, which is the slope of the curve separating liquid and solid stability regions in the pressure-temperature equilibrium diagram of Fig. 3, is a function of the enthalpy of fusion, always positive, of temperature T, and of the change of volume ΔV_f resulting from melting. For most substances, ΔV_f is positive, the solid being more dense than the liquid. The slopes of the curves of the $P(T)$ phase diagrams are therefore positive. An important exception is water, where the change of volume is negative (water is more dense than ice). The slope of the curve which separates the areas liquid water-solid water is therefore negative. (Problem No. 1).

In the solid-vapour and liquid-vapour equilibria, the volume of the condensed phase can in general be neglected with respect to that of the vapour; (V_l and V_s are very small relative to V_v). On the other hand, if the vapour is considered as an ideal gas ($V_v = RT/P$), (1.75) may be written:

$$\frac{dP}{P\ dT} = \frac{d\ \ln P}{dT} = \frac{\Delta H_t}{RT^2} \tag{1.78}$$

ΔH_t represents the enthalpy of sublimation or vaporisation. If these enthalpies are considered constant for small intervals of temperature, the variation of the saturated vapour pressure with temperature is:

$$\log P = -\frac{\Delta H_t}{RT} + \mathit{constant} \tag{1.79}$$

The constant can be determined from knowledge of the pressure at a given temperature; for example, $P = 1$ atm at the boiling temperature. For greater intervals of temperature, ΔH_t varies with the temperature, according to (1.23):

$$\Delta H_t^T = \Delta H_t^{T_t} + \int_T^{T_t} \left| C_{p(s\ \mathrm{or}\ l)} - C_{p(vap.)} \right| dT \tag{1.23}$$

If the difference $C_{p(s\ \mathrm{or}\ l)} - C_{p(vap.)}$ is assumed constant, the transformation enthalpy at temperature T will be:

$$\Delta H_t^T = \Delta H_t^{T_t} + \Delta C_p(T_t - T) = (\Delta H_t^{T_t} + \Delta C_p T_t) - \Delta C_p T \tag{1.80}$$

Replacing ΔH_t^T by its value in (1.78), and integrating, one obtains:

$$\ln P = -\left|\frac{\Delta H_t^{T_t} + \Delta C_p T_t}{RT}\right| \left(\frac{\Delta C_p}{R}\right) \ln T + \mathit{constant}$$

or

$$\log P = AT^{-1} + B \log T + C \tag{1.81}$$

This is the form most frequently used and the coefficients A, B, and C are found in reference tables.

The three curves $P = f(T)$ which represent the equilibrium solid-liquid, solid-vapour, and liquid-vapour, intercept at one point (the triple point) and divide the phase diagram in three areas in which only one phase is stable. At the phase limits, two phases co-exist in equilibrium. A study of the variance shows that it is not possible to have an area with three phases in equilibrium.

The slopes dP/dT of the two curves which represent the equilibria solid-vapour and liquid-vapour at the triple point (T and P fixed) depend only on the enthalpies of change of state (see (1.75)). In passing from solid to the vapour, two routes may be followed: direct sublimation, or melting and evaporation of the liquid. Energetically, the two processes are equivalent and the sublimation enthalpy is equal to the sum of the enthalpies of fusion and vaporisation:

$$\Delta H_s = \Delta H_f + \Delta H_v \tag{1.82}$$

As all three enthalpies are positive, ΔH_s is greater than ΔH_v; therefore the equilibrium curve for solid-vapour has a larger slope than the equilibrium curve for liquid-vapour; the angle of the two curves is then less than π at the triple point (Fig. 3).

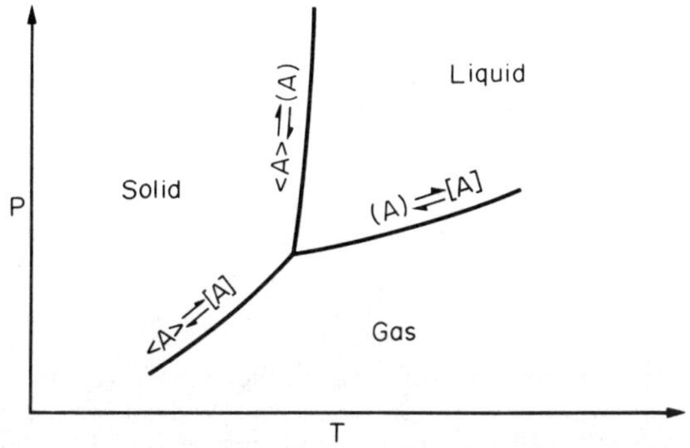

Fig. 3. Solid-liquid-vapour equilibrium

GENERAL READING

Darken, L., and R. Gurry, *Physical Chemistry of Metals*, McGraw-Hill, N.Y. (1953).

Kubaschewski, P., E.Ll. Evans and C.B. Alcock, *Metallurgical Thermochemistry*, 5th Ed., Pergamon Press, London (1979).

Prigogine, I., and R. Defay, *Chemical Thermodynamics*, Vol. 1, Longmans Green, London (1952).

Rossini, F.D., *Chemical Thermodynamics*, J. Wiley & Sons, N.Y. (1950).

Souchay, P., *Chimie Physique: Thermodynamique Chimique*, 3rd Ed., Masson et Cie., Paris (1968).

Chapter 2
SOLUTIONS

INTRODUCTION

A solution is a mixture of two or more substances, so intimately mixed that it is impossible to distinguish the properties of the individual components. It forms a single phase, as distinct from heterogeneous systems which consist of several phases. Examples of solutions are the perfectly miscible mixtures of gases, liquids, and solids represented by nitrogen and oxygen, water and alcohol and austenite.

The composition of a solution can be expressed by the number of moles, n_i, by molar fraction, N_i, or by weight per cent (% i):

$$N_i = \frac{n_i}{\Sigma n_i} \qquad (\%\ i) = \frac{m_i}{\Sigma m_i} \times 100$$

where m_i is the mass of component i in solution. The thermodynamic properties of a solution are significantly different from those of the pure constituents.

2.1. CHEMICAL POTENTIAL OF A COMPONENT IN SOLUTION

2.2.1. DEFINITION OF CHEMICAL POTENTIAL

As has been demonstrated, thermodynamic functions vary with temperature, pressure, volume, and the number of moles of the different components. Thus, for a solution which contains c components: 1, 2, 3, ..., c, the change in free energy may be written:

$$dG = \left(\frac{\partial G}{\partial T}\right)_{P,n_i} dT + \left(\frac{\partial G}{\partial P}\right)_{T,n_i} dP + \sum_{i=1}^{i=c} \left(\frac{\partial G}{\partial n_i}\right)_{T,P,n_j} dn_i \qquad (2.1)$$

The term $(\partial G/\partial n_i)_{T,P,n_j}$ represents the change in free energy when dn_i moles of i are added at constant pressure and temperature to such a large volume of solution that the molar fraction N_j of the other components does not change. It is the partial molar free energy of the component i in solution, and is represented by $\bar{G}_i$. It is also defined as the chemical potential μ_i of i in solution, a more general term, since it can be shown that:

$$\mu_i = \left(\frac{\partial U}{\partial n_i}\right)_{S,V,n_j} = \left(\frac{\partial H}{\partial n_i}\right)_{S,P,n_j} = \left(\frac{\partial F}{\partial n_i}\right)_{T,V,n_j} = \left(\frac{\partial G}{\partial n_i}\right)_{T,P,n_j} \quad (2.2)$$

At constant temperature and pressure, (2.1) can be simplified to give:

$$dG = \Sigma\mu_i \, dn_i = \Sigma\bar{G}_i \, dn_i \quad (2.3)$$

The free energy change dG depends, according to (2.3) on the number of moles n_i of the components; therefore it is an extensive function, and:

$$G(\nu n_i) = \nu G(n_i) \quad (2.4)$$

ν = stochiometric coeff. of a reaction

From this relationship, the following properties can be deduced:

(a) The integral free energy of the solution is obtained by integration of (2.3):

$$G = \Sigma\mu_i n_i = \Sigma\bar{G}_i n_i \quad (2.5)$$

(b) From the differentiation of (2.5) we obtain:

$$dG = \Sigma\mu_i \, dn_i + \Sigma n_i \, d\mu_i = \Sigma\bar{G}_i \, dn_i + \Sigma n_i \, d\bar{G}_i \quad (2.6)$$

Comparing (2.3) and (2.6), it is possible to deduce the Gibbs-Duhem equation which relates the chemical potentials of the components of an homogeneous phase:

$$\Sigma n_i \, d\mu_i = \Sigma n_i \, d\bar{G}_i = 0 \quad (2.7)$$

Integrating this relation makes it possible to determine the chemical potential of a second component in a binary or ternary solution, knowing the chemical potential of one of the components (see Chapter 3).

(c) Since G is an extensive function, i.e. proportional to the number of moles of components, it can be deduced from (2.5) that μ_i (or $\bar{G}_i$) is an intensive function, independent of the number of moles of components of the solution; therefore:

$$\mu_i(\nu n_i) = \mu_i(n_i) \quad \text{or} \quad \bar{G}_i(\nu n_i) = \bar{G}_i(n_i) \quad (2.8)$$

In particular, if ν takes the value $1/\Sigma n_i$, then:

$$\mu_i(n_i) = \mu_i(N_i) \quad \text{or} \quad \bar{G}_i(n_i) = \bar{G}_i(N_i) \quad (2.9)$$

At constant temperature and pressure, the chemical potential of a component (or its partial free energy) depends only on the composition of the mixture, and not on its quantity.

Because G is an extensive function, and μ_i (or $\bar{G}_i$) an intensive function, for the sake of simplicity it is usual to take such a number of moles of each component that $\Sigma n_i = 1$. The integral free energy of a mole of solution will then be:

$$G = \Sigma \mu_i N_i = \Sigma \bar{G}_i N_i \tag{2.5a}$$

If the number of moles in the solution is different from one, it will be necessary to multiply the integral free energy given by (2.5a) by Σn_i (number of moles of the solution). (2.7) is valid even when n_i is substituted by N_i.

2.1.2. PARTIAL AND INTEGRAL MOLAR FREE ENERGY OF MIXING

When 1 mole of A is isothermally dissolved in a large volume of solution, according to:

$$A \rightleftarrows \underline{A} \tag{2.10}$$

the change in free energy of the solution is:

$$\Delta G = \bar{G}_A - G_A \tag{2.11}$$

The difference of free energy ΔG between A in solution, and A as a pure liquid, taken as the reference state, is the "partial molar free energy of mixing" of A and is expressed by the symbol $\bar{G}_A^M$:

$$\bar{G}_A^M = \bar{G}_A - G_A^o \tag{2.12}$$

$\bar{G}_A^M$ represents the change of energy or work which a mole of pure A can make available when it goes into solution at a certain concentration, at constant pressure and temperature.

The integral free energy of mixing G^M is defined from (2.5a) and (2.12) by:

$$G^M = \Sigma \bar{G}_i^M N_i = \Sigma \bar{G}_i N_i - \Sigma G_i^o N_i \tag{2.13}$$

where $\bar{G}_i N_i$ represents the free energy of solution, while $G_i^o N_i$ represents the free energy of the pure components before entering into solution.

2.1.3. PARTITION OF A COMPONENT BETWEEN VARIOUS PHASES: EQUILIBRIUM CRITERION

Two phases, α and β, in contact with a common component A, held at the same temperature and pressure, form a closed system which may evolve by transference of A from one phase to the other:

$$\underline{A}_\alpha \underset{b}{\overset{a}{\rightleftarrows}} \underline{A}_\beta \tag{2.14}$$

If a quantity dn_A is transferred from phase α to phase β, the change in free energy of the closed system will be:

$$dG = dG_\beta + dG_\alpha = (\bar{G}_{A_\beta} - \bar{G}_{A_\alpha})\, dn_A = (\mu_{A_\beta} - \mu_{A_\alpha})\, dn_A \qquad (2.15)$$

In order that the transference of A from α to β shall be spontaneous, dG must be negative, that is, the potential of A in phase α must be higher than in phase β. If dG were positive (or the potential in phase β higher than that in phase α), transference would be spontaneous in the direction of b. If dG is zero, there would be equilibrium. In this case:

$$\bar{G}_{A_\beta} = \bar{G}_{A_\alpha} \qquad \text{or} \qquad \mu_{A_\beta} = \mu_{A_\alpha}$$

i.e. the chemical potential is the same in both phases.

Knowledge of the chemical potential of an element in a solution is important in metallurgy because it allows determination of the conditions under which that element is transferred to or from a metal which is to be refined by contact with a liquid (slag) or gaseous phase. In steelmaking, one must consider the metal, the slag, the refractories and the atmosphere over the slag/metal bath as phases in the system. For removal of an impurity from the metal, the chemical potential of that element must be lower in the slag than in the metal.

2.1.4. VARIANCE OF A SYSTEM

Solution of the mathematics of a system which includes n unknowns or variables requires n independent equations. With less than n equations (for example, $n' < n$), the values of $(n - n')$ unknowns may be fixed arbitrarily and the values of the others derived from these.

In a closed thermodynamic system, the unknowns are the molar fractions of c independent constituents, distributed among F phases in equilibrium at temperature T and pressure P, but there are established relations between the thermodynamic functions of the components in the different phases. The variance of the system is defined as the number of variables which may be arbitrarily fixed, so that the system is defined. Thus, the variance is equal to the number of unknowns, less the number of equations which relate these unknowns among themselves.

With c independent components in F phases, there will be $F(c + 2)$ variables if the temperature and pressure of each phase are considered.

The temperature, pressure, and chemical potentials of the different components A, B, C, ..., are the same in all phases $\alpha, \beta, \gamma, \ldots$; therefore there are $(c + 2)(F - 1)$ equations of the type:

$$\left.c+2 \text{ lines}\ \left\{\begin{array}{l} \mu_{\underline{A}_\alpha} = \mu_{\underline{A}_\beta} = \mu_{\underline{A}_\gamma} = \quad \cdots \\ \mu_{\underline{B}_\alpha} = \mu_{\underline{B}_\beta} = \mu_{\underline{B}_\gamma} = \quad \cdots \\ \text{-----------------------} \\ T_\alpha = T_\beta = T_\gamma \quad \cdots \\ P_\alpha = P_\beta = P_\gamma \quad \cdots \end{array}\right.\right\} \text{(F − 1) independent equalities on each line}$$

In addition, there are F relations (one for each phase) of the type:

$$\Sigma N_i = 1$$

The variance of the system is equal to:

$$v = F(c + 2) - (c + 2)(F - 1) - F = c + 2 - F \qquad (2.16)$$

If, instead of taking the independent components c, one considers the components of each phase C, (C > c), it will be necessary to consider the reactions between the independent constituents, and write r reactions of the type:

$$C_\alpha + C'_\beta \rightleftharpoons C''_\gamma$$

for which there are r equations of the form:

$$\Delta G_{react.} = 0$$

The result, however, is identical, since $c = C - r$.

When the gaseous phase is left out of consideration, the variance will be less by one unit:

$$v = c + 1 - F \qquad (2.16a)$$

Example

Consider a steel containing Fe, O, Mn, and Si in equilibrium with a liquid slag of FeO, MnO, and SiO_2 in a silica crucible. What is the variance of the system, and what can be deduced from its study? If it is assumed that the gaseous phase plays no part, the variance is therefore:

$$v = c + 1 - F \qquad \text{or} \qquad v = C - r + 1 - F$$

Independent constituents:	$c = 4$:	Fe, Mn, Si, O.
Constituents:	$C = 7$:	Fe, Mn, Si, O, MnO, FeO, SiO_2.
Reactions:	$r = 3$:	Fe + O, Mn + O, Si + 2 O → FeO, MnO, SiO_2.
Phases:	$F = 3$:	Metal, Slag, Crucible.

The variance is 2. If the temperature is fixed and one concentration (that of O, Fe, Mn, Si, MnO or FeO) the system is completely defined. The potential of SiO_2 in the slag is fixed by the silica crucible. That the gaseous phase is not considered does not mean it is not present. All of the constituents are represented in small amounts in the gaseous phase in equilibrium with the steel. If the gas phase is considered, the value of the variance is:

$$v = 4 + 2 - 4 = 2 \tag{2.17}$$

which is the same as the value arrived at earlier. If, in addition to the temperature, the concentrations of two constituents are fixed instead of one, one phase must disappear. This would occur if, for example, the concentrations of MnO and FeO in the slag were fixed; (the crucible would disappear by solution in the slag).

2.2 ACTIVITY AND MASS ACTION LAW

2.2.1. ACTIVITY OF A SUBSTANCE IN SOLUTION

Consider pure liquid A and a liquid solution AB in two crucibles at the same temperature. The gaseous phases over A and AB contain, respectively, vapour of pure A and of the mixture A and B. If p_A^o is the pressure of pure A taken as the standard state, and P_A the partial pressure of A over the solution, since the partial free energies in the vapour and the corresponding phases are the same, then:

$$G^o_{(A)} = G_{[A]\,p^o_A} \qquad \text{and} \qquad \bar{G}_{(A)} = \bar{G}_{[A]\,p_A} \tag{2.18}$$

The values of the partial free energies may be expressed as functions of the vapour pressures according to:

$$\bar{G}_{[A]\,p_A} = RT \ln p_A \tag{2.19}$$

and

$$G_{[A]\,p^o_A} = RT \ln p^o_A \tag{2.20}$$

The partial free energy of mixing of A in solution is the difference $(\bar{G}_A - G^o_A)$ according to (2.12); therefore:

$$\bar{G}^M_A = \bar{G}_A - G^o_A = RT \ln \frac{p_A}{p^o_A} \tag{2.21}$$

The activity of a substance in solution is defined as the ratio between its vapour pressure over the solution and the vapour pressure of the pure substance at the same temperature; that is:

$$a_A = \frac{p_A}{p^o_A} \tag{2.22}$$

and, thus:

$$\bar{G}_A^M = RT \ln a_A \tag{2.23}$$

When A exists in two or more phases in equilibrium, it is evident that the activity of A is the same in all phases, since the partial free energy of mixing is the same in all phases.

2.2.2. FREE ENERGY OF A COMPLEX REACTION

The reversible reaction:

$$\nu_1\underline{A}_1 + \nu_2\underline{A}_2 + \ldots + \nu_r\underline{A}_r \underset{b}{\overset{a}{\rightleftarrows}} \nu_1'\underline{A}_1' + \nu_2'\underline{A}_2' + \ldots + \nu_s'\underline{A}_s' \tag{2.24}$$

where ν_1, ν_2, etc., are the stoichiometric coefficients of the reaction and where each one of the reagents A_r and products A_s may form a separate phase, or a liquid, solid, or gaseous solution, may be separated into several reactions:

(a) Reactions between pure substances:

$$\nu_1 A_1 + \nu_2 A_2 + \ldots + \underset{2}{\overset{1}{\rightleftarrows}} \nu_1'A_1' + \nu_2'A_2' + \ldots + \nu_s'A_s' \tag{2.25}$$

The free energy of this reaction will be the standard free energy ΔG^o.

(b) Dissolution of reactants and products in the phases which make up the system:

$$\nu_r A_r \rightarrow \nu_r \underline{A}_r$$

The free energy of this reaction is the partial free energy of mixing of the components A_r given by (2.23), multiplied by the number of moles ν_r of component A_r that is:

$$\nu_r \bar{G}_{A_r}^M = \nu_r RT \ln a_{A_r} = RT \ln a_{A_r}^{\nu_r} \tag{2.26}$$

The free energy of (2.24) will then be:

$$\Delta G = \Delta G^o + \Sigma\nu_s'\bar{G}_{A_s'}^M - \Sigma\nu_r\bar{G}_{A_r}^M \tag{2.27}$$

and using activities, one obtains:

$$\Delta G = \Delta G^o + RT \ln \frac{\Pi\, a_{A_s'}^{\nu_s'}}{\Pi\, a_{A_r}^{\nu_r}} = \Delta G^o + RT \log K \tag{2.27a}$$

Depending on the sign of ΔG, (2.24) will go in direction a ($\Delta G<0$), in direction b ($\Delta G>0$), or will be in equilibrium ($\Delta G = 0$). K is equal to the product of the activities of the products $\left(\Pi\, a_{A_s'}^{\nu_s'}\right)$ of the reaction, divided by the product of the activities of the reactants $\left(\Pi\, a_{A_r}^{\nu_r}\right)$ with the appropriate exponents. At the commencement of the reaction, when the reagents are put into contact, K is nil, since the products have not formed and the free energy of reaction is therefore infinitely negative. As reaction proceeds, the free energy tends toward zero.

2.2.3. EQUILIBRIUM CONSTANT AND MASS ACTION LAW

At equilibrium, the free energy of (2.24) is zero, and therefore:

$$\Delta G^o = -RT \ln \left[\frac{\Pi\, a_{A_s'}^{\nu_s'}}{\Pi\, a_{A_r}^{\nu_r}}\right]_{eq.} \tag{2.28}$$

The equilibrium constant K_e which is dependent only on temperature, can be expressed in the form:

$$K_e = \left[\frac{\Pi\, a_{A_s'}^{\nu_s'}}{\Pi\, a_{A_r}^{\nu_r}}\right]_{eq.} \tag{2.29}$$

The expression of the equilibrium constant K_e as a function of ΔG^o, at a given temperature, is:

$$\log K_e = -\frac{\Delta G^o}{4.575\ T} \quad \text{with } \Delta G^o \text{ in cal mole}^{-1}$$

$$= -\frac{\Delta G^o}{19.14\ T} \quad \text{with } \Delta G^o \text{ in joules mole}^{-1} \tag{2.30}$$

This relation is the exact thermodynamic form of the Law of Mass Action, originally given as a function of the mole fraction N_i of each component. To retain this law in the original form it is necessary to multiply the mole fractions by the activity coefficients $\gamma_i = a_i/N_i$.

For perfect gases, pure or mixed, the value of a_i is substituted by p_i in atmospheres, if the standard state for gases is taken as 1 atm pressure.* In liquid and solid solutions, activities and mole fractions are related to each other, but their relation is not obvious, *a priori*, and must be determined experimentally.

EXAMPLES

In the blast furnace manganese is distributed between the slag, where it is manganese oxide, and the pig iron, where it is in liquid solution. In the presence of coke, the Mn oxide is reduced:

$$\underline{(MnO)}_{slag} + \underline{(C)} \rightleftarrows \underline{(Mn)}_{pig\ iron} + [CO]$$

At equilibrium:

$$\log \frac{a_{\underline{(Mn)}_m} \cdot p_{[CO]}}{a_{\underline{(MnO)}_s} \cdot a_{\underline{(C)}}} = \frac{-\Delta G^o}{4.575\ T}$$

This equation gives only one relationship between activities. To obtain the mole fraction of Mn in the pig iron and of MnO in the slag, additional information is necessary. In general, one assumes that $a_{(C)} = 1$, since the pig iron is saturated with carbon, and $p_{(CO)} = 1$ atm. In addition, it is necessary to know:

(1) the material balance for Mn;

(2) the activity coefficient γ_{Mn} of Mn in the pig iron;

(3) the activity coefficient of MnO in the slag determined experimentally.

2.2.4. CHANGE OF THE EQUILIBRIUM CONSTANT WITH TEMPERATURE

From the Clausius-Clapeyron equation (1.67), it can be deduced that:

$$\frac{\partial \log K_e}{\partial (1/T)} = - \frac{\Delta H^o}{4.575} \tag{2.31}$$

or

$$\frac{\partial \log K_e}{\partial T} = + \frac{\Delta H^o}{4.575\ T^2} \tag{2.32}$$

The equilibrium constant in an endothermic reaction $(\Delta H^o > 0)$ increases with temperature, and decreases with increase of temperature in an exothermic reaction $(\Delta H^o < 0)$. This law of variation of K_e is in agreement with Le Chatelier's Principle, which states that an increase of temperature displaces the equilibrium in the direction of absorbing heat.

*To be rigorously correct, the activity should really be substituted by the fugacity f_i and not by pressure p_i. However, in high temperature metallurgy p_i is, in most cases, a good approximation to f_i.

In the previous example, the reduction of MnO by carbon is endothermic; that is, the reaction enthalpy is positive. *A priori*, it may be foreseen that the higher the temperature, the greater the activity of Mn in the pig iron; this is what actually happens in the blast furnace.

2.3. OTHER PARTIAL AND INTEGRAL QUANTITIES OF MIXING

2.3.1. GENERALISATIONS OF PREVIOUS RESULTS

The results found for the free energy function may be extended to all the extensive functions (H, S, V, F, U, ...). For example, at constant pressure and temperature the change in an extensive quantity, X, is expressed as a function of the change of the number of moles of the components by:

$$dX = \sum_i \left(\frac{\partial X}{\partial n_i}\right)_{n_j,T,P} . \, dn_i \tag{2.33}$$

The term $\left(\frac{\partial X}{\partial n_i}\right)_{T,P,n_j}$ is the partial molar quantity of component i and is expressed by the symbol $\bar{X}_i$; therefore:

$$dX = \Sigma \bar{X}_i \, dn_i \tag{2.34}$$

The integral function for one mole of solution will then be equal to:

$$X = \Sigma \bar{X}_i N_i \tag{2.35}$$

The partial quantities of mixing are:

$$\bar{X}_i^M = \bar{X}_i - X_i^o \tag{2.36}$$

and the integral quantities of mixing, for one mole of solution:

$$X^M = \Sigma \bar{X}_i^M N_i = \Sigma \bar{X}_i N_i - \Sigma X_i^o N_i \tag{2.37}$$

For example, in a binary solution AB the integral volume of mixing is equal to:

$$V^M = N_A \bar{V}_A^M + N_B \bar{V}_B^M \tag{2.38}$$

or

$$V^M = \left(N_A \bar{V}_A + N_B \bar{V}_B\right) - \left(N_A V_A^o + N_B V_B^o\right) \tag{2.39}$$

The first term on the right-hand side represents the volume of the solution, and the second the volume of the pure components A and B, before going into solution.

The relations between thermodynamic quantities described in Chapter 1 are also valid. For example:

$$\bar{G}_i^M = \bar{G}_i - G_i^o = \bar{H}_i - H_i^o - T\left[\bar{S}_i - S_i^o\right] = \bar{H}_i^M - T\bar{S}_i^M \qquad (2.40)$$

and

$$G^M = H^M - TS^M \qquad (2.41)$$

The changes in the partial free energy of mixing with temperature and pressure are:

$$\left[\frac{\partial \bar{G}_i^M}{\partial T}\right]_{P,N_j} = -\bar{S}_i^M \qquad (2.42)$$

and

$$\left[\frac{\partial \bar{G}_i^M}{\partial P}\right]_{T,N_j} = \bar{V}_i^M \qquad (2.43)$$

The Clausius-Clapeyron law may then be written:

$$\frac{\partial\left(\bar{G}_i^M/T\right)_{P,N_j}}{\partial(1/T)} = \bar{H}_i^M \qquad (2.44)$$

2.3.2. GRAPHIC DETERMINATION OF PARTIAL FUNCTIONS OF MIXING

When the integral functions of mixing in a binary solution are known over the whole range, the corresponding partial functions may be obtained graphically by drawing the tangent to the curve $X = f(N_i)$ at the specified composition. The partial functions are the ordinates of the tangent at $N_A = 1$ and $N_B = 1$.

Equation (2.34) applied to solution AB may be written:

$$dX = \bar{X}_A\, dN_A + \bar{X}_B\, dN_B \qquad (2.45)$$

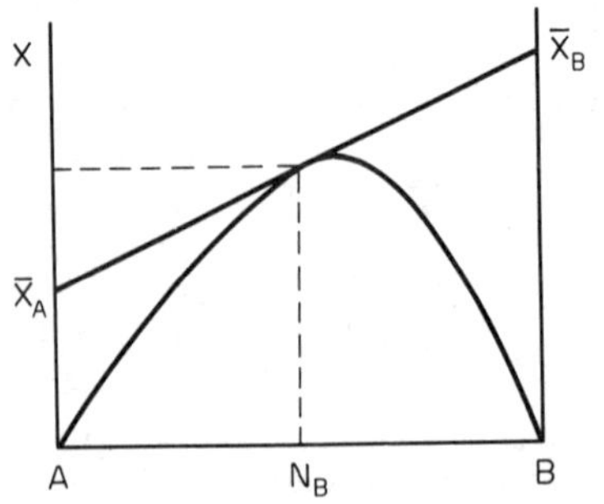

Fig. 4. Determination of partial functions of mixing from integral functions

Multiplying each member of (2.45) by N_A/dN_B, one obtains:

$$\frac{N_A dX}{dN_B} = \frac{\bar{X}_A N_A\ dN_A}{dN_B} + \frac{\bar{X}_B N_A\ dN_A}{dN_B} \tag{2.46}$$

In a binary solution, $N_A + N_B = 1$, $dN_A = -\ dN_B$, and $X = \bar{X}_A N_A + \bar{X}_B N_B$, then:

$$(1 - N_B)\frac{dX}{dN_B} = -\bar{X}_A N_A + \bar{X}_B(1 - N_B) = -X + \bar{X}_B \tag{2.47}$$

or
$$\bar{X}_B = X + (1 - N_B)\frac{dX}{dN_B} \tag{2.48}$$

(2.48) is the equation of the tangent to the curve at N_B and X in Fig. 4. The slope is (dX/dN_B). The partial quantity $\bar{X}_B$ is the ordinate of the tangent at $N_B = 1$.

2.4. IDEAL, REAL AND REGULAR SOLUTIONS

2.4.1. IDEAL SOLUTION

An ideal solution follows Raoult's law, i.e. the activity of a component i of the solution is equal to its mole fraction at all compositions. The ratio of the partial pressure p_i of i in the vapour phase in equilibrium with the solution, to the pressure p_i^o of i over pure liquid i, is also equal to the mole fraction:

$$a_i = \frac{p_i}{p_i^o} = N_i \tag{2.49}$$

The partial free energy of mixing of i is given by:

$$\bar{G}_i^M = \bar{G}_i - G_i^o = RT\ \ln(p_i/p_i^o) = RT\ \ln N_i \tag{2.50}$$

In an ideal solution, the partial and integral volumes of mixing and the partial and integral enthalpies of mixing are nil:

$$\bar{H}_i^M = \frac{\partial\left(\bar{G}_i^M/T\right)}{\partial(1/T)} = \frac{\partial(R\ \ln N_i)}{\partial(1/T)} \tag{2.51}$$

Since $R\ \ln N_i$ is not a function of temperature, $\bar{H}_i^M$ is nil.

Likewise:
$$H^M = \Sigma N_i \bar{H}_i^M = 0 \tag{2.52}$$

The partial volume of mixing is equal to:

$$\bar{V}_i^M = \left[\frac{\partial \bar{G}_i^M}{\partial P}\right]_T = \frac{\partial(RT\ \ln N_i)}{\partial P} \tag{2.53}$$

and since $RT \ln N_i$ is independent of pressure, $\bar{V}_i^M$ is nil, and:

$$V^M = \Sigma N_i \cdot \bar{V}_i^M = 0 \qquad (2.54)$$

The partial and integral entropies may be obtained from (2.42) and (2.37):

$$\bar{S}_i^M = -\frac{\partial \bar{G}_i^M}{\partial T} = -\frac{\partial RT \ln N_i}{\partial T} = -R \ln N_i \qquad (2.55)$$

and

$$S^M = -R\,\Sigma N_i \ln N_i \qquad (2.56)$$

In ideal solutions, the bond strengths between the atoms of A and B, A and A, and B and B are so similar that A and B are randomly distributed and the behaviour is ideal. In this respect, these solutions resemble a perfect gas. In fact, few solutions are ideal at all ranges of concentration. In metallurgy, for example, MnO-FeO solutions are considered to be ideal. However, a component of a solution tends to follow Raoult's law when its molar fraction is near to one.

2.4.2. EXCESS FUNCTIONS OF A REAL SOLUTION

In a real binary solution, A-B, the interaction between A and B is different from that between A and A, or B and B. There is attraction or repulsion between A and B, and there will not be a random distribution of the components of the solution.

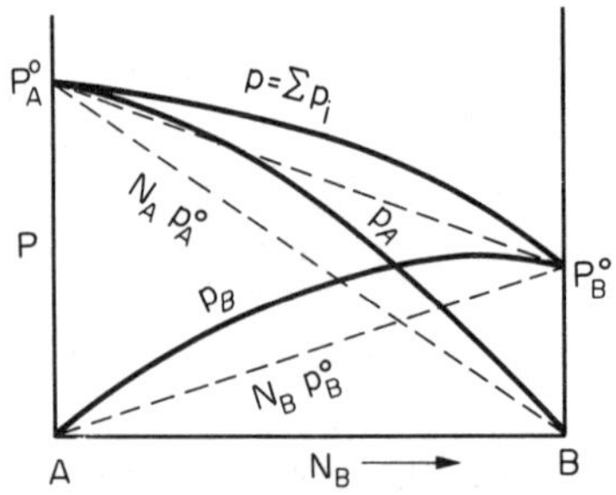

Fig. 5. Equilibrium vapour pressure over a real binary solution

The partial pressure of the component i in the gaseous phase in equilibrium with the solution is not proportional to the mole fraction of i in the solution and, consequently, a_i is different from N_i. The difference is expressed by the introduction of the activity coefficient γ_i:

$$a_A = \frac{p_A}{p_A^o} = \gamma_A N_A \qquad (2.57)$$

γ_A will be greater or smaller than one, depending on whether the partial pressure of A is larger or smaller than it would be in equilibrium with an ideal solution of the same composition. The partial molar free energy of mixing is then equal to:

$$\bar{G}_A^M = RT \ln a_A = RT \ln N_A + RT \ln \gamma_A \tag{2.58}$$

The term $RT \ln N_A$ represents the value that $\bar{G}_A^M$ would have in an ideal hypothetical solution of the same composition at the same temperature and pressure, and it is the "ideal partial free energy of mixing, $\bar{G}_A^{id.}$. The term $RT \ln \gamma_A$ is the excess partial free energy $\bar{G}_A^{ex.}$:

$$\bar{G}_A^{ex.} = \bar{G}_A^M - \bar{G}_A^{id.} \tag{2.59}$$

$$\bar{G}_A^{ex.} = RT \ln \gamma_A \tag{2.60}$$

The other partial and integral excess quantities are defined by:

$$\bar{G}_A^{ex.} = \bar{H}_A^{ex.} - T\bar{S}_A^{ex.} \tag{2.61}$$

$$G^{ex.} = \Sigma N_i \bar{G}_i^{ex.} \tag{2.62}$$

The sign of $\bar{G}_i^{ex.}$ and $G^{ex.}$ depends on the value of γ_i; they are positive when the activity coefficients are greater than one, and negative when these coefficients are less than one. In Fig. 6, γ_A and γ_B are both greater than one, so that $\Delta G^{ex.}$ is positive.

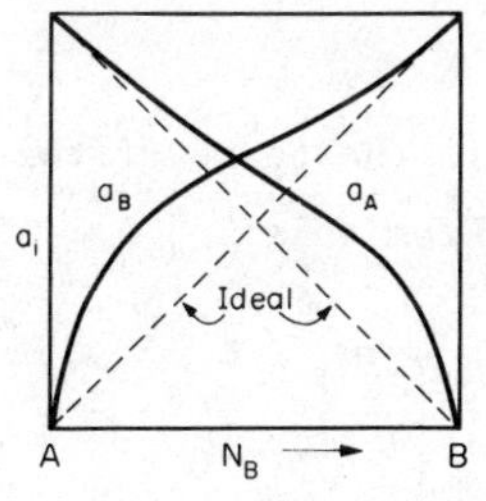

Fig. 6. Activities versus molar fractions in a real solution

Since the partial and integral ideal enthalpies and volumes of mixing are nil, we have:

$$\bar{V}_i^M = \bar{V}_i^{ex.} \tag{2.63}$$

and

$$V^M = V^{ex.} \tag{2.64}$$

Similarly:
$$\bar{H}_i^M = \bar{H}_i^{ex.} \tag{2.65}$$

and
$$H^M = H^{ex.} \tag{2.66}$$

2.4.3. GRAPHIC REPRESENTATION OF THE INTEGRAL FREE ENERGY OF MIXING OF A BINARY SYSTEM

The integral or partial free energy of mixing in real binary solutions can be separated, as has been seen before, into two terms:

$$G^M = G^{id.} + G^{ex.} \tag{2.67}$$

$$G^{id.} = \Sigma \bar{G}_i^{id.} N_i = RT\, \Sigma N_i \ln N_i \tag{2.68}$$

$$G^{ex.} = \Sigma \bar{G}_i^{ex.} N_i = RT\, \Sigma N_i \ln \gamma_i \tag{2.69}$$

From these and the second law of thermodynamics, we can deduce:

(1) The slope of the free energy curve as a function of the molar fractions, $G^M = f(N_A, N_B)$, is infinitely negative when the molar fraction of one of the components tends toward zero. From previous calculations:

$$G^M = RT\left[N_A \ln N_A + N_A \ln \gamma_A + N_B \ln N_B + N_B \ln \gamma_B\right] \tag{2.70}$$

If the free energy of mixing is differentiated with respect to one of the components, for example, N_A, then:

$$\frac{\partial G^M}{\partial N_A} = RT\left[\ln N_A + \ln \gamma_A - \ln N_B - \ln \gamma_B + \frac{N_A \partial \ln \gamma_A}{\partial N_A} + \frac{N_B \partial \ln \gamma_B}{\partial N_A}\right] \tag{2.71}$$

When N_A tends towards zero, all the terms of this equation have a finite value, except $\ln N_A$, which is $-\infty$. From this:

$$\left(\frac{\partial G^M}{\partial N_A}\right)_{N_A \to 0} \to -\infty \tag{2.72}$$

The plot of $G^M = f(N_A, N_B)$ is then asymptotic to the axis of the ordinates when N_A and N_B tend to zero.

(2) Of the two states, the more stable is that which has the more negative free energy. In Fig. 7, assume that the integral free energy of mixing is known at T over the whole range of composition.

If solutions α and β are mixed in such proportions that the final mixture has the composition N_A^γ, it can be seen from Fig. 7 that this must be homogeneous, because the integral free energy of mixing $G^M_{\gamma'}$, is more negative than G^M_{γ} for the heterogeneous mixture of the same composition at γ'.

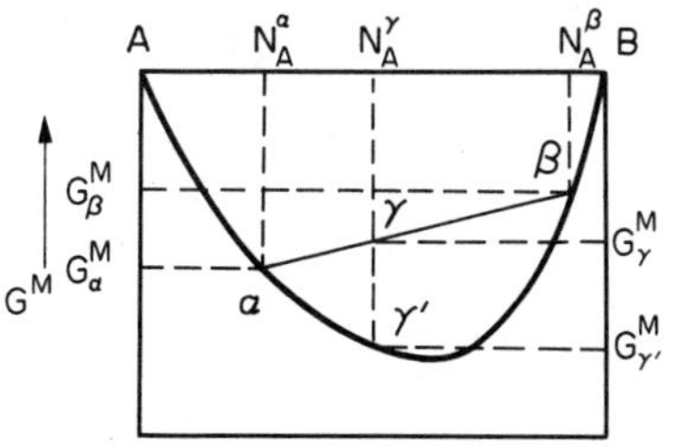

Fig. 7. Integral free energy of mixing versus composition

Consider a solid or liquid solution, and define the pure liquid as the standard state. If the solution AB is liquid over the whole range of composition at the given temperature, the integral free energy of mixing is zero when $N_A = 1$ and $N_B = 1$. In this case:

$$G^M = N_A\left(\bar{G}_A - G^o_A\right) + N_B\left(\bar{G}_B - G^o_B\right) \tag{2.73}$$

when $N_A = 1$, $N_B = 0$ and $\bar{G}_A = G^o_A$; G^M is then zero. On the contrary, if the solution is solid over the whole range of composition, when $N_A = 1$, $\bar{G}_{<A>} = G^o_{<A>}$, and the integral free energy of mixing is:

$$G^M_{N_A=1} = \bar{G}_{<A>} - G^o_{(A)} = G^o_{<A>} - G^o_{(A)} = -\Delta G_{f_A} \tag{2.74}$$

The free energy of melting of A, ΔG_{f_A}, is a function of temperature:

$$\Delta G_{f_A} = \Delta H_{f_A} \frac{T_{f_A} - T}{T_{f_A}} \tag{2.75}$$

At a temperature T, below the solidus line, the free energy of melting of A is positive, and therefore the integral free energy of mixing is negative at $N_A = 1$. For some other composition, the integral free energy of mixing of the solid will also be more negative than that of the liquid, as shown in Fig. 8; which is to be expected, since the solid is the stable form.

The change of standard state (solid instead of liquid) is obtained by displacing the diagrams $G^M_{liq.}$ and $G^M_{sol.}$ in such a way that the straight line $G^o_{<A>} - G^o_{<B>}$ joins the origins of the ordinates.

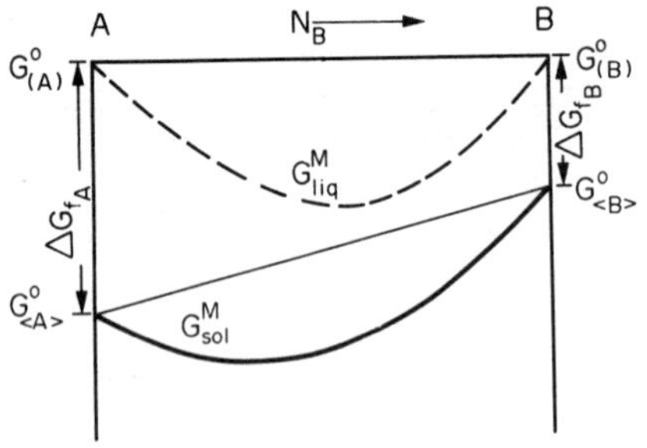

Fig. 8. G^M versus N_i. (The solid is more stable than the liquid)

(3) The G^M curve in Fig. 9 presents two minima and one maximum. In this case, between α and β, two solutions of composition N_A^{α} and N_A^{β} exist in equilibrium, because G^M for two solutions α and β (straight line $\alpha - \beta$) is more negative than G^M for one solution (broken line $\alpha - \beta$). The partial free energy of mixing of each component is the same in both phases, whatever the relative quantities of each. The values are obtained by extrapolation up to $N_A = 1$ and $N_B = 1$ of the tangent to the G^M curve at α and β.

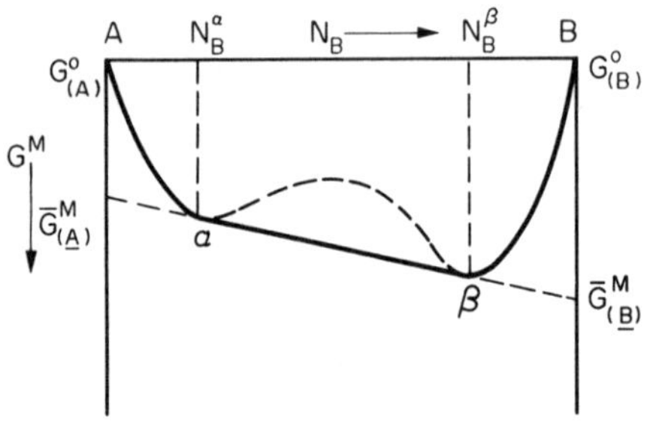

Fig. 9. G^M versus N_A and N_B for two phases in equilibrium

EXAMPLES

In the examples below, the general forms of curves $G^M = f(N_A, N_B)$ corresponding to different types of phase diagrams, will be deduced. The standard states are the pure liquids at the specified temperatures:

(i) Complete Miscibility in the Liquid and Solid

Figures 10a, b and c show how the integral free energy of a solution changes with N_A and N_B in a system of this type. In Figs. 10a', b' and c' the general forms of the corresponding activity curves are represented.

At temperature T_1 the solution is liquid over the whole range of composition. G^M is the sum of the ideal free energy of mixing $G^{id.}$ and the excess free energy $G^{ex.}$. In Fig. 10a, $G^{ex.}$ is positive, corresponding to γ_A and γ_B being in excess of one (Fig. 10a').

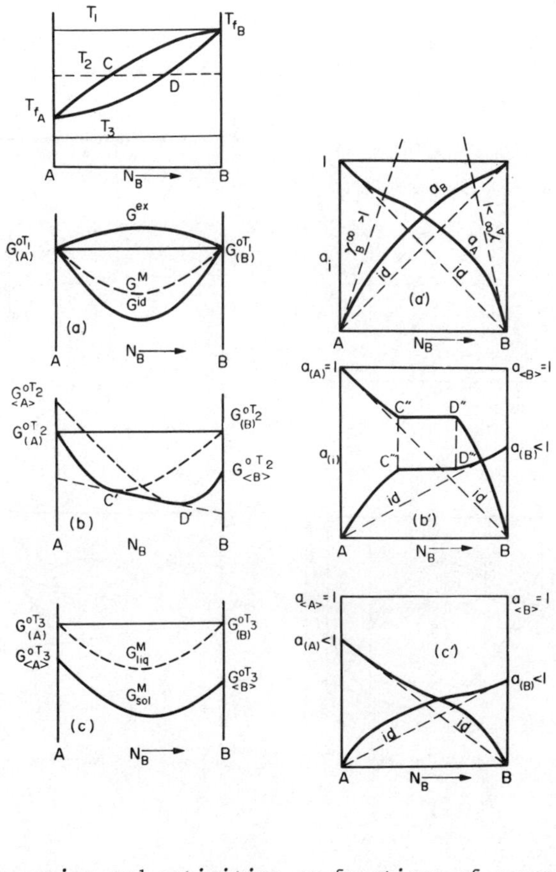

Fig. 10. Free energies and activities as functions of concentration in a complete solubility diagram in the solid and liquid states

At T_2, the solid-liquid equilibrium between C and D is represented by the tangent at C' and D' to the G^M curves for solid and liquid (Fig. 10b) and the horizontal portion of the curves of activities of Fig. 10b'. The partial free energy of mixing of pure solid B referred to the liquid standard state is negative (Fig. 10b). Since $\bar{G}_B^M$ is equal to RT ln a_B, the activity of pure solid B is then less than one, due to the standard state chosen (Fig. 10b').

At T_3, the system is solid at all concentrations of A and B. The free energy of mixing of the solid is more negative than that of the liquid. The activities of A and B, referred to the standard liquid state, change continuously, and are less than one over the whole range of composition (Fig. 10c').

(ii) Eutectic Phase Diagram

Figure 11 shows how the integral free energy of mixing varies as a function of the concentration of A and B when the solubility is total in the liquid state, and partial in the solid state.

The two tangents to the curves G^M_{solid} and G^M_{liquid} which exist when there are two phases in equilibrium at temperatures above that of the eutectic (Fig. 11b), turn into one unique tangent at the eutectic temperature where there are three phases in equilibrium (Fig. 11c), and below (Fig. 11d).

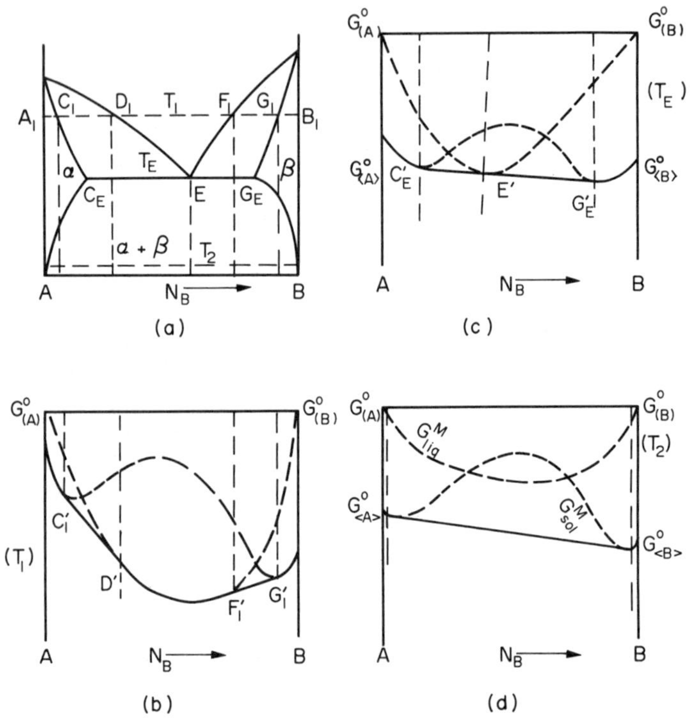

Fig. 11. Integral free energy of mixing as a function of mole fractions in an eutectic diagram

At T_2, the solubility of A in B and of B in A in solid solutions β and α are very small, and the curve G^M of the solid is a straight line joining $G^o_{<A>}$ and $G^o_{<B>}$. However, because the curve G^M is tangential to the vertical axes $N_A = 1$, and $N_B = 1$, the phases in equilibrium are A with a few atoms of B, and B with a few atoms of A. A diagram with absolutely zero solubility cannot exist.

(iii) Monotectic Diagram

The equilibrium between two liquids at a temperature above that of the monotectic is indicated by the straight line C'D' in the plot of $G^M(N_B)$ in Fig. 12b and the horizontal C"D" in the plots of $a_A(N_A)$ and $a_B(N_B)$ in Fig. 12c.

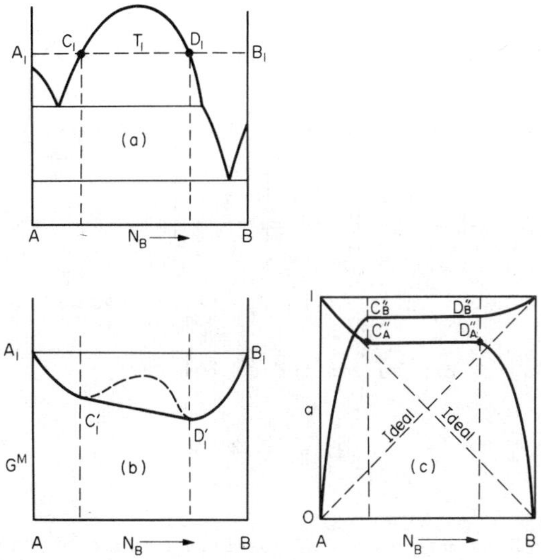

Fig. 12. Free energy of mixing and activities as functions of mole fractions in a monotectic diagram

2.4.4. REGULAR SOLUTIONS: HILDEBRAND'S MODEL

No interactions occur between the components of ideal solution. In regular solution, interactions between components are restricted to those between neighbouring atoms. The solution behaves as if completely random AB pairs are formed, without action on neighbouring pairs. This corresponds to zero values of the excess entropy and volume, as in an ideal solution, but with a value of enthalpy of mixing different from zero. This enthalpy corresponds to the enthalpy of formation of pairs, and may be positive or negative, depending on the type of union between the components. The fact that the excess entropy is zero has the following consequences:

(a) The entropy of mixing is the same as the ideal entropy:

$$S^M = S^{id.} = \Sigma N_i \bar{S}_i^M = - R \; \Sigma N_i \ln N_i \tag{2.76}$$

(b) From this relation and $H^M = G^M + T\,S^M$, it can be seen that:

$$H^M = RT \left[\Sigma N_i \ln a_i - \Sigma N_i \ln N_i \right] = RT \, \Sigma N_i \ln \gamma_i \tag{2.77}$$

and

$$\bar{H}_i^M = RT \ln \gamma_i \tag{2.78}$$

In a regular solution, as in any solution, the excess partial and integral enthalpies are equal to the partial and integral enthalpies of mixing, since the ideal enthalpies are nil. In addition, if (2.69) and (2.77) are compared, it can be seen that in a regular solution the enthalpies of mixing are also equal to the excess free energies:

$$\bar{H}_i^{ex.} = \bar{H}_i^M = \bar{G}^{ex.} \quad \text{and} \quad H^{ex.} = H^M = G^{ex.} \tag{2.79}$$

(c) The variation of excess free energy with temperature is nil:

$$\left[\frac{\partial \bar{G}_i^{ex.}}{\partial T} \right] = \left[\frac{\partial (RT \ln \gamma_i)}{\partial T} \right] = - \bar{S}_i^{ex.} = 0 \tag{2.80}$$

From this it can be deduced that $RT \ln \gamma_i$ is independent of temperature.

(d) A function alpha, α, can be defined as follows:

$$\alpha_i = \frac{\ln \gamma_i}{(1 - N_i)^2} \tag{2.81}$$

It can be demonstrated by use of the Gibbs-Duhem equation that this function α is independent of the composition in a regular solution, and has the same value for the two components of a binary solution. Taking into consideration relations (2.77) and (2.81), the integral enthalpy of mixing can be written:

$$H^M = RT(N_A \ln \gamma_A + N_B \ln \gamma_B) = RT \, \alpha \left[N_A(1 - N_A)^2 + N_B(1 - N_B)^2 \right] \tag{2.82}$$

and since $(1 - N_A) = N_B$ and $(1 - N_B) = N_A$:

$$H^M = RT \, \alpha N_A N_B = b N_A N_B \tag{2.83}$$

The term b, equal to $RT \, \alpha$ and therefore to $(RT \ln \gamma_i)/(1 - N_i)^2$, is independent of temperature, according to (2.80), and the composition of the solution, (since α does not change with the mole fraction). In a regular solution, the curve of the integral enthalpy of mixing (or of the excess integral enthalpy) is a parabola symmetrical with respect to $N_A = N_B = 0.5$. However, it is not sufficient to find a parabolic relation of this type in a binary system to conclude that it is a regular solution. In addition, the coefficient b must be independent of T.

2.5. DILUTE SOLUTIONS

2.5.1. HENRY'S LAW

In dilute solutions, it is found experimentally that, in general, the partial pressure of a solute i is proportional to its mole fraction:

$$p_i = k' \cdot N_{i(N_i \to 0)} \tag{2.84}$$

Since the ratio p_i/p_i^o is equal to the activity a_i, it can be written:

$$a_i = \frac{k'}{p_i^o} N_{i(N_i \to 0)} = KN_{i(N_i \to 0)} \tag{2.85}$$

This is the expression of Henry's Law which is represented in Fig. 13 by the straight portion of the activity plot when N_i tends towards zero.

2.5.2. SELECTION OF STANDARD STATE

For dilute solutions, the standard state may be either the pure species, liquid or solid, or the infinitely dilute solution. In the first case, the activity is equal to the mole fraction when this tends towards one (Raoult's Law):

$$a_i = N_{i(N_i \to 1)}$$

In the second instance, the standard state is based on Henry's Law, and the activity is equal to the mole fraction when this tends towards zero:

$$a_i = N_{i(N_i \to 0)}$$

Generally, the letter f is used to designate the activity coefficient when the infinitely dilute state is chosen as the standard state. Thus, the activity of component i of a solution is given by:

$$a_i^{\infty} = f_i N_i$$

when the standard state is the infinitely dilute solution (Henry's Law), and:

$$a_i = \gamma_i N_i$$

when the pure species is taken as the standard state (Raoult's Law).

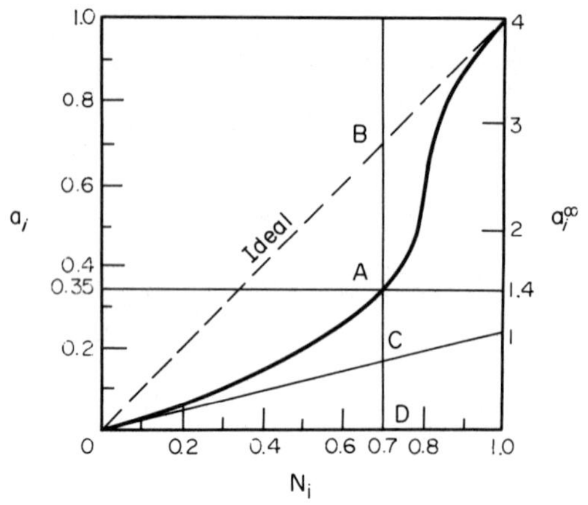

Fig. 13. Activity diagram showing the change of scale on changing the standard state.

(At point A, $N_i = 0.7$, $a_i = 0.35$, $a_i^\infty = 1.4$, $\gamma_i^\infty = \frac{CD}{OD} = 0.25$, $\gamma_i = \frac{DA}{DB} = 0.5$, $f_i = \frac{DA}{DC} = 2$)

At infinite dilution, f_i is unity, according to the above definition, and γ_i is equal to γ_i^∞. Therefore the following relation exists between the two coefficients:

$$\frac{\gamma_i}{\gamma_i^\infty} = f_i \tag{2.86}$$

If, for a given mole fraction N_i, the activity a_i of component i referred to the pure species is compared with the activity a_i^∞ referred to an infinitely dilute solution, we obtain:

$$\frac{a_i}{a_i^\infty} = \frac{\gamma_i N_i}{f_i N_i} = \gamma_i^\infty \tag{2.87}$$

The partial free energy of mixing of component i in solution will be different, depending on the standard state selected. Taking the pure species as the standard state:

$$\bar{G}_i^M = \bar{G}_i - G_i^o = RT\ (\ln \gamma_i + \ln N_i) \tag{2.88}$$

When the infinitely dilute solution is the standard state:

$$\bar{G}^M_{i\infty} = \bar{G}_i - G^o_{i\infty} = RT\ (\ln f_i + \ln N_i) \tag{2.89}$$

The difference between $\bar{G}_i^M$ and $\bar{G}_{i\infty}^M$ is therefore equal to:

$$\bar{G}_i^M - \bar{G}_{i\infty}^M = G_{i\infty}^o - G_i^o = RT \ln \frac{\gamma_i}{f_i} = RT \ln \gamma_i^\infty \tag{2.90}$$

EXAMPLE

The activity coefficient of a component i in an infinitely dilute solution is 0.25 ($\gamma_i^\infty = 0.25$). When the mole fraction of i is 0.7 ($N_i = 0.7$), its activity referred to the pure substance is 0.35. From this data, the following can be deduced (Fig. 13):

The activity coefficient referred to the pure species:

$$\gamma_i = \frac{a_i}{N_i} = 0.5$$

The activity coefficient referred to the infinitely dilute solution:

$$f_i = \frac{\gamma_i}{\gamma_i^\infty} = 2$$

The activity of i referred to the infinitely dilute solution:

$$a_i^\infty = f_i N_i = 1.4$$

The partial free energy of mixing of i referred to the pure species:

$$\bar{G}_i^M = RT \ln 0.35$$

The partial free energy of mixing of i referred to the infinitely dilute solution:

$$\bar{G}_{i\infty}^M = RT \ln 1.4$$

The difference in free energy due to the change of standard state:

$$\Delta G_i^o = \bar{G}_i^M - \bar{G}_{i\infty}^M = RT \ln 0.25$$

Since γ_i^∞ is less than one, the partial free energy of mixing referred to the infinitely dilute state is greater than the other.

In metallurgy, and particularly in iron and steel making, it is frequently most convenient to choose as the standard state the infinitely dilute solution expressed in weight %, instead of as a mole fraction. This is defined as follows:

$$\frac{a_i}{(\%\ i)} = 1 \quad (\text{when } (\%\ i) \to 0) \tag{2.91}$$

For weight (% i) different from zero:

$$a_i = f_i(\%\ i) \tag{2.92}$$

By comparing the activity of i referred to the pure species a_i, with the activity of i referred to the infinitely dilute solution expressed in weight %, a_i^∞, at the same composition, one obtains:

$$\frac{a_i}{a_i^\infty} = \frac{\gamma_i N_i}{f_i(\%\ i)} = \gamma_i^\infty \frac{N_i}{(\%\ i)} \tag{2.93}$$

For very low concentrations of i in a solvent M, mole fraction and weight % are related by:

$$\frac{N_i}{(\%\ i)} = \frac{A_M}{100\ A_i} \tag{2.94}$$

In this formula, A_M and A_i are the atomic weights of M and i. Substituting this value in (2.93), gives:

$$\frac{a_i}{a_i^\infty} = \gamma_i^\infty \frac{A_M}{100\ A_i} \tag{2.95}$$

The difference in partial free energy of mixing due to the change in standard state is:

$$G^o_{i^\infty} - G^o_i = RT \ln \gamma_i^\infty \frac{A_M}{100\ A_i} \tag{2.96}$$

2.5.3. SIEVERT'S LAW

Diatomic gases (N_2, O_2, H_2) dissolve in liquid metals according to:

$$H_2 \rightleftarrows 2(\underline{H}) \tag{2.97}$$

The constant K for this reaction is equal to:

$$K = a^2_{(\underline{H})}/p_{H_2} \tag{2.98}$$

In general, the atomic fraction of a gas dissolved in a metal is small, and its activity is proportional to its atomic fraction (Henry's Law). (2.98) may then be written:

$$N_{(H)} = K\sqrt{p_{H_2}} \tag{2.99}$$

The atomic fraction of a diatomic gas in a metallic solution is therefore proportional to the square root of the partial pressure of that gas in equilibrium with the solution. This result is Sievert's Law. For example, for oxygen, this law holds until saturation, which takes place when an oxide separates from the metal.

2.5.4. DILUTE SOLUTIONS WITH VARIOUS COMPONENTS

In industrial practice, it is not usual to obtain the pure metal, but an alloy which contains impurities or alloying elements in small amounts. The thermodynamic behaviour of one solute may be changed by the presence of others. Thus, for the solute 1:

$$\ln \gamma_1 = f(N_1, N_2, N_3, \ldots, N_1) \tag{2.100}$$

Developing this function as a Taylor series, for small concentrations of all solutes, one obtains:

$$\ln \gamma_1 = \ln \gamma_1^{\infty} + N_i \frac{\partial \ln \gamma_1}{\partial N_1} + N_2 \frac{\partial \ln \gamma_1}{\partial N_2} + \ldots + \tfrac{1}{2}\left[N_1^2 \frac{\partial^2 \ln \gamma_1}{\partial N_1^2} + N_1 N_2 \frac{\partial^2 \ln \gamma_1}{\partial N_1 N_2} + \ldots \right] \tag{2.101}$$

Since N_1, N_2, N_3, etc., are small, the second and higher order terms can be neglected. The differentials of $\ln \gamma_1$: $(\partial \ln \gamma_1/\partial N_1)$, $(\partial \ln \gamma_1/\partial N_2)$, etc., have constant values up to concentrations of some per cent. In this form $\ln \gamma_1$ is a linear function of the mole fractions of the various solutes present in the solution:

$$\ln \gamma_1 = \ln \gamma_1^{\infty} + N_1\varepsilon_1^1 + N_2\varepsilon_1^2 + N_3\varepsilon_1^3 + \ldots + N_i\varepsilon_1^i \tag{2.102}$$

The coefficients ε_i^j are the interaction or Wagner parameters, and are defined by:

$$\varepsilon_i^j = \left(\frac{\partial \ln \gamma_i}{\partial N_j} \right)_{N_j \to 0} \tag{2.103}$$

It can be shown that $\varepsilon_i^j = \varepsilon_j^i$ by utilising the relationships which exist between $\ln \gamma_i$, $\bar{G}_i^{ex.}$ and $G^{ex.}$:

$$\bar{G}_i^{ex.} = RT \ln \gamma_i \tag{2.104}$$

and, by definition:

$$\bar{G}_i^{ex.} = \frac{\partial G^{ex.}}{\partial N_i} \tag{2.105}$$

By simultaneous differentiation of $G^{ex.}$ with respect to N_i and N_j, we have:

$$\frac{\partial^2 G^{ex.}}{\partial N_i \partial N_j} = \frac{\partial \bar{G}_i^{ex.}}{\partial N_j} = \frac{\partial \bar{G}_j^{ex.}}{\partial N_i} \tag{2.106}$$

and

$$\frac{\partial \ln \gamma_i}{\partial N_j} = \frac{\partial \ln \gamma_j}{\partial N_i} \tag{2.107}$$

or

$$\varepsilon_i^j = \varepsilon_j^i \tag{2.108}$$

If the infinitely dilute solution is chosen as the standard state, the term $\ln \gamma_i^{\infty}$ disappears, and (2.102) can be written:

$$\ln f_i = N_1 \varepsilon_i^1 + N_2 \varepsilon_i^2 + \ldots + N_j \varepsilon_i^j \tag{2.109}$$

Using concentrations expressed in weight %, and decimal logarithms, the interaction parameter e_i^j is defined by:

$$e_i^j = \frac{\partial \log f_i}{\partial (\% \; j)} \tag{2.110}$$

The activity coefficient of a component 1 is then expressed as a function of these parameters and the weights % of solutes:

$$\log f_i = e_1^1 \; (\% \; 1) + e_1^2 \; (\% \; 2) + \ldots + e_1^i \; (\% \; i) \tag{2.111}$$

The relationship between the interaction parameters of j upon i and of i upon j are, in this case:

$$e_i^j = \frac{A_i}{A_j} e_j^i \tag{2.112}$$

The relationship between e_i^j and ε_i^j is, if Fe is the solvent:

$$e_i^j = \frac{56}{2.3 \; A_j \; 10^2} \varepsilon_i^j = \frac{0.2425}{A_j} \varepsilon_i^j \tag{2.112a}$$

A_i and A_j are the atomic weights of the solutes i and j respectively.

When 1 is the only solute, (2.102) is simplified to:

$$\ln \gamma_1 = \ln \gamma_1^{\infty} + N_1 \varepsilon_1^1 \tag{2.113}$$

The activity coefficient γ_1 varies with the molar fraction N_1 of solute, and Henry's Law does not hold even in dilute solutions if ε_1^1 is not zero.

The values of the interaction parameters ε_i^j and e_i^j for various solutes in liquid iron are given in Tables 20 and 21 (Chapter 7).

Another form of expressing the interaction parameters is by plotting the values of the logarithms of the activity coefficients versus the weight per cent of these solutes (see *Examples* in Chapter 7, Figs. 113, 114, 123, etc.). The slopes of the curves for $(\%\ j) = 0$ are the interaction parameters

$$e_i^j = \left(\frac{\partial \log f_i}{\partial (\%\ j)}\right)_{(\%\ j)=0}$$

2.5.5. DEPARTURE FROM RAOULT'S LAW IN DILUTE SOLUTIONS

It has been seen earlier that a binary solution $M - i$ does not follow Henry's Law when the interaction parameter ε_i^i of a solute on itself is not zero. By integrating the Gibbs-Duhem equation, it can be proved that when the solute i does not follow Henry's Law, the solvent does not follow Raoult's Law. For a simple binary system, $M - i$, the Gibbs-Duhem relation is written:

$$N_i \ d \ln a_i + N_M \ d \ln a_M = 0 \qquad (2.114)$$

or

$$N_i(d \ln N_i + d \ln \gamma_i) + N_M(d \ln N_M + d \ln \gamma_M) = 0 \qquad (2.115)$$

Because $d \ln N_i = dN_i/N_i$, $d \ln N_M = dN_M/N_M$ and $(dN_i + dN_M) = 0$, (2.115) can be simplified, and the Gibbs-Duhem equation may be expressed as:

$$N_i \ d \ln \gamma_i + N_M \ d \ln \gamma_M = 0 \qquad (2.116)$$

Using the value of $\ln \gamma_i$ given by (2.113) for a binary solution containing only one solute, we find:

$$d \ln \gamma_i = \varepsilon_i^i \ dN_i \qquad (2.117)$$

On the other hand, if $\ln \gamma_M$ is assumed to be a linear function of N_M when the values of N_M are near to one (see Fig. 14), then:

$$\frac{d \ln \gamma_M}{dN_M} = -\frac{\ln \gamma_M}{1 - N_M} = \textit{constant} \qquad (2.118)$$

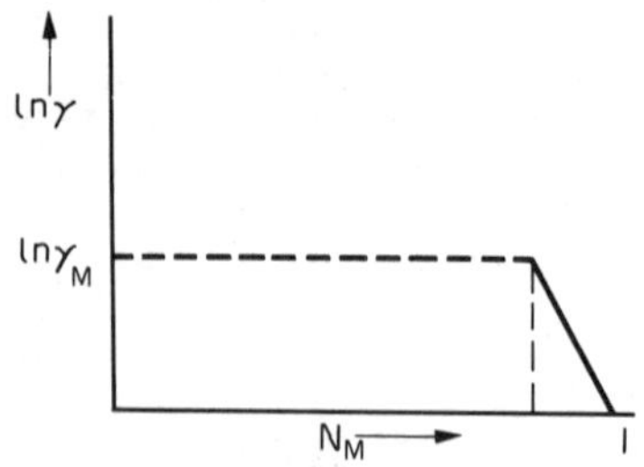

Fig. 14. ln γ_M versus N_M for the values of N_M near to 1

Replacing d ln γ_i and d ln γ_M, in (2.116), and integrating between $N_M = 1$ and N_M, $N_i = 0$ and N_i, this becomes:

$$\varepsilon_i^i \int_0^{N_i} N_i dN_i = \frac{\ln \gamma_M}{1 - N_M} \int_1^{N_M} N_M dN_M \tag{2.119}$$

or

$$\varepsilon_i^i N_i^2 = \frac{N_M^2 - 1}{1 - N_M} \ln \gamma_M = -(1 + N_M) \ln \gamma_M \tag{2.120}$$

and therefore

$$\ln \gamma_M = -\frac{1}{1 + N_M} \varepsilon_i^i N_i^2 \simeq -\tfrac{1}{2} \varepsilon_i^i N_i^2 \tag{2.121}$$

When the interaction parameter ε_i^i is different from zero, ln γ_M is zero only for $N_i = 0$, and Raoult's Law does not hold. However, it must be noted that the approach of ln γ_M to zero is second-order with respect to N_i, while the approach of ln γ_i to ln γ_i^∞ is first-order.

It can be shown, for a solution with more than one solute, that the activity coefficient of the solvent is a function of the mole fraction N_1, N_2, ..., N_i, N_j, ..., N_n of all the solutes:

$$\ln \gamma_M = -\left[\frac{1}{1 + N_M}\right] \sum_{i,j}^{n} \varepsilon_i^j N_i N_j \tag{2.122}$$

2.5.6. LOWERING OF THE MELTING POINT OF A SUBSTANCE BY ADDITION OF A SOLUTE

Consider a simple binary system M - i (Fig. 15) at a temperature, T, lower than the melting temperature T_f of M. From the liquidus to the solidus curves there is equilibrium between liquid and solid:

$$\langle \underline{M} \rangle \rightleftarrows (\underline{M}) \tag{2.123}$$

This equilibrium can be separated into three independent equilibria:

$$\langle \underline{M} \rangle \rightleftarrows \langle M \rangle \qquad \Delta G = -\bar{G}^{M}_{\langle M \rangle} = -RT \ln a_{\langle M \rangle} \tag{2.124}$$

$$\langle M \rangle \rightleftarrows (M) \qquad \Delta G_{f_M} = \Delta H_{f_M} \frac{T_f - T}{T_f} \tag{2.125}$$

$$(M) \rightleftarrows (\underline{M}) \qquad \Delta G = \bar{G}^{M}_{(M)} = RT \ln a_{(M)} \tag{2.126}$$

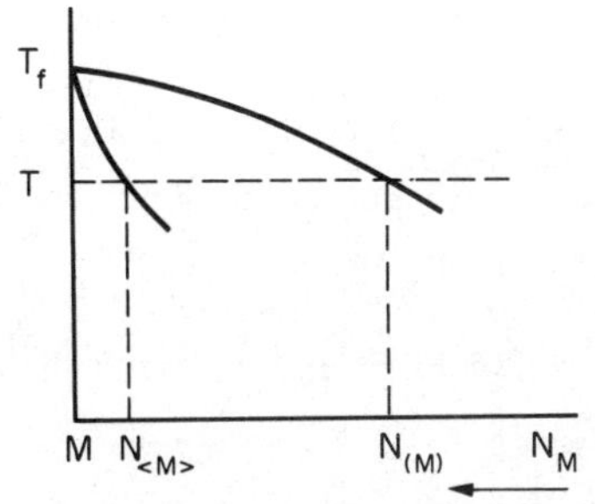

Fig. 15. Section of the phase diagram M - i near to $N_M = 1$

The sum of the free energies of these reactions is zero, since the free energy change of (2.123) is nil. Therefore:

$$\ln \frac{a_{(M)}}{a_{\langle M \rangle}} = -\frac{\Delta H_f (T_f - T)}{RT \cdot T_f} \tag{2.127}$$

Consider the cooling of a liquid solution M - i rich in M when there is no solubility of i in solid M; then the activity of the solvent M in the solid is unity. Because the liquid is a dilute solution, lowering of the melting point ($\Delta T = T_f - T$) is small, and the temperature T, when crystallisation of M starts, is close to the crystallisation temperature of the pure solvent $M (T \simeq T_f)$. (2.127) can then be written:

$$\Delta T = -\frac{RT_f^{\,2}}{\Delta H_f} \ln a_{(M)} = -\frac{RT_f^{\,2}}{\Delta H_f} \left[\ln N_{(M)} + \ln \gamma_{(M)} \right] \tag{2.128}$$

In a binary solution, $\ln \gamma_M$ is given by (2.121). The term $\ln N_{(M)}$ is equal to $\ln(1 - N_i)$, and when the solute fraction is small, the value of $\ln(1 - N_i)$ is near to $-N_i$. Replacing $\ln \gamma_M$ and $\ln N_M$ by their values, (2.128) becomes:

$$\Delta T = \frac{RT_f^{\,2}}{\Delta H_f} \left[N_i + \tfrac{1}{2} \varepsilon_i^i N_i^{\,2} \right] \tag{2.129}$$

When Henry's Law is followed in dilute solutions, i.e. when ε_i^i is zero, the lowering of the melting point is proportional to the mole fraction of the solute.

This is Raoult's Cryoscopic Law.

The proportionality constant $RT_f^2/\Delta H_f$ depends only on the properties of the solvent. By differentiating (2.129) with respect to N_i, we have:

$$\frac{d\Delta T}{dN_i} = \frac{RT_f^2}{\Delta H_f}\left[1 + \varepsilon_i^i N_i\right] \tag{2.130}$$

$RT_f^2/\Delta H_f$ represents, then, the slope of the tangent to the liquidus at the melting point of M. A value of 2.3 T_f has been established by Trouton for the enthalpy of melting of the metals. The slope of the tangent to the liquidus at $N_i = 0$ and $T = T_f$ is then proportional to the melting point of pure M. The lowering of the melting point by n solutes or impurities can be deduced from (2.122) and (2.128).

$$\Delta T = -\frac{RT_f^2}{\Delta H_f}\left(\ln N_M - \frac{1}{1 + N_M}\sum_{i,j}^{n} \varepsilon_i^j N_i N_j\right) \tag{2.131}$$

As previously, in very dilute solutions $\ln N_M$ may be replaced by:

$$-\sum_{i=1}^{n} N_i$$

2.6. PROPERTIES OF BINARY SOLUTIONS

Real Solutions	*Ideal Solutions*
$G^M = H^M - TS^M$	$H^M = 0, \quad V^M = 0$
$G^M = G^{id.} + G^{ex.}$	$S^{ex.} = 0, \quad \bar{S}_i^{ex.} = 0$
$X = \Sigma N_i \bar{X}_i$	$\bar{S}^{id.} = -R \ln N_i$
$X = (G, H, S, \ldots)^{M,id.,ex.}$	$S^{id.} = -R\,\Sigma N_i \ln N_i$
$\bar{H}_i^{id.} = 0$	$G^M = G^{id.} = -TS^{id.}$
$\bar{H}_i^M = \bar{H}_i^{ex.}$	$\bar{G}_i^M = \bar{G}_i^{id.} = RT \ln N_i$
$\bar{S}_i^{id.} = -R \ln N_i$	$G^{ex.} = 0, \quad \bar{G}_i^{ex.} = 0$
$\bar{G}_i^{id.} = -T\bar{S}_i^{id.} = RT \ln N_i$	$a_i = N_i$
$\bar{G}_i^M = RT \ln a_i$	$\gamma_i = 1$
$\bar{G}_i^{ex.} = RT \ln \gamma_i$	
$a_i = \gamma_i N_i$	

Regular Solutions	*Dilute Solutions*
$\bar{S}_i^{ex.} = 0$	Standard state: pure body
$\bar{S}_i^M = \bar{S}_i^{id.} = -R \ln N_i$	$a_i = \gamma_i N_i$
$\bar{G}_i^{ex.} = \bar{H}_i^{ex.} = \bar{H}_i^M$	$a_i = N_i$ when $N_i \to 1$ (Raoult's Law)
$\bar{G}_i^{ex.} = RT \ln \gamma_i = b(1 - N_i)^2$	$\gamma_i \to \gamma^\infty$ when $N_i \to 0$ (Henry's Law)
	$\ln \gamma_i = \ln \gamma_i^\infty + \sum_{j=1}^{n} \varepsilon_i^j N_j$
$G^{ex.} = H^M = bN_i(1 - N_i)$	Standard state: dilute solution
b independent of N_i and T	$a_i = f_i N_i$
$b = \dfrac{RT \ln \gamma_A}{(1 - N_A)^2} = \dfrac{RT \ln \gamma_B}{(1 - N_B)^2}$	$a_i = N_i$ when $N_i \to 0$
	$f_i = \gamma_i / \gamma_i^\infty$
	$f_i \to 1$ when $N_i \to 0$
	$\ln f_i = \sum_{j=1}^{n} \varepsilon_i^j N_j$

GENERAL READING

Darken, L., and R. Gurry, *Physical Chemistry of Metals*, McGraw-Hill, N.Y. (1953).

Kubaschewski, O., E.Ll. Evans and C.B. Alcock, *Metallurgical Thermochemistry*, 5th Ed., Pergamon Press, London (1979).

Kunii, D and Levenspill, *Fluidisation Engineering*, John Wiley and Sons Inc. New York (1969).

Mackowiak, J., *Physical Chemistry for Metallurgists*, Allen and Unwin, London (1965).

Prigogine, I., and R. Defay, *Chemical Thermodynamics*, Vol. 1, Longmans Green, London (1952).

Richardson, F.D., *Physical Chemistry of Melts in Metallurgy*, Academic Press, London (1974).

Souchay, P., *Chimie Physique: Thermodynamique Chimie*, 3rd Ed., Masson et Cie., Paris (1968).

Wagner, C., *Thermodynamics of Alloys*, Addison-Wesley, London (1952).

Chapter 3

EXPERIMENTAL AND BIBLIOGRAPHICAL METHODS

INTRODUCTION

Knowledge of the values of the thermodynamic functions is of fundamental importance in extractive metallurgy; this makes it possible to find the best conditions of operation and to determine the highest efficiencies which may be obtained in a metallurgical process. The values which are found in specialized literature are far from being definitive and systematic, and they are being improved as experience is gained. The objective of this chapter is to describe laboratory methods and calculations used to determine thermodynamic functions, and also to indicate where to find and how to use the data available in the literature.

3.1. EXPERIMENTAL DETERMINATION OF THERMODYNAMIC DATA

Laboratory research has as its goal in this domain the determination of two principal types of information; the standard values of thermodynamic functions, and the activities of components in solutions. The methods used in both instances are very similar and they will be studied together. The principal categories of the methods are:

Heterogeneous equilibria.
Electrochemical methods.
Calorimetric methods.

3.1.1. HETEROGENEOUS EQUILIBRIA

The methods used for determination of the activity of a component in solution may serve in some instances for determination of the free energy of a reaction.

A. DETERMINATION OF THE ACTIVITY OF A COMPONENT IN SOLUTION

These methods consist basically of the establishment of equilibrium between a phase α containing component C at unknown activity, with another β, solid, liquid, or gaseous, in which the activity of C has previously been determined. At equilibrium, the activity of C is the same in both phases, and it is then only necessary to determine the quantity of C in α to know its activity as a function of the mole fraction.

Two types of method can be considered: in the first, the component C is distributed between various phases without reacting with any of them. These are based on the measurement of vapour pressure, or the partition of a component between two liquid phases. In the second, equilibrium is established between a phase which contains component C and another in which C is present as a compound, after having reacted with a second component. These are the metal-slag-atmosphere equilibria.

(A) *EQUILIBRIUM WITHOUT REACTION BETWEEN PHASES*

(i) Vapour Pressure Measurements

This method consists of comparing the vapour pressure p_C^T of C in equilibrium with an alloy, with the vapour pressure p_C^{oT} in equilibrium with pure C at the same temperature. By definition:

$$a_C^T = \frac{p_C^T}{p_C^{oT}} \tag{3.1}$$

This method applies with the following constrictions:

One of the components of the solution must have a vapour pressure which can be measured at the temperature of the experiment, and for the concentrations considered. The pressure limits depend on the method of measurement used, and they range from 10^{-12} to 10^2 mm Hg.

The partial pressures of all of the components of complex solutions can be measured simultaneously by mass spectrometry. Otherwise, separate measurement is only possible when the partial pressure of the selected compound is very large compared to those of any others present.

The state of aggregation of the components of the gaseous phase must be known; that is, the molecular weight of the vapour. For example, in the equilibrium:

$$n(\underline{S}) \rightleftarrows S_n$$

depending on temperature and pressure, the sulphur vapour is in the form S, S_2, S_4, ..., or a mixture of two or more forms, which may introduce a large error in the measurement of activities. When in doubt, the molecular weights are previously determined by mass spectrometry.

Vapour pressure is measured by static and transport methods as well as those based on evaporation.

Static methods. These consist of establishing equilibrium between the vapour of component C in a solution B-C at temperature T_s, and pure C at temperature T_p, lower than T_s. If $p_C^{oT_p}$ is the vapour pressure of pure C at T_p, and $p_{\underline{C}}^{T_s}$ is the vapour pressure of C in solution B-C at T_s, when these pressures are equal, i.e.:

$$p_C^{o\,T_p} = p_{\underline{C}}^{T_s} \tag{3.2}$$

the activity of C in solution will be:

$$a_{\underline{C}} = \frac{p_{\underline{C}}^{T_s}}{p_C^{o\,T_s}} = \frac{p_C^{o\,T_p}}{p_C^{o\,T_s}} \tag{3.3}$$

It is then enough to know the variation of p_C^o with temperature to determine the activity of C in solution.

In the "dew point method", the apparatus consists of a closed vessel in which a crucible, which contains the alloy B-C at the temperature T_s, is placed. A temperature gradient is established over the crucible, and the temperature T_p at which the first drop of volatile component C condenses, is read. The activity of C in the solution is derived then from (3.3). This method is applicable only to solutions which contain a very volatile component (Mg, Zn, Hg, Ca, S, etc.).

In the balance method, two crucibles are joined by a sealed tube, and balanced on the axis of a scale, the tube serving as one of the arms. One crucible contains the alloy B-C at temperature T_s, and the other the pure component C, at temperature T_p. If the vapour pressure of C is greater in one crucible, vapour will condense in the other, until equilibrium is reached, i.e. when the vapour pressures are the same in both crucibles the scale will remain in a fixed position. The activity is calculated as in the former instance.

Transport method. In this method, the vapour of C over the alloy B-C is transported by an inert carrier gas i which, on passing over (or through the solution), is saturated with C. The partial pressure of C in the gaseous phase is equal to:

$$p_C = P \frac{n_C}{(n_C + n_i)} \tag{3.4}$$

where P is the total pressure, n_C and n_i are the numbers of moles of C and inert gas i in the gaseous phase respectively. The number of moles of inert gas is measured with a suitable instrument, and that of C is determined either by measuring the loss of weight of the sample, or by condensing the vapour of C outside the vessel. The activity of C in solution is given by (3.1). This method only applies if the carrier gas is saturated with C which must be ensured by a separate previous experiment.

Methods based on the evaporation laws. Langmuir's formula allows for calculation of the evaporation rate of a component C, per unit surface and time, according to:

$$-\frac{dn_C}{dt} = \left(\frac{2\pi RT}{M}\right)^{-\frac{1}{2}} \cdot p_C \ (\text{mole cm}^{-2}\ \text{sec}^{-1}) \tag{3.5}$$

In this expression, n_C is the number of moles of C which evaporate in time t, p_C the vapour pressure of C at temperature T, and M the molecular weight of the component C in the gaseous phase. These methods require an advanced and very precise technology.

In the Langmuir method, the weight loss of a filament made of the alloy B-C heated in vacuum, is measured at temperature T. The vapour pressure of the volatile component C is proportional to the weight loss of the filament, per unit time and surface.

In Knudsen's method, a cell of an inert metal such as platinum or tantalum is used to contain the alloy. The cell has a calibrated hole in the upper part. On heating the cell to temperature T in vacuum, vapour of the volatile species C of the solution escapes through the orifice. Originally, this vapour condensed on a plate. The pressure of C was proportional to the mass deposited on the plate, per unit time, and inversely proportional to the cross-section of the hole.

In the modern version, the cell is coupled to a mass spectrometer. The vapour above the cell is ionised by electron bombardment. Ions of different species are separated according to the mass/charge ratio by a magnetic or high frequency field. The currents generated by the discharge of these ions on their respective plates are proportional to the vapour pressures of the constituents. This advanced method makes it possible to determine the activities of several elements in an alloy simultaneously.[17-19]

In the torsion-effusion method, a small cylinder suspended by means of a torsion wire is placed horizontally in a furnace. The cylinder has two calibrated holes, placed symmetrically and opposite each other on the sides with respect to the centre. The B-C alloy is placed in the cylinder, which is heated in vacuum. The vapour of the volatile species C escapes through the holes, producing a torsional moment, which causes the cylinder to turn a certain angle. The vapour pressure of C in solution is proportional to this angle.

(ii) Partition Method

When a component C is dissolved in two non-miscible components, A and B, it separates between the two phases $\alpha(A + C)$ and $\beta(B + C)$. In equilibrium, the activity of C is the same in the two phases:

$$a_C^{\alpha} = a_C^{\beta} \qquad \text{or} \qquad \gamma_C^{\alpha} N_C^{\alpha} = \gamma_C^{\beta} N_C^{\beta} \tag{3.6}$$

Knowing the value of the activity of C in one phase, it is possible to determine its activity in the other, for small concentrations of C in A and B. At greater concentrations of C, the solubilities of A in B and B in A increase, as may be seen in Fig. 16, where the equilibrium lines α-β between phases are shown. Beyond a certain concentration of C, the solutions α and β must be considered as forming a ternary equilibrium and, at the limit, there is only one liquid phase.

The partition coefficient K is defined as the ratio between the mole fractions of C in the two phases in equilibrium:

$$K = \frac{N_C^{\alpha}}{N_C^{\beta}} \tag{3.7}$$

From (3.6) we can see that $N_C^{\alpha}/N_C^{\beta} = \gamma_C^{\beta}/\gamma_C^{\alpha}$, and:

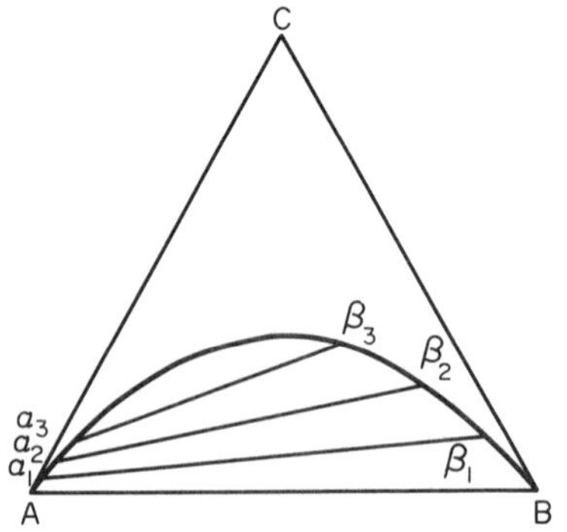

Fig. 16. Ternary phase diagram with a miscibility gap

$$K = \frac{\gamma_C^{\beta}}{\gamma_C^{\alpha}} \tag{3.8}$$

The activity coefficients γ_C^{β} and γ_C^{α}, and therefore the partition coefficient K, are functions of the temperature and the concentrations of C in the two phases, and must be determined for each one of them. When the component C follows Henry's Law in both phases, the activity coefficients γ_C^{α} and γ_C^{β} are constants, and the partition coefficient K is not a function of the concentrations, but only of the temperature:

$$K_T = \frac{\overset{\infty}{\gamma}{}_C^{\beta}}{\overset{\infty}{\gamma}{}_C^{\alpha}} \tag{3.9}$$

The partition method for the determination of activities applies only to systems of non-miscible or only slightly miscible components such as Fe-Ag, Fe-Pb, Al-Pb, Zn-Pb, etc.

(B) EQUILIBRIUM OF TWO OR MORE PHASES WITH REACTION BETWEEN THE COMPONENTS

These equilibria are those found most often in extractive metallurgy, particularly in iron and steel making where, generally, three phases (and occasionally two) are in equilibrium:

A metallic phase containing M.

An oxide phase (slag) where M is present as MX_2 or MO.

A gaseous phase, containing a component which reacts with M.

Consider the equilibrium between metal, slag and gas:

$$2(\underline{M})_m + [O_2] \rightleftarrows 2(\underline{MO})_s \tag{3.10}$$

The standard free energy ΔG^o is related to the equilibrium constant K_e by:

$$\Delta G^o = - RT \ln K_e = - RT \ln \frac{a^2{}_{(MO)_s}}{p_{O_2} \times a^2{}_{(\underline{M})_m}} \tag{3.11}$$

To determine the activity of M in the metallic phase, it is necessary to know:

The standard free energy of (3.10).

The oxygen pressure in the gas phase.

The activity of MO in the slag.

Slag is not necessarily formed when a metallic phase is in the presence of a reactive gas. For very low partial pressures, the gas dissolves, and it is not until it reaches saturation that the slag appears. Up to saturation, the activity of the gas in the metal increases with its partial pressure in the gas phase (Henry's Law or Sievert's Law for diatomic gases). Above saturation, the activity of the gas in the metal does not change as long as the slag phase formed contains a single compound.

Metal-atmosphere and metal-slag-atmosphere equilibria of interest in metallurgy are considered below.

(i) Metal-Atmosphere Equilibrium

This type of equilibrium makes it possible to measure the activity of a non-reactive gas (H_2, N_2, etc.) and of an element which has a volatile compound in solution in a metal. Table 1 shows several elements whose activity in solution can be measured by metal-gas equilibrium.

For example, the activity of carbon in liquid metal in equilibrium with a known mixture of CO and CO_2 may be calculated from:

$$\Delta G_T^o = - RT \log \frac{p_{CO}^2}{a_{\underline{C}} \cdot p_{CO_2}} \tag{3.12}$$

The activity of carbon is the only unknown in this equation.

Gaseous mixtures are obtained by simple mixing, by saturation of a gas with a volatile solid or liquid (H_2, H_2O), or by reaction of a gas with a solid (CO_2-CO with carbon: Boudouard's equilibrium). The mixture is passed slowly over a crucible containing the metal at the desired temperature. Equilibrium is reached when there is no change with time of the compositions of the liquid and of the gas.

(ii) Metal-Slag-Atmosphere Equilibrium

These equilibria make it possible to determine the activity of an element M in the metallic phase, as well as that of its compound MO in the slag. For example, the activity of silicon in steel can be determined from the equilibrium between the metal, slag containing SiO_2, and a mixture of H_2O and H_2:

TABLE 1 Elements whose Activities can be Determined by means of the Metal-Atmosphere Equilibrium

Element	Mixture of Gases used	Reaction considered	Standard Free Energy of Reaction
H_2	-	Direct Measurement	-
C	$CO - CO_2$	$[CO_2] + <C> \rightleftarrows 2[CO]$	$\Delta G^o = 40{,}800 - 41{\cdot}71\,T$
	$H_2 - CH_4$	$2[H_2] + <C> \rightleftarrows [CH_4]$	$\Delta G^o = -16{,}520 + 12{\cdot}25\,T\log T - 15{\cdot}62\,T$
O	$CO - CO_2$	$2[CO] + [O_2] \rightleftarrows 2[CO_2]$	$\Delta G^o = -135{,}000 + 41{\cdot}5\,T$
	$H_2 - H_2O$	$2[H_2] + [O_2] \rightleftarrows 2[H_2O]$	$\Delta G^o = -117{,}800 + 26{\cdot}5\,T$
S	$H_2 - H_2S$	$2[H_2] + [S_2] \rightleftarrows 2[H_2S]$	$\Delta G^o = -43{,}160 + 23{\cdot}6\,T$
	$O_2 - SO_2$	$2[O_2] + [S_2] \rightleftarrows [SO_2]$	$\Delta G^o = -173{,}240 + 34{\cdot}62\,T$
N_2	-	Direct Measurement	
	$H_2 - NH_3$	$[N_2] + 3[H_2] \rightleftarrows 3[NH_3]$	$\Delta G^o = -24{,}100 + 53{\cdot}4\,T$

$$(\underline{Si})_{metal} + 2[H_2O] = <SiO_2>_{slag} + 2[H_2] \qquad (3.13)$$

The standard free energy of this reaction is equal to:

$$\Delta G^o = -RT \log \frac{p^2_{H_2} \cdot a_{<SiO_2>slag}}{p^2_{H_2O} \cdot a_{(\underline{Si})metal}} \qquad (3.14)$$

The activity of pure silica is one if the reaction is carried out in a silica crucible. The variance of this system is $3(T, \underline{P}, p_{H_2}/p_{H_2O})$. For different values of the ratio p_{H_2}/p_{H_2O}, at a given total pressure and temperature, the activity of silicon can be determined over a large range of concentration.

In a second stage, it is possible to determine the activity of silica in a complex slag, which may also contain alumina, lime, magnesia, etc. The equilibrium is then written:

$$\underline{(Si)}_{metal} + 2\left[\underline{H_2O}\right] = \underline{(SiO_2)}_{slag} + 2\left[\underline{H_2}\right] \tag{3.15}$$

$$\Delta G^o = - RT \log \frac{a_{\underline{(SiO_2)}_s} \cdot p^2_{H_2}}{a_{\underline{(Si)}_m} \cdot p^2_{H_2O}} \tag{3.15a}$$

As the activity of silicon in the metallic phase has been determined previously as a function of its mole fraction, the activity of silica in the slag can be readily calculated. The system Fe-Si, $CaO-SiO_2-Al_2O_3$, H_2-H_2O has a variance of 5. Compared with the previous system ($v = 3$), it may be noted that it is possible to fix two more variables, such as the concentration of Al_2O_3 and CaO in the slag, which makes it possible to determine the activity of SiO_2 for any composition in the ternary diagram $SiO_2-Al_2O_3-CaO$.

In iron and steel making, the gas phase, separated from the molten metal by the slag, is often neglected, and only the metal-slag equilibrium is considered, such as:

$$\underline{(Si)}_m + 2\underline{(FeO)}_s = \underline{(SiO_2)}_s + 2\underline{(Fe)}_m \tag{3.16}$$

or

$$\underline{(Si)}_m + 2\underline{(O)}_m = \underline{(SiO_2)}_s \tag{3.17}$$

To determine the activity of SiO_2 in the slag, from (3.17) it is necessary to know the activities of both the silicon and oxygen dissolved in the metal.

B. DETERMINATION OF THE STANDARD FREE ENERGY OF FORMATION OF A COMPOUND

The previous examples indicate how the activity of a component in a system in equilibrium can be determined, knowing the activity of the other components and the standard free energy of the reaction. Conversely, it is possible to determine the standard free energy of a reaction when the activities of all the constituents in equilibrium are known. The values of ΔG^o for reactions between pure condensed compounds and a mixture of gases with assumed ideal behaviour are the easiest to determine. For example, in the determination of the standard free energy of the reaction:

$$\langle FeO \rangle + [\underline{H_2}] = \langle Fe \rangle + [\underline{H_2O}] \tag{3.18}$$

the activities of FeO and Fe (separate phases) are equal to one, and those of the gases are equal to the partial pressures. It is sufficient to measure these partial pressures at equilibrium in order to calculate the value of ΔG^o at the temperature of the experiment:

$$\Delta G^o = -RT \ln \frac{p_{H_2O}}{p_{H_2}} \tag{3.19}$$

Because the error introduced by assuming that ΔG^o varies linearly with T is generally less than the uncertainty in basic data, when the value of ΔG^o is known at two or more temperatures, the relationship between ΔG^o and T can be obtained by drawing the best straight line representing:

$$\Delta G^o = \Delta H^o - T\Delta S^o \tag{3.20}$$

In addition, $\ln \left(p_{H_2O}/p_{H_2}\right)$ can be plotted against $1/T$ and the equation of the straight line through the values will be:

$$\ln K = -\frac{\Delta H^o}{R} \cdot \frac{1}{T} + \frac{\Delta S^o}{R} \tag{3.21}$$

Reactions (3.18) may be separated into the two reactions of formation of pure H_2O and FeO from their elements:

$$H_2 + \tfrac{1}{2}O_2 \rightleftarrows H_2O \tag{3.22}$$

$$\langle Fe\rangle + \tfrac{1}{2}O_2 \rightleftarrows \langle FeO\rangle \tag{3.23}$$

If the free energy of formation of water vapour is known, then from Hess's Law it is possible to calculate ΔG^o for the formation of iron oxide:

$$\Delta G^o_{FeO} = \Delta G^o_{(3.18)} - \Delta G^o_{H_2O} \tag{3.24}$$

3.1.2. ELECTROCHEMICAL METHODS

These methods can be used to determine the standard free energy of formation of a compound as well as the activity of a component in solution.

A cell of the type $M/MX_n/X_2$ contains ions M^{n+} and X^- in the salt or oxide electrolyte MX_n. The electrodes are metal M and a gas X_2, and the reactions at each electrode are:

$$M^{n+} + ne \rightleftarrows M \qquad E_{M^{n+}/M} = E^o_{M^{n+}/M} - \frac{RT}{nF} \log \frac{1}{a_{M^{n+}}} \tag{3.25}$$

and

$$\frac{n}{2}X_2 + ne \rightleftarrows nX^- \qquad E_{X_2/X^-} = E^o_{X_2/X^-} - \frac{RT}{nF} \log \left(a^n_{X^-}/p^{n/2}_{X_2}\right) \tag{3.26}$$

In (3.25) and (3.26), F is Faraday's constant (96,500 coulomb, or 96,500/4.18 = 23,060 cal volt^{-1} equivalent) and n is the number of electrons exchanged in the reactions. The total reaction of the cell is the difference between (3.26) and (3.25), i.e.:

$$M + \frac{n}{2}X_2 \rightleftarrows M^{n+} + nX^- \qquad (3.27)$$

and the electromotive force, when operating under reversible conditions, is equal to:

$$E = E_{X_2/X^-} - E_{M^{n+}/M} = E^o - \frac{RT}{n\mathcal{F}} \log \frac{a_{X^-} \leftarrow a_{M^{n+}}}{p_{X_2}^{n/2}} \qquad (3.28)$$

This e.m.f. is related to the free energy of (3.27) by:

$$E = -\frac{\Delta G}{n\mathcal{F}} \qquad (3.29)$$

If the pure salt in equilibrium with its ions is taken as the standard state, we have:

$$MX_n \rightleftarrows M^{n+} + nX^- \qquad (3.30)$$

$$a_{MX_n} = a_{X^-}^n \cdot a_{M^{n+}} \qquad (3.31)$$

The equilibrium (3.27) between the electrodes and the ions can be replaced by:

$$\frac{n}{2}X_2 + M \rightleftarrows MX_n \qquad (3.32)$$

and the free energy change for (3.32) is:

$$\Delta G = -n\mathcal{F}E = \Delta G^o + RT \log \frac{a_{MX_n}}{a_M \cdot p_{X_2}^{n/2}} \qquad (3.33)$$

When MX_n and M are pure and the gas pressure at the electrode is 1 atm, the free energy of reaction (3.32) is the standard, and equal to $-n\mathcal{F}E^o$. Then it is enough to measure the e.m.f. of the cell to determine the standard free energy of formation of the salt.

If the metal M or the electrolyte MX_n in (3.32) are in solution, the e.m.f. E and the free energy ΔG are different from the standard values, and are given by (3.33). Depending on whether M or MX_n are in solution we may have either an electrode or an electrolyte concentration cell.

ELECTROLYTE CONCENTRATION CELLS

The electrolyte concentration cell makes it possible to determine the activity of a salt MX_n in solution in another salt M'_n. To this effect, the e.m.f.s of the two following cells are compared:

(a) $(M)/(MX_n)/X_2$ and (b) $(M)/MX_n)_{M'X_{n'}}/X_2$

In (a), the pure salt is formed from the elements:

$$\frac{n}{2}X_2 + (M) \rightleftarrows (MX_n) \rightleftarrows M^{n+} + nX^- \tag{3.34}$$

and

$$E_{(a)} = E^o = -\frac{\Delta G^o}{nF} \tag{3.35}$$

In (b), the salt MX_n formed, is dissolved in $M'X_n$, according to:

$$\frac{n}{2}X_2 + (M) \rightleftarrows (\underline{MX_n}) \rightleftarrows M^{n+} + nX^- \tag{3.36}$$

From (3.33) the e.m.f. of this reaction is a function of the activity of MX_n:

$$E_{(b)} = \frac{-\Delta G_{(b)}}{nF} = E^o - \frac{RT}{nF} \log a_{(\underline{MX_n})} \tag{3.37}$$

From the difference between $E_{(a)}$ and $E_{(b)}$ the activity of the salt in solution can be obtained directly:

$$E_{(a)} - E_{(b)} = \frac{RT}{n} \log a_{(\underline{MX_n})} \tag{3.38}$$

The activity of a salt in solution could also be determined directly from the e.m.f. of a single cell:

(c) $(M)/(MX_n)/(\underline{MX}_n)_{M'X_{n'}}/(M)$

In this case, the global reaction which takes place would correspond to the solution of the salt MX in $M'X_{n'}$ through the interface:

$$(MX_n) \rightleftarrows (\underline{MX_n}) \tag{3.39}$$

and the e.m.f. of the cell would be given by (3.38), but it is different because there is an additional e.m.f. generated at the interface which is difficult to measure or calculate, and which may introduce an appreciable error. For this reason, it is more accurate to use two cells.

ELECTRODE CONCENTRATION CELLS

In this type of cell, the e.m.f. of two cells whose metal electrodes are different, one being pure M and the other the solution $(\underline{M})_{M'}$, are compared. The two cells

$(M)/MX_n/X_2$ and $(\underline{M})_{M'}/MX_n/X_2$, may be combined, and since the gas electrode and the electrolyte are in common, it is possible to do away with the gas electrode and measure the e.m.f. of the resultant cell directly:

(d) $(M)/MX_n/(\underline{M})_{M'}$ or $(M)/M^{n+}/(\underline{M})_{M'}$

The reactions which take place in the cell are:

$$\text{Anode:} \qquad (M) \rightleftarrows M^{n+} + ne^- \tag{3.40}$$

$$\text{Cathode:} \qquad M^{n+} + ne^- \rightleftarrows (\underline{M}) \tag{3.41}$$

$$\text{The total reaction:} \qquad (M) \rightleftarrows (\underline{M}) \tag{3.42}$$

produces an e.m.f. E which corresponds to the partial free energy of mixing of (M):

$$E = \frac{-\bar{G}^M_{\underline{M}}}{nF} = -\frac{RT}{nF}\log a_{(\underline{M})} \tag{3.43}$$

The activity of M in solution or the partial free energy of mixing can be determined by measuring the e.m.f. of the cell.

By changing the temperature, it is possible to determine the partial entropy of mixing of M:

$$\bar{S}^M_{\underline{M}} = -\frac{d\bar{G}^M_{\underline{M}}}{dT} = nF\frac{dE}{dT} \tag{3.44}$$

An electrode, capable of measuring the oxygen potential of slag, metal or gas at up to steelmaking temperatures, can be constructed[20, 21] using a tube of zirconia stabilised by lime as the electrolyte through which the $O^=$ ions pass. The oxygen potential inside the tube is fixed, either by air, oxygen, or a mixture of a metal and its oxide (Ni/NiO) which serves as a reference electrode. The e.m.f. between the interior reference and the exterior is measured by two platinum or iridium wires in contact with the inner and outer faces of the tube and the exterior oxygen potential can be calculated from the value.

For example, by measuring the oxygen potential of a metal in equilibrium with its pure oxide:

$$M + (\underline{O}) = (MO)$$

$$\Delta G^o = -nFE^o$$

it is possible to calculate ΔG^o for the oxide formation. When MO is in solution, as in a slag, the difference between E^o and the measured e.m.f. is related to the activity of MO

$$E = \frac{\Delta G}{nF} = E^o - \frac{RT}{nF}\log\frac{1}{a_{MO}} \qquad \text{(see 3.25)}$$

C. LIMITATIONS OF ELECTROCHEMICAL METHODS

The following conditions must be met if the measurement of cell e.m.f.s is to be accurate:

The electrolyte conductivity must be entirely ionic, or have only a small electronic contribution.

The valency n of the metallic ion in the salt must be unique. It must be known, since it is essential for calculation of the free energy of formation of the salt and the activity of the salt in the electrolyte or of the metal in the electrode.

In electrode concentration cells, the metal of the salt forming the electrolyte must be less noble than the other, otherwise a displacement reaction will take place:

$$n'(M^{n+}) + n(\underline{M}') \rightleftarrows n'(\underline{M}) + n\left(M'^{n'+}\right) \tag{3.45}$$

Under these circumstances, the composition of the electrolyte will change at the electrode-electrolyte interface, resulting in errors and instability in the e.m.f. values. These problems diminish the greater the difference between the standard free energies of formation of the two salts MX_n and $M'X_{n'}$, and the greater the concentration of M in the electrode. For example, if the difference in standard free energy of formation of two salts is 7000 cal mole^{-1} (29,000 joules mole^{-1}), the relative error in the activity of M is below 10% when its mole fraction in the electrode is greater than 0.1.

3.1.3. CALORIMETRIC METHODS

The enthalpies and internal energies can be determined calorimetrically and the thermodynamic values which derive from them can be calculated. For example, at constant pressure it is possible to measure:

<u>The standard enthalpy of a pure component</u>: At temperature T, H_T^o, i.e. the heat necessary to raise its temperature from 298 K to T K. The specific heat at constant pressure C_p, and the standard entropy S_T^o are equal, respectively to:

$$C_p = \left(\frac{\partial H}{\partial T}\right)_p \qquad \text{and} \qquad S_T^o = S_{298}^o + \int_{298}^{T} \frac{C_p}{T}\, dT$$

<u>The standard enthalpy of reaction or formation of a compound ΔH_{298}^o</u>: The standard enthalpy and entropy at temperature T can be derived from the equations:

$$\Delta H_T^o = \Delta H_{298}^o + \int_{298}^{T} \Delta C_p\, dT \tag{3.46}$$

$$\Delta S_T^o = S_{298}^o + \int_{298}^{T} \frac{\Delta C_p}{T}\, dT \tag{3.47}$$

where ΔS^o_{298} and ΔC_p are equal, respectively, to $\Sigma S^{o\ prod.}_{298} - \Sigma S^{o\ react.}_{298}$ and $\Sigma C_p^{prod.} - \Sigma C_p^{react.}$. The standard free energy of the reaction ΔG^o_T can be expressed as a function of the temperature by:

$$\Delta G^o_T = \Delta H^o_T - T\Delta S^o_T \tag{3.48}$$

The enthalpy of transformation ΔH_t: At the temperature of transformation, the entropy of transformation ΔS_t is equal to $\Delta H_t/T_t$.

The integral enthalpy of mixing H^M of a solution M-M': If a quantity of M at the temperature of the environment is added to liquid M' in a crucible inside a calorimeter at temperature T, the enthalpy difference measured by the calorimeter is the sum of: (a) the enthalpy to heat the metal M from the environment temperature to T; (b) the enthalpy of fusion of M; and (c) the integral enthalpy of mixing of the solution. By deducting the first two, it is possible to find the third, H^M. It is also possible to find the partial enthalpies of mixing of M and M' by the graphical method previously described.

3.2. METHODS OF CALCULATION OF THERMODYNAMIC ACTIVITIES

The limitations imposed on laboratory methods are very often an obstacle to experimental determination of the activities of components in solution. For this reason, estimates and calculations are necessary in order to obtain the missing data. Thus, for instance, it is possible to use phase diagrams, apply solution models, or integrate the Gibbs-Duhem equation for binary or ternary solutions.

3.2.1. ACTIVITY OF A COMPONENT OF A BINARY SOLUTION AT ONE POINT OF LIQUIDUS OF A PHASE DIAGRAM

In the phase diagram M-M' in Fig. 17, liquid of composition L is in equilibrium with solid of composition S.

$$<\underline{M}> = (\underline{M}) \tag{3.49}$$

The activity of M is the same in both phases when it is referred to the same state. If pure solid M is chosen as the standard state for the solid solution, and pure liquid M for the liquid solution, the activities are related (see (2.127)) by:

$$\ln a_{(\underline{M})} = \ln a_{<M>} - \frac{\Delta H_f (T_f - T)}{RT\ T_f} \tag{3.50}$$

It is necessary to know the activity of M in the solid in order to calculate its activity in the liquid. Many phase diagrams indicate zero or very low solubility in the solid state. In the case of zero solubility, the activity of M in the solid $a_{<M>}$ is unity, and the relationship (3.50) can be simplified. With very low solubility, it may be assumed that Raoult's Law applies and the activity may be replaced by the mole fraction $N_{<M>}$ of M in the solid solution.

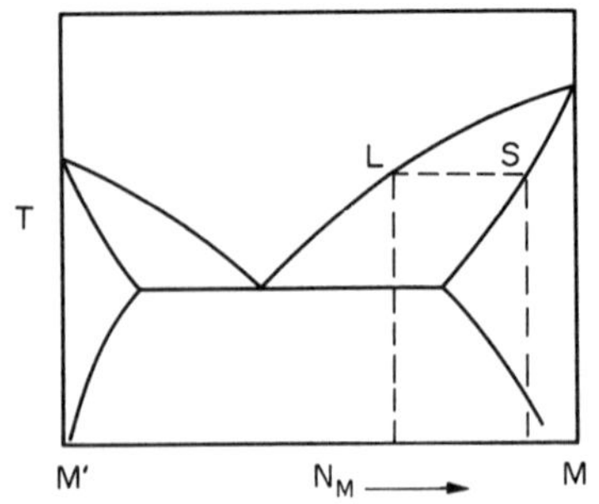

Fig. 17. Binary phase diagram

3.2.2. ACTIVITY OF A COMPONENT OF A SOLUTION AT A TEMPERATURE ABOVE THE LIQUIDUS

The activity of a component M at a given concentration in a solution at a high temperature can be calculated from that at a lower temperature (for instance, that of the liquidus) by integration of the Clausius-Clapeyron relationship:

$$\frac{\partial \ln a_{(\underline{M})}}{\partial(1/T)} = \frac{\bar{H}_M^M}{R} \tag{3.51}$$

The partial enthalpy of mixing of M can either be obtained by calorimetry or it can be calculated by means of a model that is adjusted to the particular case of the solution being studied. Examples of the simplest and most frequently used models are given below.

IDEAL SOLUTIONS

In ideal solutions, the partial and integral enthalpies of mixing are nil, and the activity of a component of the solution is equal to its molar fraction at any temperature.

REGULAR SOLUTIONS: HILDEBRAND MODEL

The integral enthalpy of mixing in regular solutions (see (2.83)) can be expressed in the form:

$$H^M = bN_{(\underline{M})} \cdot N_{(\underline{M}')} \tag{3.52}$$

b is a constant independent of temperature and composition. On the other hand, the partial enthalpy of mixing is equal to:

$$\bar{H}_M^M = RT \ln \gamma_{(\underline{M})} = b(1 - N_{(\underline{M})})^2 \tag{3.53}$$

Since $\bar{H}_M^M$ is independent of temperature, the Clausius-Clapeyron equation (3.52) may be easily integrated. Integration between T_L (liquidus) where the activity of M, $a_{(\underline{M})}^{T_L}$, is known (3.51), and the temperature T at which it is necessary to know the activity $a_{(\underline{M})}^{T}$ leads to:

$$\ln a_{(\underline{M})}^{T} = \ln a_{(\underline{M})}^{T_L} + \frac{\bar{H}_{(M)}^{M}}{R} \left(\frac{1}{T} - \frac{1}{T_L}\right) \qquad (3.54)$$

By replacing $\bar{H}_{(\underline{M})}^{M}$ by $RT_L \ln \gamma_{(M)}$ and substituting $a_{(\underline{M})}/N_{(\underline{M})}$ for $\gamma_{(\underline{M})}$, it is possible to write:

$$\ln a_{(\underline{M})}^{T} = \ln a_{(\underline{M})}^{T_L} + \frac{T_L - T}{T} \left[\ln a_{(\underline{M})}^{T_L} - \ln N_{(\underline{M})}\right] \qquad (3.55)$$

By simplifying (3.55) and replacing $\ln a_{(\underline{M})}^{T_L}$ by the value in (3.50), the activity of M at T referred to the pure liquid is obtained:

$$\ln a_{(\underline{M})}^{T} = \frac{T_L}{T} \ln a_{<\underline{M}>}^{T_L} - \frac{T_f - T_L}{RT\,T_f} \Delta H_f + \frac{T - T_L}{T} \ln N_{(\underline{M})} \qquad (3.56)$$

In this expression, the only unknown is the activity of M in the solid at T_L i.e. $a_{<\underline{M}>}^{T_L}$; but it is possible to make simplifications when the solubility of M is low, or nil.

OTHER MODELS OF SOLUTIONS

Van Laar and Kuprowski have proposed other models of solutions. For instance, in Kuprowski's model, instead of having, as in regular solutions,

$$\frac{RT \ln \gamma_{(\underline{M})}}{(1 - N_{(\underline{M})})^2} = b \qquad \text{or} \qquad \ln \gamma_{(\underline{M})} = \frac{b'}{T} (1 - N_{(\underline{M})})^2$$

we have:

$$\ln \gamma_{(\underline{M})} = \frac{\beta}{T^n} (1 - N_{(\underline{M})})^m \qquad (3.57)$$

The coefficient β and the exponents n and m are determined experimentally.

Another model having the structure found for regular solution ($H^M = bN_M N_{M'}$) has been proposed by Guggenheim for a binary solution MM'. In this model, b is a complex function, symmetrical with respect to the mole fractions of M and M':

$$H^M = N_{(\underline{M})} N_{(\underline{M}')} \sum_{i=0}^{n} B_i (N_{(\underline{M})} - N_{(\underline{M}')})^i \qquad (3.58)$$

The coefficients B_0, B_1, B_2, etc., are experimentally determined. When they are nil, except B_0, the solution is regular. The solution in which the two first coefficients are different from zero has been called "sub-regular" by Hardy. In this case:

$$H^M = N_{(\underline{M})}N_{(\underline{M}')}\left[B_0 + B_1(N_{(M)} - N_{(M')})\right] = (A_1N_{(\underline{M})} + A_2N_{(\underline{M}')})N_{(\underline{M})}N_{(\underline{M}')} \tag{3.59}$$

and the coefficients A_1 and A_2 are independent of temperature and the mole fractions of the components.

Models of ternary solutions can be studied in the chapter by Ansara[1], p. 403, of the *Proceedings of the Symposium on Metallurgical Chemistry*, 1971, edited by Kubaschewski and published by H.M.S.O. in 1972.

3.2.3. INTEGRATION OF THE GIBBS-DUHEM EQUATION IN A BINARY SYSTEM

Laboratory methods generally allow measurement of the activity of only one component of a solution. Knowledge of the activity of the second component is, at times, essential. For instance, in an SiO_2 - CaO slag in equilibrium with iron, the activity of SiO_2 can be determined by means of equilibrium with silicon in the metal. The activity of calcium in the iron phase is too low to be measured, and so it is not possible to determine experimentally the activity of lime in the slag. In order to study desulphurisation and dephosphorisation it is essential to know this. It must be calculated by graphical integration of the Gibbs-Duhem equation.

For a binary solution, the Gibbs-Duhem equation can be expressed in the form:

$$N_A d\mu_A + N_B d\mu_B = 0 \tag{3.60}$$

The graphical integration of this equation can be carried out directly if the chemical potential of one of the components is known. Similar relationships between activities or activity coefficients may be utilised:

(A) CALCULATION FROM ACTIVITIES

The chemical potentials can be expressed as functions of the activities in the form:

$$\mu_i = \mu_i^o + RT \ln a_i \tag{3.61}$$

or, for constant temperature:

$$d\mu_i = RT \; d \ln a_i \tag{3.62}$$

Inserting these values in (3.60), we obtain:

$$N_A \; d \ln a_A + N_B \; d \ln a_B = 0 \tag{3.63}$$

If the activity of B is known over the whole concentration range, it is possible to integrate (3.63) for the boundary conditions $N_A = 1$ and N_A, so obtaining:

$$(\ln a_A)_{N_A=N_A} = (\ln a_A)_{N_A=1} - \int_{N_A=1}^{N_A=N_A} \left(\frac{N_B}{N_A}\right) d \ln a_B \qquad (3.64)$$

As the activity of the pure component A ($N_A = 1$) is unity, the term $(\ln a_A)_{N_A} = 1$ is nil. In general, there is not a mathematical expression to represent $\ln a_B$ as a function of N_B, and it is necessary to construct a graphical integration. For this purpose, the curve N_B/N_A versus $-\ln a_B$ is plotted (Fig. 18). The value of $\ln a_A$ for a certain value of the mole fraction N_A is given by the area contained between the curve, the axis of the abscissa, and the value of $-\ln a_B$ which corresponds to N_A (hatched area of Fig. 18).

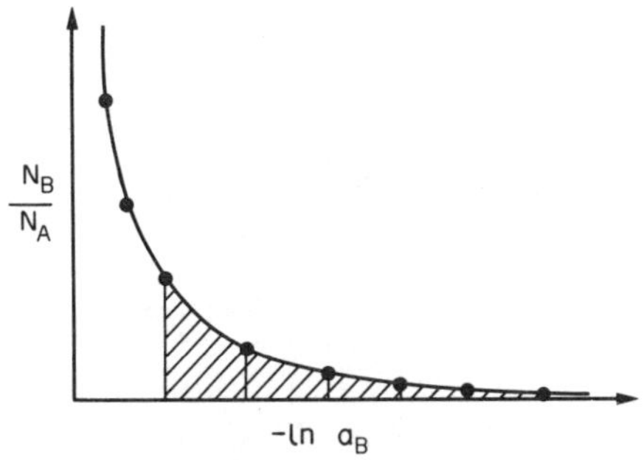

Fig. 18. Graphical integration of the Gibbs-Duhem equation by using the activities of the components

Nevertheless, when $N_A = 1$, $-\ln a_B$ has an infinite value, and the curve approaches the axis of the abscissa asymptotically. In this case, the area which represents the value $(\ln a_A)_{N_A}$ cannot be determined accurately.

(B) CALCULATION FROM ACTIVITY COEFFICIENTS

In (3.63), a_A and a_B are replaced by $\gamma_A N_A$ and $\gamma_B N_B$. The Gibbs-Duhem equation may then be written in the form given in (2.115), and simplified to:

$$N_A \, d \ln \gamma_A + N_B \, d \ln \gamma_B = 0 \qquad (3.65)$$

This new form of the Gibbs-Duhem relationship is integrated between $N_A = 1$ and N_A, according to:

$$(\ln \gamma_A)_{N_A} = -\int_{N_A=1}^{N_A} \left(\frac{N_B}{N_A}\right) d \ln \gamma_B \tag{3.66}$$

The value of $\ln \gamma_A$ at the composition N_A is given by the area between the curve, the axis of the abscissa, and the value of $-\ln \gamma_B$ corresponding to N_A. This area can be determined by dividing it into a number of elemental areas. In this case, the difficulties of the first method do not exist since γ_B (and $\ln \gamma_B$) always have a finite value.

It can be seen that the value of $\ln \gamma_B$ remains constant in the range in which the component B follows Henry's Law. This is represented in this case by the vertical portion of the curve for $N_B/N_A \simeq 0$. In this range, $\ln \gamma_A = 0$, γ_A is unity, and component A follows Raoult's Law (Figs. 19 and 20).

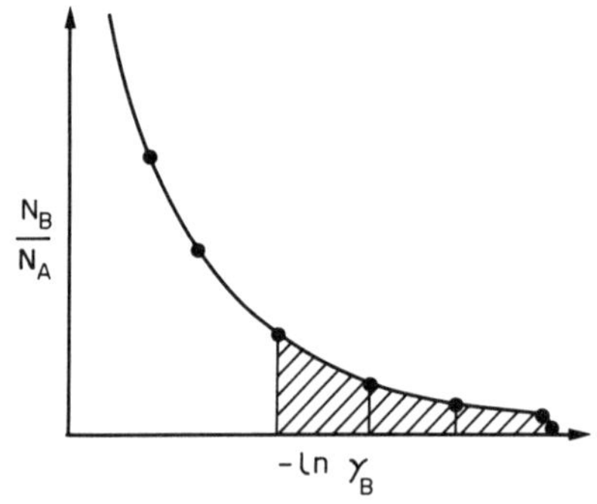

Fig. 19. Integration of the Gibbs-Duhem equation utilising the activity coefficient

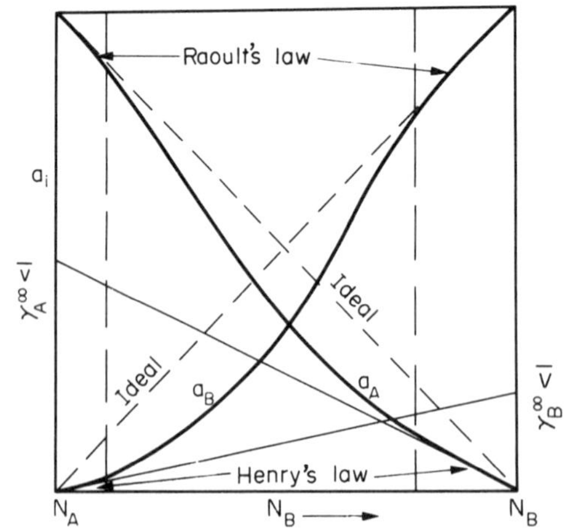

Fig. 20. Activity diagram for a binary system (γ_A and $\gamma_B < 1$). The vertical dotted lines indicate the limits of validity of Raoult's and Henry's Laws

$\ln \gamma_B$ varies continuously between $\ln \gamma_B^\infty$ (Henry's Law) and zero (Raoult's Law). If $\ln \gamma_B^\infty$ is positive (i.e. $\gamma_B^\infty > 1$), $\ln \gamma_B$ can only decrease, and $d \ln \gamma_B$ is negative. The integral (3.66) is then positive, and $\gamma_A > 1$. Similarly, it can be seen that $\ln \gamma_A$ is negative when $\ln \gamma_B$ is negative. In Fig. 20, γ_B and γ_A are less than one.

(C) CALCULATION FROM THE FUNCTION α

By dividing each item of (3.65) by dN_A, we obtain

$$\frac{N_A \, d \ln \gamma_A}{dN_A} + \frac{N_B \, d \ln \gamma_B}{dN_A} = 0 \tag{3.67}$$

Thus

$$\int d \ln \gamma_B = - \int \frac{N_A}{N_B} \frac{d \ln \gamma_A}{dN_A} \, dN_A \tag{3.68}$$

Integrating by parts ($\int UdV = UV - \int VdU$) gives

$$\Big[\log \gamma_B \Big]_{N_B=1}^{N_B} = - \left[\frac{N_A}{N_B} \log \gamma_A \right]_{N_A=0}^{N_A} + \int_{N_A=0}^{N_A} \frac{\ln \gamma_A}{N_B^2} \, dN_A \tag{3.69}$$

where $\ln \gamma_A / N_B^2$ is defined (2.81) as function α. If we assume that $\ln \gamma_A$ is known over the whole range of concentration, the second term of the right-hand side of (3.69) represents the area between the axis and the vertical lines corresponding to the values of N_A, as shown in Fig. 21. (The curve is drawn from Problem No. 9.) The first term can be calculated from $\ln \gamma_A$ and the value of N_A/N_B. It is obvious from (3.69) that when α is a constant (see 2.4.4. Regular Solutions) the graphical solution is not necessary, because $\ln \gamma_A = \alpha N_B^2$ and $\ln \gamma_B = \alpha N_A^2$.

3.2.4. INTEGRATION OF THE GIBBS-DUHEM EQUATION IN A TERNARY SYSTEM

In a ternary solution, the activities of two components can be calculated from the activities of the third by the methods of Darken[2], Wagner[3], or Schuhmann[4]. A brief account of the last of these is given below.

In the ternary system ABC, according to the Gibbs-Duhem equation,

$$N_A \partial \mu_A + N_B \partial \mu_B + N_C \partial \mu_C = 0 \tag{3.70}$$

$$\mu_A = \left(\frac{\partial G}{\partial n_A} \right)_{N_B N_C} \quad \text{and} \quad \mu_B = \left(\frac{\partial G}{\partial n_B} \right)_{N_A N_C} \tag{3.71}$$

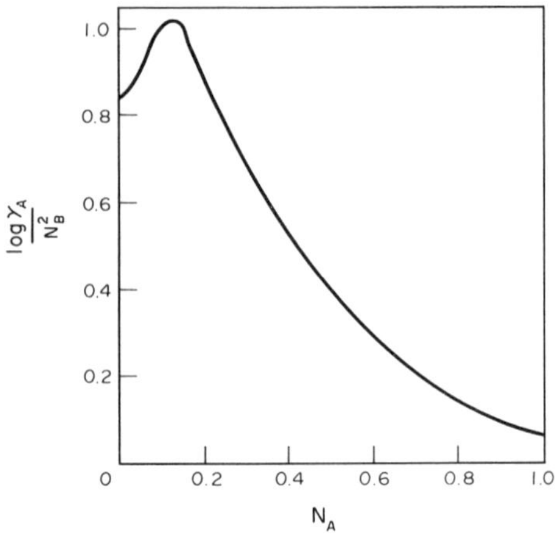

Fig. 21. Integration of the Gibbs-Duhem equation using the function α

Assuming that the potential of A is known over the whole range of concentrations in the ternary diagram (Fig. 22), from (3.71):

$$\left(\frac{\partial\mu_B}{\partial n_A}\right)_{N_BN_C} = \left(\frac{\partial^2 G}{\partial n_A \partial n_B}\right)_{N_C} = \left(\frac{\partial^2 G}{\partial n_B \partial n_A}\right)_{N_C} = \left(\frac{\partial\mu_A}{\partial n_B}\right)_{N_AN_C} \tag{3.72}$$

By definition:

$$\left(\frac{\partial\mu_A}{\partial n_B}\right)_{N_AN_C} = -\left(\frac{\partial\mu_A}{\partial n_A}\right)_{N_BN_C} \times \left(\frac{\partial n_A}{\partial n_B}\right)_{\mu_AN_C} \tag{3.73}$$

and

$$\left(\frac{\partial\mu_B}{\partial n_A}\right)_{N_BN_C} = \left(\frac{\partial\mu_B}{\partial\mu_A}\right)_{N_BN_C} \times \left(\frac{\partial\mu_A}{\partial n_A}\right)_{N_BN_C} \tag{3.74}$$

Substituting from (3.73) and (3.74) in (3.72) and simplifying:

$$\left(\frac{\partial\mu_B}{\partial\mu_A}\right)_{N_BN_C} = -\left(\frac{\partial n_A}{\partial n_B}\right)_{\mu_AN_C} = -\left(\frac{\partial N_A}{\partial N_B}\right)_{\mu_AN_C} \tag{3.75}$$

By integration at constant N_B and N_C (or N_B/N_C ratio), indicated by line AA' in Fig. 22, we obtain:

$$\left[\mu_B^{II} = \mu_B^{I} - \int_{\mu_A^{I}}^{\mu_A^{II}} \left(\frac{\partial N_A}{\partial N_B}\right)_{\mu_A N_C} d\mu_A\right]_{N_B/N_C} \tag{3.76}$$

The sum in (3.76) is of the same form as (3.64) for a binary system, with the difference that N_A/N_B has replaced $\partial N_A/\partial N_B$. As for the binary system, the integration can be carried out graphically when μ_B^{I} and the isopotential curves of the constituent A are known, as shown in Fig. 22. At the intersection of each curve with the line AA' of constant N_B/N_C ratio (for example, point P) one draws the tangent to the curve. This intersects the side AB of the triangle at T. The ratio BT/AT (which can be negative if the intercept PT falls outside the triangle ABC) is equal to $(\partial N_A/\partial N_B)_{\mu_A N_C}$. The curve $(\partial N_A/\partial N_B)_{\mu_A N_C}$ is plotted against μ_A and the area included between it, the abscissa axis, μ_A^{I} and μ_A, represents the value of $\mu_B - \mu_B^{I}$. As for binary systems, the potential μ_i can be replaced by $\ln a_i$ or $\ln \gamma_i$.

Rein and Chipman[16] have used this method to calculate the activities of CaO and Al_2O_3 in the SiO_2 - CaO - Al_2O_3 system from measurements of a_{SiO_2} (Fig. 88).

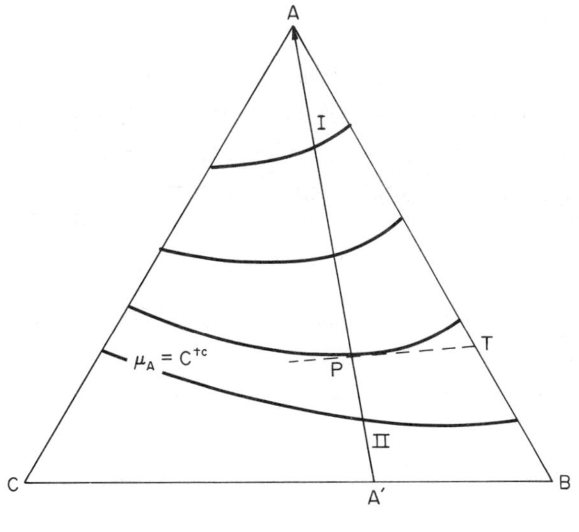

Fig. 22. Ternary diagram giving isopotential curves for A and indicating the method of determining $(\partial N_A/\partial N_B)_{\mu_B N_C}$

3.3. TABULATION OF THERMODYNAMIC DATA

Values of the thermodynamic functions are presented in the literature, usually as graphs or tables.

3.3.1. PRESENTATION OF DATA AS TABLES

Many tables of thermodynamic data are available. The best known are those published by the U.S. National Bureau of Standards for pure substances and Hultgren's compilations for solutions

(A) DATA FOR PURE SUBSTANCES

(i) The selected values of chemical thermodynamic properties and the Bulletins which serve as supplements, published by the National Bureau of Standards provide for several forms of pure substances (ions, atoms, molecules), three series of data:

The first gives the values of the functions:

$$\Delta H^o_0,\ \Delta H^o_{298},\ \Delta G^o_{298},\ \ln K,\ S^o,\ \text{and}\ C_p$$

From these, it is possible to calculate approximately (because the variations of C_p with the temperature are not given), the standard enthalpy and the standard entropy at all temperatures (equations (1.20) and (1.38)), the enthalpy and the entropy of reaction (equations (1.25) and (1.41)) and the corresponding free energy.

In the second series, temperatures of transformation (transition, fusion, boiling, sublimation) of the elements and the corresponding entropy and enthalpy changes are given.

In the third series, enthalpies and free energies of formation of compounds, ΔH^o and ΔG^o, and the enthalpy differences $H_T - H_{298}$ are given at different temperatures.

(ii) More recently, the N.B.S. has published *JANAF Thermodynamic Tables*, where, for many substances, at every 100 degrees the values of the following quantities are given:

$$C_p,\ S^o,\ \frac{G^o - H^o_{298}}{T},\ H^o - H^o_{298},\ \Delta H_{\text{formation}},\ \Delta G_{\text{formation}},\ \ln K_e$$

To calculate the free energy of a body at an intermediate temperature, one can use

$$G^o = T\left\{\frac{G^o - H^o_{298}}{T}\right\} + H^o_{298}$$

Since $(G^o - H^o_{298})/T$ is almost unchanged with temperature, it is possible to estimate the value of G^o at any temperature by linear interpolation.

(iii) Other tables of data for the pure substances, on the same models as those above (but often prior to them) are interesting in a limited field; (for example, the refractory elements and compounds, ...). These tables are in the bibliography of this chapter[5-8].

(B) DATA RELATING TO ALLOYS

The most important work concerning binary alloys is *Selected Values of the Thermodynamic Properties of Binary Alloys* by R. Hultgren, P.D. Desai, D.T. Hawkins, M. Gleiser, K.K. Kelley and Wagman[11]. This is a complete revision of *Selected Values of the Thermodynamic Properties of Metals and Alloys* by Hultgren, Orr, Anderson and Kelley. It includes the following data:

The binary phase diagram

The diagrams of integral enthalpy, integral entropy, and integral free energy of mixing of liquids at a given temperature.

Tables which give, for different compositions of the binary alloy and generally at two temperatures (one for the solid, the other for the liquid), the main partial and integral thermodynamic quantities of mixing and of excess:

G^M, H^M, S^M, G^{xs}, S^{xs}, $\bar{G}^M_{A \text{ and } B}$, $\bar{G}^{xs}_{A \text{ and } B}$, $a_{A \text{ and } B}$, and $\gamma_{A \text{ and } B}$.

This compilation is derived from many other works cited in the Bibliography[9-15], which give the binary diagrams and other data, for metals and oxides.

3.3.2. THERMODYNAMIC DATA IN THE FORM OF DIAGRAMS

The presentation of thermodynamic data as diagrams has the advantage of permitting rapid evaluation of the thermodynamic functions, without tedious comparative calculations. The more frequently used diagrams are:

(a) The Ellingham, Richardson and Jeffes diagrams.

(b) The curves of $\log K_e$ as a function of $1/T$.

(A) ELLINGHAM, RICHARDSON AND JEFFES' DIAGRAMS

In constructing these diagrams, it is necessary to keep in mind that the entropy as well as the enthalpy of formation of a compound does not change significantly with temperature, as long as there is no change of state. The diagrams of most practical interest are those for oxides, sulphides, halides, carbides, and nitrides. The oxide diagrams will be used in order to explain their use. The special points of each of the other diagrams will also be discussed. (See Figs. 23, 58, 60 and 61.)

(i) Variation of the Standard Energy with T

The standard free energy of formation of MO_2 from pure metal M and pure oxygen:

$$\langle M \rangle + [O_2] \rightleftarrows MO_2 \tag{3.77}$$

is expressed as a function of ΔH^o_T and ΔS^o_T, which are virtually constant, by:

$$\Delta G^o_T = \Delta H^o_T - T\Delta S^o_T \simeq \Delta H^o_{298} - T\Delta S^o_{298} \tag{3.78}$$

and the functions ΔG^{o}_{T} can be represented in the form of straight lines, as in Fig. 23.

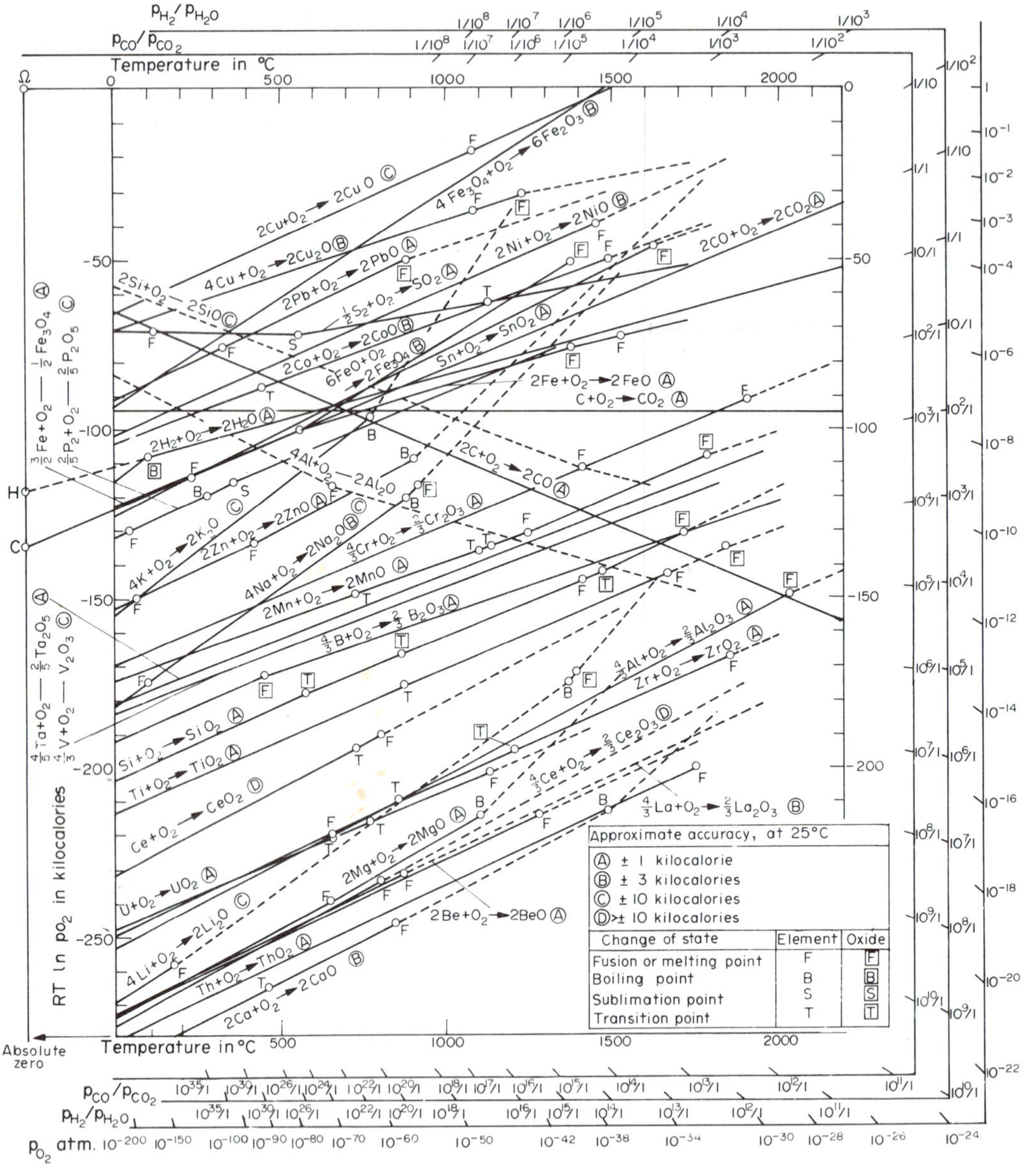

Fig. 23. Free energies of formation of some oxides
(By courtesy of Mr. Olette and Mme. Ancey-Moret)

The relative positions of the lines in the diagram make it possible to visualise the relative affinities ($A^o = -\Delta G^o$) of the metals for oxygen directly in their standard conditions. It is possible to see from the diagram that there are easily reducible oxides such as CuO, Cu_2O, PbO and Fe_2O_3 (to Fe_3O_4) as well as the oxides of Au, Ag and Hg (not shown in the diagram). There are others of greater stability such as FeO, P_2O_5, MnO, etc., and oxides, which are very difficult to reduce (refractory), such as ZrO_2, Al_2O_3, CaO, MgO, etc.

In order to obtain metal from the oxides of Au, Ag, Hg, Cu and Pb a weak reductant or a slight increase of the temperature is sufficient. This is the reason why gold, which has a very low affinity for oxygen, is found native. The line ΔG^o_T for the reaction $4Fe_3O_4 + O_2 = 6Fe_2O_3$ cuts the abscissa at 1500°C, which is that of the inversion of the formation of Fe_2O_3 in O_2 at 1 atm. Below this temperature ΔG^o for Fe_2O_3 is negative and it is stable. Above the inversion temperature, only Fe_3O_4 is stable and at it, the three phases, Fe_2O_3, Fe_3O_4 and O_2 at 1 atm are in equilibrium.

The moderately reducible oxides need stronger reductants such as carbon, hydrogen or metals, since the temperatures required to obtain the metals by thermal decomposition of the oxides are too high to be practicable. For a metal, M', to be able to reduce MO_2, the M' must have a greater affinity for oxygen than M. In this case, the standard free energy of the reaction:

$$(MO_2) + (M') \rightleftarrows (M'O_2) + (M) \tag{3.79}$$

will be negative, and the equilibrium will be displaced toward the right. In these diagrams, it is common to find that lines which represent the free energies of formation of two compounds intersect each other, as in the case of P_2O_5 and FeO. Phosphorus is a more energetic reductant than iron below 1323°C and less energetic above this temperature. This temperature is therefore the inversion temperature for:

$$(P_2O_5) + 5(Fe) \rightleftarrows 5(FeO) + 2(P)$$

and all four components of the system are in equilibrium at 1323 K.

Oxides that cannot be reduced on a commercial scale by another metal, for example, the refractory oxides, can be obtained by electrolysis either of the oxide dissolved in a fused salt (Al_2O_3), or of the fused chloride ($MgCl_2$).

(ii) Entropy of Formation

The slope of the line that represents the free energy of formation of a compound (equation (3.78))) is equal to $-\Delta S^o$ ($-\Delta S^o = d\,\Delta G^o/dT$), which is a function of the entropy of the products and the reactants:

$$\Delta S^o = \Sigma S^o_{prod.} - \Sigma S^o_{react.} \tag{3.80}$$

In general, the entropies of solids and liquids are small compared with the values for gases. According to Trouton's empirical rule, the entropies of fusion and vaporisation are approximately equal to 2.3 EU and 23 EU respectively. On the other hand, in a solid-liquid or liquid-liquid reaction, the difference of entropy between products and reactants is of minor importance. The change of entropy will then depend on the states of the reactants and products. When both are condensed, or when the number of moles of gas formed is equal to the number of moles of gas reacted, the variation of entropy is very small, and the plots of $\Delta G^o = f(T)$ are horizontal, as is the case for the formation of carbides (Fig. 58), or the oxidation of carbon to CO_2 (Fig. 23).

The change of entropy is negative when the number of moles of reactant gas is larger than the number of moles of product gas, and the slopes of $\Delta G^o = f(T)$ are positive. This is the most common case for oxide, sulphide, nitride, and halide diagrams. Because the reactions of formation of compounds from elements are written for the same number of moles of reactant gas, the lines are practically parallel for a given state (solid or liquid).

The change of entropy after reaction is positive when the number of moles of reactant gas is smaller than the number of moles of product gas; thus, the slope of the curves ΔG^o_T is negative. This is the case for the oxidation of carbon to CO and the formation of sub-oxides and sub-halides such as SiO and AlCl:

$$2\langle C\rangle + [O_2] \rightleftarrows 2[CO]$$

$$2\langle Si\rangle + [O_2] \rightleftarrows 2[SiO]$$

$$2\langle Al\rangle + [Cl_2] \rightleftarrows 2[AlCl]$$

The standard free energies of these reactions become more negative as the temperature is raised. In Fig. 23 it is apparent that the line representative of ΔG^o_T for CO formation intersects the lines of all other metals. This explains why carbon is an energetic reductant at high temperature. On the other hand, the change of entropy due to fusion or vaporisation, ΔS^o_f or ΔS^o_v, explains the changes in slope at the temperatures of fusion and vaporisation. The slopes increase if the reactants change their state with a rise of temperature, and decrease when the change of state affects the products. The change of slope is larger for vaporisation than for fusion, due to the larger change of entropy.

(iii) Enthalpy of Formation of a Compound

The enthalpy of formation of a compound, ΔH^o_T, is the value of ΔG^o_T at $T = 0^oK$. When a reactant or product is not stable at low temperature, this value can be obtained by extrapolation.

The enthalpy of change of state, fusion, or vaporisation of the metal or its oxide, can be determined graphically. For example, the enthalpy of fusion of a metal is the difference between the enthalpies of the two following reactions:

$$\langle M \rangle + [O_2] \rightleftarrows (MO_2) \qquad \langle M \rangle \text{ is solid}$$

$$(M) + [O_2] \rightleftarrows (MO_2) \qquad (M) \text{ is liquid}$$

Thus, it is enough to extrapolate the straight lines representing ΔG^o and to measure this difference at the origin of the abscissa.

Since the free energy of change of state ΔG_t is zero at the temperature of change of state, the corresponding entropy, ΔS_t, is then equal to $\Delta H_t/T_t$, which in turn is the difference in slope of the representative lines of these two states, at this temperature.

(iv) Correction for Variation of Oxygen Pressure

If for the formation of an oxide, only the oxygen is not in its standard state (p_{O_2} is different from 1 atm), the free energy of reaction is a function of the equilibrium constant which, in this particular case, is $1/p_{O_2}$:

$$\Delta G = \Delta G^o + RT \ln K_e = \Delta G^o - RT \ln p_{O_2} \tag{3.81}$$

The difference in free energy between the standard and non-standard conditions is equal to $RT \ln p_{O_2}$. This difference, proportional to $\ln p_{O_2}$, varies linearly with the temperature. The correction that has to be applied to the standard free energy when p_{O_2} is different from 1 atm is expressed as a function of temperature in Fig. 24 (line OA). The lines representative of the free energy in the non-standard conditions (BC and BD) can be obtained from the difference between $\Delta G^o_{(T)}$, $(B\text{-}T_{inv.})$, and the line which corresponds to the correction $RT \ln p_{O_2}$, (OA). If p_{O_2} is less than 1 atm, ΔG is less negative than ΔG^o, and vice versa. From Fig. 24, it can be seen that the inversion temperature is lowered as the oxygen pressure is reduced.

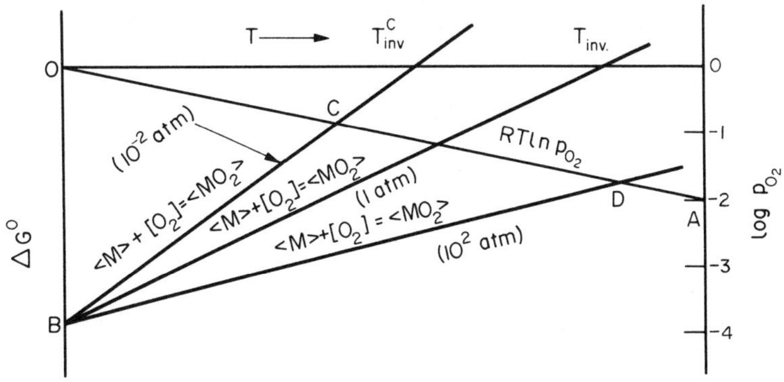

Fig. 24. ΔG vs T for pressure conditions other than standard

This modification is useful not only when the oxygen pressure is different from 1 atm, but also when a reactant or product is not in its standard state. Thus, in (3.81) and Fig. 24 $\ln p_{O_2}$ can be replaced by $-\ln K_e$, and it becomes apparent that it is thermodynamically more difficult to reduce an oxide when it is in solution ($a_{MO} < 1$) than in the pure state. Conversely, it is easier to obtain a metal in solution ($a_M < 1$) than as pure metal. Likewise, the metallothermic reduction of an oxide of a volatile metal is facilitated by the use of reduced pressure.

(v) Free Energy of Oxidation of Gases at Non-Standard Pressures

In the Ellingham diagram for oxides, the plots of ΔG^o_T for the oxidation of CO and H_2:

$$2[CO] + [O_2] \rightleftarrows 2[CO_2] \qquad \text{and} \qquad 2[H_2] + [O_2] \rightleftarrows 2[H_2O] \tag{3.82}$$

are based on all of the gases being at 1 atm pressure. For non-standard conditions, the free energy of reaction will be expressed by:

$$\Delta G = \Delta G^o + 2\ RT \ln \frac{p_{CO_2}}{p_{CO}} - RT \ln p_{O_2} \tag{3.83}$$

When the partial pressures of CO and CO_2 (or H_2 and H_2O) are different from 1 atm, that is, when the ratios p_{CO_2}/p_{CO} or p_{H_2O}/p_{H_2} are different from 1, the correction to the free energy function is equal to $2\ RT \ln (p_{CO_2}/p_{CO})$ and the plots of ΔG_T corresponding to (3.82) for any ratio of p_{CO_2}/p_{CO} or p_{H_2O}/p_{H_2} can be found by joining the point C (or H) on the ordinate of the diagram to the point on the p_{CO_2}/p_{CO} or p_{H_2O}/p_{H_2} grid indicating the specified ratio, as can be seen in Fig. 25.

This method of presentation has the advantage of indicating directly the conditions of temperature and pressure within which the reduction of an oxide by a gas is possible. For example, Fig. 25 shows that, with $p_{CO}/p_{CO_2} = 1$, the reduction of FeO to Fe by CO cannot be effected at temperatures higher than 973 K. With $(p_{CO}/p_{CO_2}) = 10^{-1}$, the maximum reduction temperature would be 573 K (if FeO was stable at that temperature).

(B) LOG K_e VS 1/T

From the expression:

$$\Delta G = \Delta G^o + RT \ln K_e \tag{3.84}$$

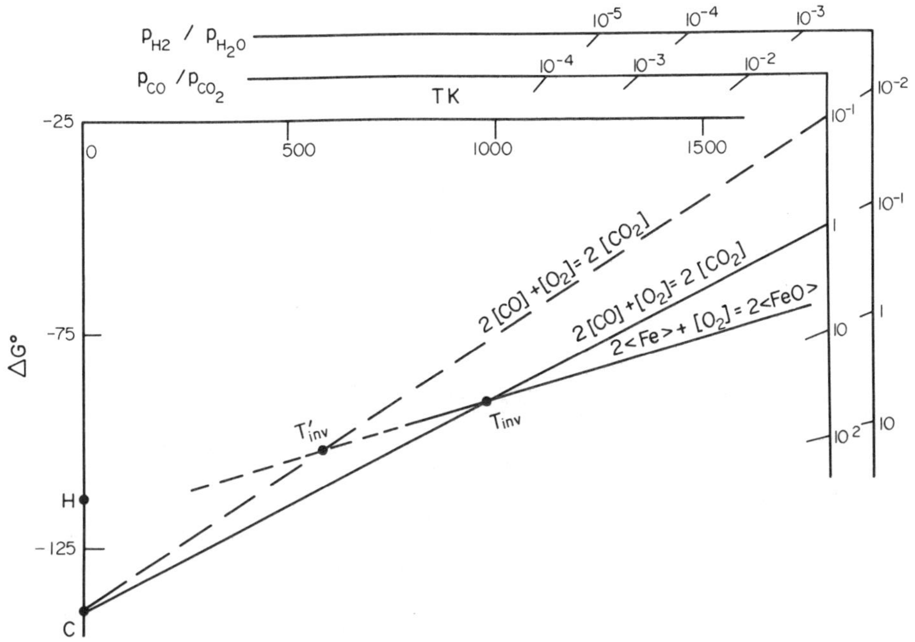

Fig. 25. Change of free energy of oxidation of CO according to the ratio CO/CO_2

the following relation can be deduced at equilibrium ($\Delta G = 0$):

$$\ln K_e = -\Delta G^o/RT = -\Delta H^o/RT + \Delta S^o/R \tag{3.85}$$

As ΔH^o and ΔS^o vary very little with the temperature, log K_e is a linear function of 1/T. The straight lines of Fig. 26 represent this function for the reduction of oxides by CO according to:

$$<MO> + [CO] \rightleftarrows <M> + [CO_2] \tag{3.86}$$

This method of presentation has the advantage of indicating the value of $\ln K_e$ at a given temperature, without any additional calculation. The ordinate at the origin of one of the straight lines represents the change of entropy for reaction (3.86). It can be seen in Fig. 26 that, for reactions of the same type, the values of ΔS^o of reaction are similar, except when the number of gaseous molecules varies as, for example, when the metal produced is volatile, as is the case with zinc.

The slopes of the straight lines:

$$\frac{d \ln K_e}{d(1/T)} = -\frac{\Delta H^o}{R} \tag{3.87}$$

indicate whether the reactions are endothermic or exothermic. A positive slope corresponds to a negative ΔH^o (exothermic reaction), and vice versa.

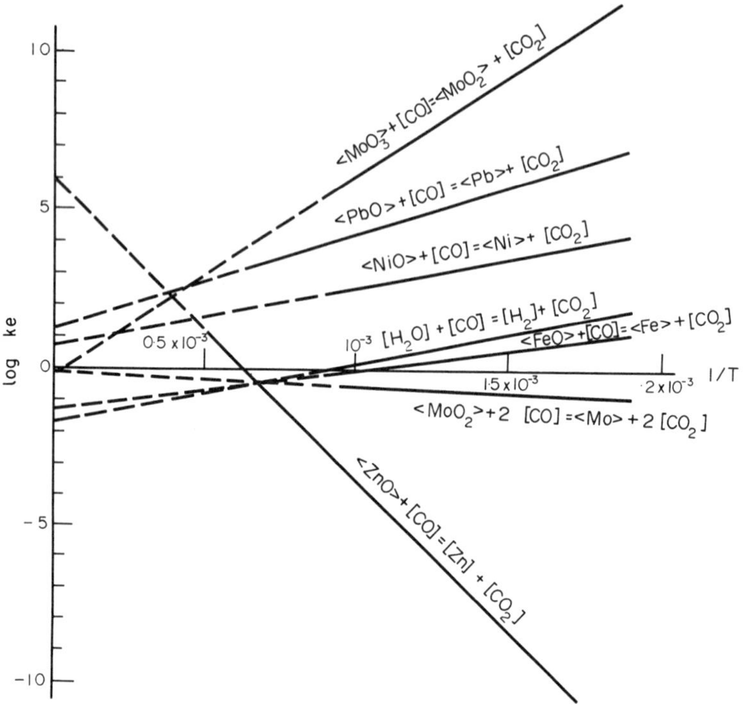

Fig. 26. Diagram of log K_e = f(1/T) for reaction (3.86)

REFERENCES

1. I. Ansara, "Prediction of Thermodynamic Properties of Mixing and Phase Diagrams in Multicomponent Systems", *Proceedings of the Symposium on Metallurgical Chemistry*, p. 403. Ed. Kubaschewski, H.M.S.O. (1972).

2. L.S. Darken, *J. Am. Chem. Soc.* 72, 2909 (1950).

3. C. Wagner, *Thermodynamics of Alloys*, pp. 19-22, Addison-Wesley, Reading, Mass. (1952).

4. R. Schuhmann, *Acta Metal.* 3, 219 (1955).

5. JANAF, *Thermochemical Tables*, 2nd Ed., NSRDS-NBS 37 (1971).

6. JANAF, *Thermochemical Tables*. The Dow Chemical Company, Midland, Mich.

7. H.L. Schick, *Thermodynamics of Certain Refractory Compounds*. New York, Academic Press (1966).

8. Barin and Knacke, *Thermochemical Properties of Inorganic Substances*. Springer Verlag, Berlin, Heidelberg (1978).

9. F.A. Shunk, *Constitution of Alloys*. II TRI B 6082-1, IIT Research Institute, Chicago, Ill., Annual Report No. 1 (1968).

10. F.A. Shunk, *Constitution of Binary Alloys*, 2nd Supplement, McGraw-Hill, N.Y. (1969).

11. R. Hultgren, P.D. Desai, D.T. Hawkins, M. Gleiser, K.K. Kelley and Wagman, *Selected Values of the Thermodynamic Properties of the Elements*. American Society for Metals, Metal Park, Ohio.

12. M. Hansen and K. Anderko, *Constitution of Binary Alloys*, 2nd Ed., McGraw-Hill, N.Y. (1958).

13. J.F. Elliott, M. Gleiser and V. Ramakrishna, *Thermochemistry for Steelmaking*, Vol. II. Addison-Wesley, Reading, Mass. (1963).

14. J.F. Elliott, *Constitution of Binary Alloys*. 1st Supplement, McGraw-Hill, N.Y. (1965).

15. E.M. Levin *et al.*, *Phase Diagrams for Ceramists*. American Ceramic Society (1956, 1959, 1964).

16. R.H. Rein and J. Chipman, *Trans. A.I.M.E.* 233, 415 (1965).

17. M.G. Ingraham and J. Drowart, *High Temperature Technology*, pp. 219-240, McGraw-Hill, N.Y. (1959).

18. J. Drowart, *Condensation and Evaporation of Solids*, p. 255, Gordon and Breach, N.Y. (1964).

19. C. Chatillon, A. Pattoret and J. Drowart, *High Temperatures-High Pressures* Vol. 7, pp. 119-148 (1975).

20. W.F. Caley and C.R. Masson, *Metal-Slag-Gas Reactions and Processes*, pp. 140, 154, Ed. Z.A. Foroulis and W.W. Smelter. The Electrochemical Society (1975).

21. J. Fouletier and G. Vitter, *Applications of Solid Electrolytes*, Ed. T. Takahashi and A. Kozawa, Japanese Electrochemical Society Press Inc. (1980).

GENERAL READING

Bockris, J.O'M., J.L. White and J.D. Mackenzie, *Physico-chemical Measurements at High Temperatures*, Academic Press, N.Y. (1959).

Bonnier, E., *Methods de Laboratoire de Determination des Activites*, E.N.S. Electrochimie et Electrometallurgie de Grenoble.

Hume-Rothery, W., J.W. Christian and W.B. Pearson, *Metallurgical Equilibrium Diagrams*, Institution of Physics, London (1952).

Kubaschewski, O. and J.A. Catterall, *Thermochemical Data of Alloys*, Pergamon Press, London (1956).

Kubaschewski, O., E.Ll. Evans and C.B. Alcock, *Metallurgical Thermochemistry*, Pergamon Press, London (1967).

Chapter 4

HETEROGENEOUS REACTIONS IN METALLURGY, KINETICS AND CATALYSIS

INTRODUCTION

The direction of a reaction under given conditions and the state of the system at equilibrium can be calculated from thermodynamic data, but this does not provide any information on the mechanism or velocity of the reaction. This is obtained from an examination of the kinetics.

In addition to the variables (T, P, V, and N_i) which control the direction of a reaction and the equilibrium state, reaction kinetics depend on many other factors. These include the form in which reactants and products are present, and the presence of a catalyst or external energy which initiates or accelerates reaction.

Generally, two types of kinetics can be distinguished: homogeneous and heterogeneous. In the first, the gaseous or liquid reactants and products form a single homogeneous phase. The reactions in which solid reactants and products are involved do not correspond to this definition, since there may be several solid phases, or a single phase with a concentration gradient. In the second, the reactants form two or more phases, and the reactions are confined to the surfaces of contact between phases.

Metallurgical reactions are generally heterogeneous, and the two phases may be:

Two solids: Carbon in contact with an iron plate diffuses from the surface toward the centre of the plate until the iron is saturated.

Two liquids: Slag-metal, slag-matte, etc., in which the products are distributed between the phases after reaction at the contact surface.

A fluid and a solid: Reductant gas-metal oxide, oxidising gas-metal sulphide, etc.

This chapter is mainly concerned with the study of solid-gas and liquid-liquid reactions. The first portion is an analysis of the kinetics of transformation of a solid particle by reaction with a gas. This serves as the base for the study, in the second section, of the kinetics of reaction within a particulate bed. In the third section, these results are extended to the influence of catalysts and in the fourth, the interchanges between two fluid phases will be studied.

4.1. TRANSFORMATION OF A PARTICLE BY A GAS

A study of the kinetics of transformation of a solid particle by a gas is aimed at determination of the relationship between the rate of transformation and:

(a) The characteristics of the particle, such as its chemical and mineral composition, micrographic structure, porosity, shape, and dimensions. The effects of several of these factors are difficult to estimate and express as kinetic equations. For this reason they are included in the constants which are determined experimentally. Porosity changes the transformation kinetics of a solid completely; therefore a separate study is necessary for the cases of dense and porous particles. Two forms of particle will be examined: flat, and spherical.

(b) The temperature that influences the velocities of reaction and diffusion; two phenomena that affect reaction kinetics.

(c) The characteristics of the gases; for example, their partial pressures or concentrations. The change in composition of the gaseous reactant as it passes over a particle, decreases as the rate of flow increases. So as not to introduce a supplementary variable which is difficult to control, it will be assumed that the gas flow is large enough that the composition is effectively unchanged on passing over the particle, and that a variation of the gas flow will not introduce a change in the composition of the gas surrounding the particle.

4.1.1. PHENOMENA IN HETEROGENEOUS KINETICS

At high temperature, a reaction between a gas and a solid, of the type:

$$[A] + <B> \underset{b}{\overset{a}{\rightleftarrows}} [C] + <D> \qquad (4.1)$$

occurs in several elementary steps, which are:

(a) diffusion of the reactants towards the interface;

(b) chemical reaction at the interface;

(c) diffusion of the products away from the interface.

Apart from diffusion and reaction, the accumulation of gaseous reactants and products in the pores as well as the adsorption of gaseous reactant upon the reaction area, should be considered. It will be seen later that the accumulation velocity, which corresponds to Fick's second law, may be neglected when chemical reaction occurs. Similarly, the adsorption of gaseous reactant upon the reaction surface, which will be studied in heterogeneous catalysis, can be ignored at high temperature.

(A) REACTION VELOCITY

In a reversible reaction like (4.1), the velocity of reaction, r_r, is the difference between the rates of the forward and reverse reactions. These velocities can be assumed to be proportional to the area of the reaction surface (S_i) assuming also first order reactions in both directions and the partial pressures of the gaseous reactant and product at the surface (p_i and p_i'). With these hypotheses, the reaction velocity can be written:

$$r_r = -\frac{dn_A}{dt} = S_i(kp_i - k'p_i') \tag{4.2}$$

In this expression, k and k' are the specific velocity constants of reaction (4.1) in the directions a and b respectively. At equilibrium, the reaction velocity r_r is zero, and the partial pressures are those at equilibrium, p^* and p'^*. Consequently:

$$kp^* = k'p'^* \qquad \text{or} \qquad \frac{p'^*}{p^*} = \frac{k}{k'} = K \tag{4.3}$$

K is the equilibrium constant of (4.1). Since:

$$p_i + p_i' = P = p^* + p'^* = p^*(1 + K) \tag{4.4}$$

(4.2) may be written:

$$r_r = kS_i\left(p_i - \frac{k'}{k}p_i'\right) = kS_i\left[p_i\left(1 + \frac{1}{K}\right) - \frac{P}{K}\right] \tag{4.5}$$

Finally, replacing P by its value as a function of p^*, given by (4.4), we obtain:

$$r_r = kS_i\left(1 + \frac{1}{K}\right)(p_i - p^*) \tag{4.6}$$

If, instead of the partial pressures, the concentrations or mole fractions are considered, for a perfect gas where $C_i = p_i/RT$ and $N_i = p_i/P$, then:

$$r_r = kRT\left(1 + \frac{1}{K}\right)S_i(C_i - C^*) = BS_i(C_i - C^*) \tag{4.7}$$

With $$B = kRT\left(1 + \frac{1}{K}\right)$$

and $$r_r = kP\left(1 + \frac{1}{K}\right)S_i(N_i - N^*) \tag{4.8}$$

C_i and N_i are the concentration and molar fraction of the gaseous reactant at the interface, C^*, and N^* the concentration and molar fraction at equilibrium.

Variation of constants with temperature. The equilibrium constant K of reaction (4.1) varies, as has been seen in the thermodynamic sections, according to:

$$K = e^{-\frac{\Delta G^o}{RT}} = K_o e^{-\frac{\Delta H^o}{RT}} \tag{4.9a}$$

and the theory of absolute reaction velocities indicates that the specific reaction velocity constant, k, varies with temperature according to the equation:

$$k = k_o T e^{-\frac{E}{RT}} \tag{4.9b}$$

in which E is the activation energy of the reaction.

Accumulation of gaseous reactants. As reaction progresses, the reaction interface recedes, leaving in the solid product pores which are filled with gas, giving rise to an accumulation of reactants. The amount of gas accumulated in the pores is small in comparison with that which reacts, as can be seen from this example: The reduction of wustite with hydrogen takes place according to:

$$<FeO> + H_2 \rightleftarrows <Fe> + H_2O$$

Assuming that wustite is dense, and that the iron sponge produced has a porosity $\varepsilon = 0.3$ (30%), the reduction of 1 mole of wustite (the volume of which is approximately 13 cm^3), requires 22,400 cm^3 of H_2. The maximum amount of H_2 which may be accumulated in the pores is 13×0.3, i.e. 3.9 cm^3, about 0.02% of the amount that has reacted. The accumulation velocity, equal to the amount of gas accumulated per unit of time, can be neglected in this case.

(B) DIFFUSION VELOCITY

According to Fick's first law, the flux of a substance through a surface, S, in the direction x is proportional to the concentration gradient of this substance in that direction:

$$r_d = \frac{dn}{dt} = - SD \frac{dC}{dx} \tag{4.10}$$

The diffusion coefficient D has the dimension of a unit area per unit time.

Diffusion can take place in the solid or gaseous phase. In the case of diffusion in the solid, the reactant diffuses through the crystal lattice towards the reaction interface. In the case of diffusion in the gas, both reactants and products diffuse in countercurrent through the mixture which exists in the pores of the solid. In this chapter, the analysis is concentrated on those cases where gaseous diffusion is the only significant mode of transfer of reactants and products.

Variation of the diffusion coefficient of a gas with temperature and pressure in a solid. The diffusion coefficient, D, of a mixture of ideal gases in an open system is proportional to $T^{3/2}$, and inversely proportional to P. For a mixture of real gases, the following expression may be used:

$$D_1 = \frac{D_o}{p} \left[\frac{T}{T_o} \right]^n$$

The diffusion coefficient in normal conditions of temperature ($T \sim 273$ K) and pressure ($p = 1$ atm), D_o, has a value between 0.1 and 1 $cm^2\ sec^{-1}$. For example:

$$D_{H_2-H_2O} = 0.75\ cm^2\ sec^{-1}, \qquad D_{CO-CO_2} = 0.14\ cm^2\ sec^{-1}.$$

The exponent n (dimensionless), varies between 1.75 and 2. Generally, it is taken as equal to 1.8.

It can also be shown that in a porous solid of open porosity ε and a pore size much greater than the mean free path λ of the molecules, the diffusion coefficient D is proportional to the porosity and inversely proportional to the sinuosity coefficient of the pores, s, defined as the route effectively travelled by a gas molecule in passing from one point to another inside the porous solid, distant one unit of length. When the porous solid is a group of spheres, the sinuosity coefficient is equal to $\sqrt{2}$. Consequently:

$$D = D_1 \frac{\varepsilon}{s} = D_1 \frac{\varepsilon}{\sqrt{2}}$$

Note that the diffusion coefficient, D, in a porous solid is always smaller than the coefficient D_1 outside it.

4.1.2. TRANSFORMATION OF A DENSE OR NON-POROUS SOLID PARTICLE BY A FLUID

In the transformation of a non-porous particle by a fluid, the product of reaction may be:

A dense product: the high temperature oxidation of iron with oxygen forms dense wustite, which impedes the passage of oxygen. Diffusion, in this case, is through the solid.

A porous product: for example, the reduction of iron ore with hydrogen. The gases diffuse through the pores and the boundary layer that surrounds the solid. Diffusion through the boundary layer will be neglected.

A soluble or volatile product: the solution of a solid metal in a molten solution. Diffusion occurs through the boundary layer of fluid that surrounds the solid. This case will be studied separately (see section 4.1.3).

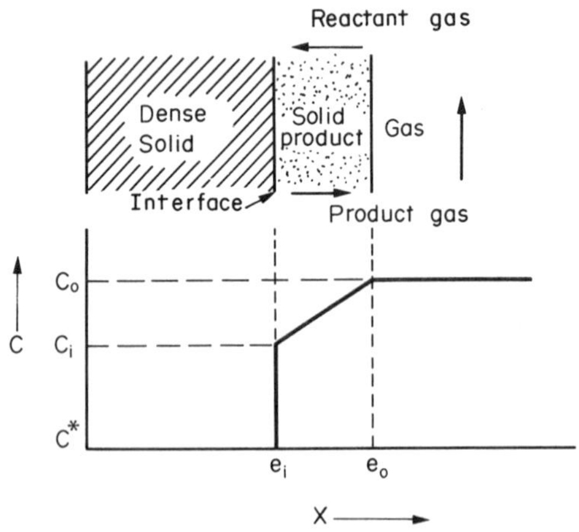

Fig. 27. Transformation of a dense solid particle by a gas

As fluid cannot penetrate the non-porous solid reactant, the reaction interface will be that of the non-reacted solid. The surface is simple and geometrically similar to the surface of the initial reactant. Chemical reaction and diffusion are well defined and consecutive, because one takes place at the interface and the other through the layer of product which surrounds this surface.

The reaction velocity is proportional to $(C_i - C^*)$, according to (4.7), and it attains the maximum, r_{mr}, when $C_i = C_o$, while the diffusion velocity, when the concentration gradient is linear, as in the case of reaction with a flat particle (Fig. 27), is proportional to $(C_o - C_i)$, and the maximum, r_{md} occurs when $C_i = C^*$. The two velocities are represented in Fig. 28 as functions of the concentration C_i of the reactant at the interface.

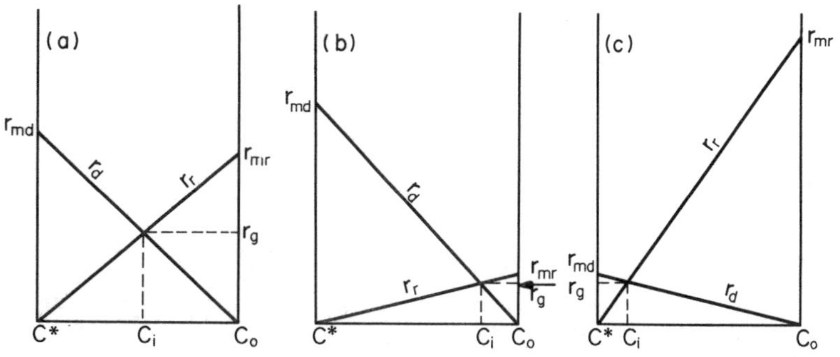

Fig. 28. Graphical representation of the global velocity, r_g, and the concentration of gaseous reactant, C_i, at the reaction interface: (a) mixed regime; (b) chemical regime; (c) diffusional regime

The two velocities r_d and r_r are not independent; the amount of gas which diffuses to the reaction interface cannot be larger than that which reacts, and vice versa; so r_d and r_r must be equal. The overall velocity r_g, is obtained by making r_r and r_d equal, as is indicated in Fig. 28(a). Nevertheless, when the two velocities are very different, the lowest maximum velocity may be considered, as a good approximation to the overall velocity (Fig. 28(b) and (c)). Then it can be said that the lower velocity controls the rate of the process. When chemical reaction velocity is the slower, the regime is "chemical", and when the diffusion velocity is the slower, it is "diffusional".

(A) FUNCTIONS THAT DESCRIBE THE PROGRESS OF THE TRANSFORMATION

The progress of a transformation can be described by the fraction of reactant transformed, F, or else by the relative penetration, f.

(i) The fraction of reactant transformed is the ratio between the volume transformed, $V_o - V_i$, and the initial volume, V_o:

$$F = \frac{V_o - V_i}{V_o} \tag{4.11}$$

for a flat particle:

$$F = \frac{e_o - e_i}{e_o} \tag{4.12}$$

for a sphere:

$$F = \frac{R_o^3 - R_i^3}{R_o^3} \tag{4.13}$$

In these relationships, V_o, e_o, and R_o are, respectively, the initial volume, the initial thickness of the flat particle, and the initial radius of the sphere, and V_i, e_i, and R_i the volume, thickness, and radius of the residual solid reactant.

(ii) The relative penetration is defined as the fraction of the thickness of the flat particle or of the radius of the sphere transformed:

for a flat particle:

$$f = \frac{e_o - e_i}{e_o} = F \tag{4.14}$$

for a sphere:

$$f = \frac{R_o - R_i}{R_o} = 1 - (1 - F)^{1/3} \tag{4.15}$$

(iii) Relationship between r_g, F, and f: The overall reaction velocity, r_g (in moles of solid transformed per unit of time), is proportional to the change of the volume of reactant per unit of time:

$$r_g = -q\frac{dV_i}{dt} \tag{4.16}$$

The proportionality coefficient, q, is the number of moles of reactant per unit of volume. From (4.11) to (4.16), a relationship between r_g, dF/dt, or df/dt can be established:

For a flat particle:

$$\frac{dF}{dt} = \frac{df}{dt} = -\frac{1}{V_o}\cdot\frac{dV_i}{dt} = \frac{r_g}{qe_oS_o} \tag{4.17}$$

for a sphere:

$$\frac{dF}{dt} = -\frac{1}{V_o}\cdot\frac{dV_i}{dt} = \frac{r_g}{qV_o} \tag{4.18}$$

and as

$$dV_i = 4\pi R_i^2 dR_i = S_i dR_i :$$

$$\frac{df}{dt} = -\frac{1}{R_o}\cdot\frac{dR_i}{dt} = -\frac{1}{R_oS_i}\cdot\frac{dV_i}{dt} = \frac{r_g}{qR_oS_i} \tag{4.19}$$

(B) DEDUCTION AND APPLICATION OF A GENERAL KINETIC LAW

When the velocities of reaction and diffusion are of the same order, the concentration C_i at the interface has a value between C_o and C*. The overall velocity is equal to both velocities, as indicated in Fig. 28(a), i.e.:

$$BS_i(C_i - C^*) = -SD\frac{dC}{dx} \tag{4.20}$$

(i) Flat Particle

For a flat particle, the reaction interface, S_i, and the surface, S, through which the gas diffuses, are equal to S_o. On the other hand, since the flux of gas between e_o and e_i is conservative, the concentration gradient of the reactive gas is constant (Fig. 27) and:

$$-\frac{dC}{dx} = \frac{C_o - C_i}{e_o - e_i} = \frac{C_o - C_i}{e_o f} \tag{4.21}$$

since $f = (e_o - e_i)/e_o$, by definition. From (4.20) and (4.21), a relationship between f and C_i can be deduced:

$$\frac{C_i - C^*}{C_o - C^*} = \frac{1}{1 + (Be_o f/D)} \tag{4.22}$$

Replacing r_g ($= r_r$) and C_i in (4.7) by the value derived from f or df/dt according to (4.17) and (4.22), we find:

$$qe_o\frac{df}{dt} = \frac{B(C_o - C^*)}{1 + (Be_o \,.\, f/D)} \tag{4.23}$$

or

$$\frac{1}{B}\cdot\frac{df}{dt} + \frac{e_o}{D}\cdot f\cdot\frac{df}{dt} = \frac{C_o - C^*}{qe_o} \tag{4.24}$$

Integrating this between f=0 and f, t=0 and t gives:

$$\frac{1}{B}\cdot f + \frac{e_o}{2D}\cdot f^2 = \frac{(C_o - C^*)t}{qe_o} \tag{4.25}$$

Chemical regime. If the overall velocity is equal to the maximum reaction velocity, r_{mr}, when this is much less than that of diffusion (Fig. 28b), then the regime is "chemical". This occurs when the product of reaction is very porous and allows the reactant gas to diffuse almost freely. The gas concentration at the interface is then approximately equal to C_o (Fig. 29). The overall velocity, r_g, equal to the maximum velocity of reaction, r_{mr}, is given by (4.7) when $C_i = C_o$. Solving this equation, makes it possible to relate the relative penetration, f, to time. In a simpler form, (4.25) may be used. In a chemical regime, B is small and D large, and (4.25) may be simplified to:

$$f = \frac{B(C_o - C^*)t}{qe_o} = \frac{t}{\tau_{chem.}} \tag{4.26}$$

where $$\tau_{chem.} = \frac{qe_o}{B\left(C_o - C^*\right)} \tag{4.27}$$

$\tau_{chem.}$ is equal to the total time necessary for the complete transformation of the solid reactant in a chemical regime. Therefore the relative penetration is proportional to the reaction time.

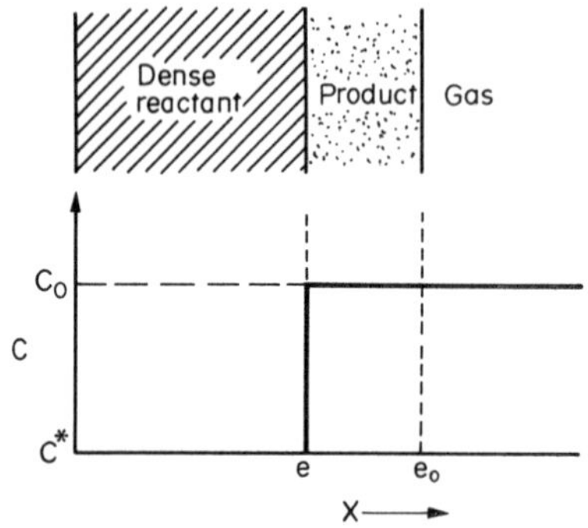

Fig. 29. Concentration gradient of gaseous reactant across the layer of product for a chemical regime

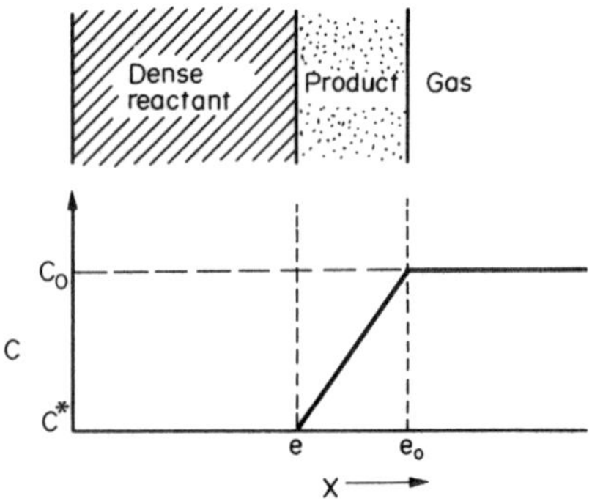

Fig. 30. Concentration gradient of gaseous reactant across the layer of product for a diffusion regime

Diffusion regime. This regime predominates when the chemical reaction velocity is much greater than the diffusion rate (Fig. 28(C)), i.e. when the reaction product is a dense solid. In this case, the concentration at the interface will be the equilibrium concentration C^*, as can be seen in Fig. 30. The overall velocity, equal to the maximum diffusion velocity, is the flux of reactant gas through the layer of product according to (4.10):

$$r_g = r_{md} = DS\,\frac{C_o - C^*}{e_o - e_i} \tag{4.28}$$

The solution to this equation is given by (4.25) in which D is considered small in comparison with B. Under these conditions, (4.25) is simplified:

$$f^2 = \frac{2D\left(C_o - C^*\right)}{qe_o^{\,2}}\,t = \frac{t}{\tau_{diff.}} \tag{4.29}$$

with $$\tau_{diff.} = \frac{qe_o^{\,2}}{2D\left(C_o - C^*\right)} \tag{4.30}$$

$\tau_{diff.}$ is the time for complete transformation in a diffusional regime. The relative penetration is then proportional to the square root of the time.

Mixed Regime. In a mixed regime, when neither reaction velocity nor diffusion velocity predominates, it is not possible to simplify equation (4.25). The overall velocity of transformation is therefore neither the maximum reaction velocity r_{mr} nor the maximum diffusion velocity r_{md}. By means of Fig. 28 it is possible to deduce that:

$$r_g = \left[\frac{C_o - C_i}{C_o - C^*}\right] r_{md} \qquad \text{and} \qquad r_g = \left[\frac{C_i - C^*}{C_o - C^*}\right] r_{mr}$$

In the same way, the total transformation time in a mixed regime, $\tau_{mix.}$, can be related to the total transformation time in a chemical or diffusional regime:

$$\tau_{diff.} = \left[\frac{C_o - C_i}{C_o - C^*}\right] \tau_{mix.} \tag{4.31}$$

and

$$\tau_{chem.} = \left[\frac{C_i - C^*}{C_o - C^*}\right] \tau_{mix.} \tag{4.32}$$

$\tau_{mix.}$ can be calculated from the sum of (4.31) and (4.32):

$$\tau_{mix.} = \tau_{diff.} + \tau_{chem.} \tag{4.33}$$

Inserting this value in equation (4.25), and rearranging, we obtain:

$$\frac{\tau_{chem.}}{\tau_{mix.}} f + \frac{\tau_{diff.}}{\tau_{mix.}} f^2 = \frac{t}{\tau_{mix.}} \tag{4.34}$$

The characteristic curves for the fraction of transformed reactant $F(t)$ or the relative penetration $f(t)$ as a function of time are plotted in Fig. 31.

(ii) Sphere

In the case of a sphere, the surface S through which the gas diffuses, is a function of the radius R. The gas flow between R_i and R_o, which is conservative ($\phi = r_d$ = constant), at a given time, can be expressed by:

$$r_d = - SD \frac{dC}{dR} = - 4\pi DR^2 \frac{dC}{dR} \tag{4.35}$$

Integrating (4.35) between $R = R_i$ and R_o, $C = C_i$ and C_o, gives:

$$r_d = \frac{4\pi DR_oR_i}{R_o - R_i} (C_o - C_i) \tag{4.36}$$

Equating the chemical reaction and diffusion velocities from (4.7) and (4.36), we have:

$$4\pi B R_i^2 (C_i - C^*) = \frac{4\pi D R_o R_i (C_o - C_i)}{R_o - R_i} \tag{4.37}$$

Replacing R_i by its value as a function of $f (R_i = R_o(1 - f)$ by definition), after simplifying, the following relationship is found between f and C_i:

$$B(1 - f)(C_i - C^*) = \frac{D(C_o - C_i)}{R_o f} \tag{4.38}$$

or

$$\frac{C_i - C^*}{C_o - C^*} = \frac{1}{1 + R_o \frac{B}{D} f(1 - f)} \tag{4.39}$$

With the help of (4.19) $(df/dt = r_g/qR_oS_i$, where $r_g = r_r = r_d)$ and (4.7), $C_i - C^*$ can be related to df/dt:

$$qR_o \frac{df}{dt} = \frac{r_g}{S_i} = B(C_i - C^*) \tag{4.40}$$

Inserting this in (4.39), we obtain:

$$qR_o \frac{df}{dt} = \frac{B(C_o - C^*)}{1 + \frac{B}{D} R_o f(1 - f)} \tag{4.41}$$

i.e.

$$\frac{1}{B}\frac{df}{dt} + \frac{R_o}{D} f(1 - f) \frac{df}{dt} = \frac{C_o - C^*}{qR_o} \tag{4.42}$$

Integration between $t=0$ and t, $f=0$ and f, gives:

$$\frac{1}{B} f + \frac{R_o}{D} f^2(3 - 2f) = \frac{C_o - C^*}{qR_o} t \tag{4.43}$$

<u>Chemical regime</u>. As has been seen in the case of a flat particle, the chemical regime predominates when the product formed is very porous. In this case, R_o/D in (4.43) is much smaller than $1/B$, and the second term of this equation can be neglected:

$$f = \frac{B(C_o - C^*)}{qR_o} t = \frac{t}{\tau_{chem.}} \tag{4.44}$$

and the total transformation time $\tau_{chem.}$ is equal to:

$$\tau_{chem.} = \frac{qR_o}{B(C_o - C^*)} \tag{4.45}$$

The relative penetration is then proportional to the reaction time in the same form as for the case of a flat particle.

Diffusion regime. When the product of reaction is a dense solid, diffusion is very slow compared with the chemical reaction; then, in (4.43), the term $(1/B)f$ can be neglected, giving:

$$f^2(3 - 2f) = \frac{D(C_o - C^*)}{qR_o^2} t = \frac{t}{\tau_{diff.}} \tag{4.46}$$

and the total time of transformation, $\tau_{diff.}$, is equal to:

$$\tau_{diff.} = \frac{qR_o^2}{D(C_o - C^*)} \tag{4.47}$$

In this case, the relative penetration is not a simple function of time.

Mixed regime. In the general case of a mixed regime, it is not possible to simplify (4.43). Defining the total time of transformation in a mixed regime in the same way as for a flat particle:

$$\tau_{mix.} = \tau_{diff.} + \tau_{chem.} \tag{4.48}$$

and inserting these values in (4.43), we have:

$$\frac{\tau_{chem.}}{\tau_{mix.}} f + \frac{\tau_{diff.}}{\tau_{mix.}} f^2(3 - 2f) = \frac{t}{\tau_{mix.}} \tag{4.49}$$

Figure 32 indicates the variation of the relative penetrations with time, in the three cases studied: chemical, diffusion, and mixed regimes.

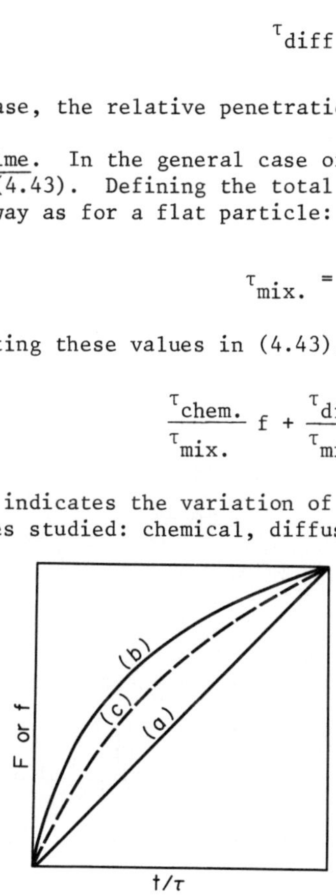

Fig. 31. Fraction of reactant transformed, F, and relative penetration, f, versus time in the transformation of a flat particle

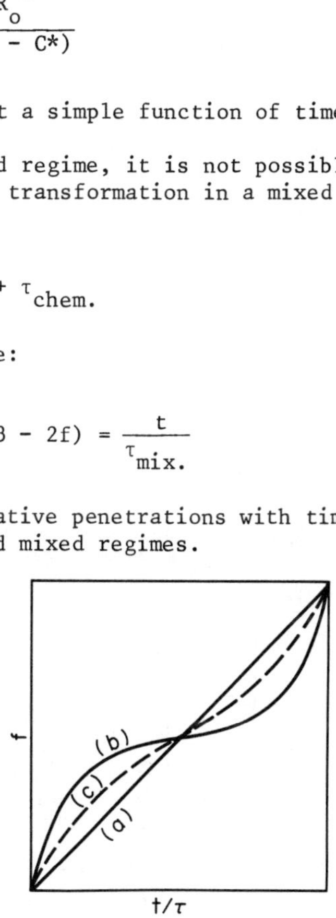

Fig. 32. Relative penetration, f, versus time in the transformation of a sphere

(a) Chemical; (b) Diffusion; (c) Mixed regime

Equations (4.33) and (4.48) show that, in a mixed regime, the total transformation time is the sum of the times that correspond to the maximum velocities of transformation in the diffusion and chemical regimes.

4.1.3. SOLUTION OF A DENSE SOLID IN A FLUID†

The solution of a soluble solid in a fluid differs from the previous case in that there is no layer of dense or porous product to impede the passage of fluid to the solid reactant interface. Nevertheless, there is always a film of fluid, the "boundary layer", which is formed adjacent to the surface and reduces the velocity of solution, since the dissolved molecules or ions have to diffuse through it toward the reactant interface. Even when agitation of the fluid is violent (turbulent regime), this layer does not disappear, and the thickness is 10^{-2} to 10^{-3} cm. In the following analysis, which is developed to demonstrate the interference of the boundary layer in the velocity of transformation, it is assumed, as before, that the fluid concentration does not change on passing over the particle, and that the boundary layer has a constant thickness, δ.

The calculations for expressing the solution of the solid are developed in the way described previously, with the exception that, in this case, the notion of relative penetration has no meaning. As previously, the reaction velocity is expressed in the form:

$$r_r = BS_i(C_i - C^*) \tag{4.7}$$

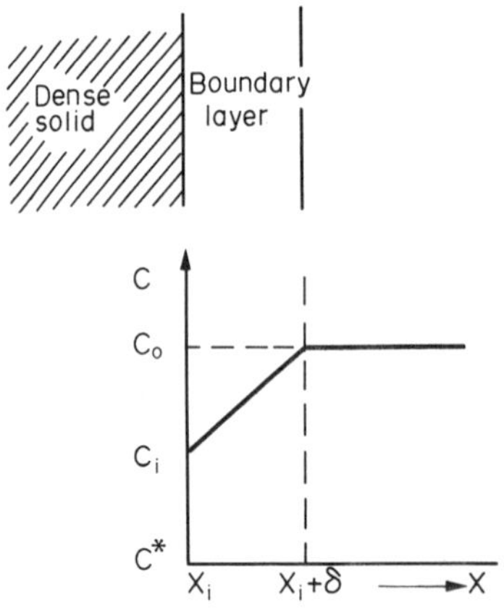

Fig. 33. Solution of a flat particle

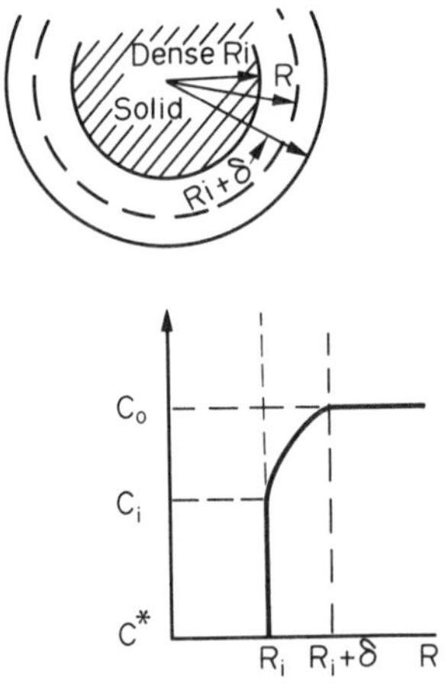

Fig. 34. Solution of a sphere

Concentration gradient in the boundary layer

†This paragraph is concerned only with particles which are solid at the temperature of solution (for example W in steel). If the solid is molten at this temperature (scrap steel), the kinetics to be considered are those of melting, followed by solution of one liquid in another.

and the diffusion of the gaseous or liquid reactant through the boundary layer by:

$$r_d = -DS\frac{dC}{dx} \qquad (4.10)$$

The overall velocity, r_g, equal to r_r and r_d, is related to the disappearance of the solid reactant by:

$$r_g = -q\frac{dV_i}{dt} = qV_o\frac{dF}{dt} \qquad (4.17) \text{ or } (4.18)$$

(i) Flat Particle

The flux through the boundary layer is conservative, therefore the concentration gradient is constant, since the surface of the particle does not change during transformation. According to Fig. 33, the value of this gradient will be:

$$-\frac{dC}{dx} = \frac{C_o - C_i}{\delta} \qquad (4.50)$$

C_o and C_i are the concentrations of reactant in the bulk solution and at the surface of the particle, respectively. Applying these values to (4.10), and equating the expressions for the velocities of reaction and diffusion, gives:

$$\frac{C_i - C^*}{C_o - C^*} = \frac{D/\delta}{B + D/\delta} \qquad (4.51)$$

Replacing C_i by its value in (4.7) or (4.10), the velocity of solution can be expressed as a function of C_o:

$$r_g = \frac{BSD/\delta}{B + D/\delta}(C_o - C^*) \qquad (4.52)$$

In this equation, the overall velocity is expressed in the same form as for a first-order reaction, notwithstanding the influence of diffusion through the boundary layer.

According to (4.52), when the concentration of reactant around the solid does not change, the velocity of solution is constant. Equating (4.17) and (4.52), and integrating between F=0 and F, t=0 and t, the fraction of solid transformed is obtained as a function of time:

$$F = \frac{e_o - e_i}{e_o} = \frac{B(D/\delta)}{e_o \cdot q(B + D/\delta)}(C_o - C^*)t \qquad (4.53)$$

The thickness decreases linearly with time.

(ii) Sphere

In the case of a sphere surrounded by a fluid (Fig. 34), the flux of reactant is conservative in the boundary layer, but the gradient is not constant, because the area involved increases with the distance from the sphere. Equating (4.7) and (4.10), gives:

$$4\pi BR_i^{\,2}\,(C_i - C^*) = -\,4\pi DR^2\,\frac{dC}{dR} \tag{4.54}$$

R_i is the instantaneous radius of the sphere, and R is a radius between R_i and $(R_i + \delta)$. Integrating (4.54) between $C = C_i$ and C_o, $R = R_i$ and $(R_i + \delta)$, we find:

$$\frac{C_i - C^*}{C_o - C^*} = \frac{(R_i + \delta)(D/\delta)}{BR_i + (R_i + \delta)(D/\delta)} \tag{4.55}$$

Replacing C_i in (4.7) by the value in (4.55), the velocity of transformation, r_g, is given as a function of R_i and C_o.

$$r_g = 4\pi BR_i^{\,2}\,\frac{(R_i + \delta)(D/\delta)}{BR_i + (R_i + \delta)(D/\delta)}\,(C_o - C^*) \tag{4.56}$$

When R_i is $> 100\ \mu$, δ can be neglected, other than in such terms as D/δ.

Since $r_g = -\,q\,\frac{dV_i}{dt} = -\,4\pi qR_i^{\,2}dR_i/dt$, (4.56) can be written:

$$\frac{BR_i + (R_i + \delta)(D/\delta)}{(R_i + \delta)}\,dR_i = -\,\frac{B(D/\delta)}{q}\,(C_o - C^*)\,dt \tag{4.57}$$

Integrating (4.57) between $R_i = R_o$ and R_i, and $t = 0$ and t, gives:

$$(B + D/\delta)(R_o - R_i) - B\delta\,\log\frac{R_o + \delta}{R_i + \delta} = \frac{B(D/\delta)}{q}\,(C_o - C^*)t \tag{4.58}$$

When R_i is large with respect to δ, (4.58) can be simplified to:

$$\frac{R_o - R_i}{R_o} = \frac{B(D/\delta)(C_o - C^*)}{qR_o(B + D/\delta)}\,t \tag{4.59}$$

Thus, R_i decreases linearly with time, and (4.59) as well as (4.53) can be put into the form:

$$\frac{X_o - X_i}{X_o} = \frac{1}{X_o q(\delta/D + 1/B)}\,(C_o - C^*)t \tag{4.60}$$

X_o and X_i represent the dimensions of the particle (thickness or radius). On the other hand, (4.26) and (4.44), which describe the transformation of a particle in a chemical regime (i.e. when diffusion across the solid transformed and/or the boundary layer is very rapid), can be written:

$$\frac{X_o - X_i}{X_o} = \frac{1}{X_o q(1/B)} (C_o - C^*)t \qquad (4.61)$$

(4.60) and (4.61) have similar forms, indicating that reaction at the interface and diffusion through the boundary layer act as two resistances in series, of values δ/D and $1/B$.

The solution of a piece of refractory metal in liquid steel is a first-order reaction, similar to (4.60). In the case of scrap, because the melting point is the same as that of the steel, there is no solution, only a question of heat absorption, fusion and diffusion.

4.1.4. TRANSFORMATION OF A POROUS PARTICLE BY A GAS

When a dense solid reacts with a gas, the reaction interface is simple and well-defined, and the two phenomena that control the overall velocity of reaction (chemical reaction and diffusion), are separated in space. When the solid is porous, the gas penetrates the pores and reacts throughout the mass of the particle and the reaction interface has no simple geometry. The phenomena of reaction and diffusion are dispersed throughout the solid in such small volumes that they can be considered as occurring in the same space.

The progress of reaction cannot be observed, except in certain cases, by movement of the interface because it is not well defined. The kinetic expression which describes the process is different from that for a dense solid.

(A) GENERAL EQUATION FOR THE CONCENTRATION GRADIENT OF A GAS REACTING WITH A POROUS SOLID

The elemental volume $dx \, . \, dy \, . \, dz$ of Fig. 35 is porous, and the gaseous reactant flows through it in a direction parallel to dx.

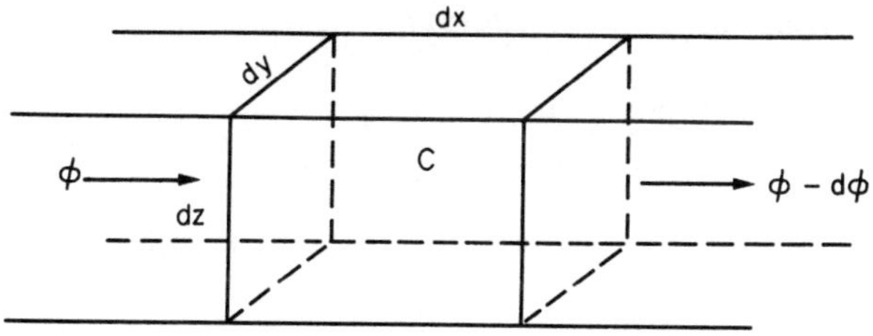

Fig. 35. Variation of reactant gas flow in an elemental volume of porous solid

The flux ϕ of reactant gas enters at one end, and $(\phi - \{d\phi/dx\}dx)$ comes out at the other. The change of flux $d\phi$ results from the reaction that consumes the gaseous reactant within the volume $dx \, . \, dy \, . \, dz$, since the amount of gas that accumulates in the pores in unit time is negligible. Consequently:

$$-\frac{d\phi}{dx} dx = r_r \qquad (4.62)$$

The flow of gas is given by Fick's first law:

$$\phi = -DS\frac{dC}{dx} = -D\frac{dC}{dx}dz\,dy \tag{4.63}$$

and if D is assumed constant:

$$-\frac{d\phi}{dx} = D\frac{d^2C}{dx^2}dz\,dy = D\frac{d^2(C - C^*)}{dx^2}dz\,dy \tag{4.64}$$

The reaction velocity, r_r, retains the form of (4.7) in which the reaction interface, S_i, is proportional to the volume dx . dy . dz. The coefficient of proportionality, s_e, is the specific surface of the porous solid, assumed constant:

$$r_r = BS_i(C - C^*) = Bs_e(C - C^*)\,dx\,dy\,dz \tag{4.65}$$

Replacing $d\phi/dx$ and r_r by their values in (4.62), gives:

$$\frac{Dd^2(C - C^*)}{dx^2} = Bs_e(C - C^*) \tag{4.65a}$$

or

$$\frac{d^2(C - C^*)}{dx^2} - \frac{C - C^*}{h^2} = 0 \tag{4.66}$$

with

$$h^2 = \frac{D}{Bs_e} = \frac{D}{kRT(1 + 1/K)s_e} \tag{4.67}$$

For a sphere, in which the flow is radial:

$$\frac{d^2(C - C^*)}{dR^2} + \frac{2}{R}\frac{d(C - C^*)}{dR} - \frac{1}{h^2}(C - C^*) = 0 \tag{4.68}$$

and, in general:

$$\nabla^2(C - C^*) - \frac{1}{h^2}(C - C^*) = 0 \tag{4.69}$$

(B) FORMS OF THE CURVES C(x) AND C(R)

By solving (4.66) and (4.68), it is possible to determine the gas concentration profile for a flat or spherical particle:

Flat particle. For a particle of thickness $2e_o$, completely surrounded by gas of composition C_o, the solution of (4.66) is of the form:

$$C - C^* = Ae^{-x/h} + Be^{x/h} \tag{4.70}$$

The boundary conditions are:

When $x = 0$, $\frac{dC}{dx} = 0$ (for reasons of symmetry)

When $x = e_o$, $C = C_o$

Under these conditions, the values of A and B can be determined:

$$B = A = \frac{C_o - C^*}{2 \cosh(e_o/h)} \tag{4.71}$$

The solution of (4.66) will then be:

$$\frac{C - C^*}{C_o - C^*} = \frac{\cosh(x/h)}{\cosh(e_o/h)} \tag{4.72}$$

Sphere. The general solution of (4.68) is:

$$C - C^* = \frac{A}{R} \sinh(R/h) + \frac{B}{R} \cosh(R/h) \tag{4.73}$$

the boundary conditions being:

When $R = R_o$, $C = C_o$

When $R = 0$, $\frac{dC}{dR} = 0$ (for reasons of symmetry)

From these conditions, A and B can be calculated:

$$B = 0, \qquad A = \frac{R_o(C_o - C^*)}{\sinh(R_o/h)}$$

The solution of (4.68) is then:

$$\frac{C - C^*}{C_o - C^*} = \frac{R_o \sinh(R/h)}{R \sinh(R_o/h)} \tag{4.74}$$

Figure 36 indicates how the concentration of reactant gas varies with the depth of penetration into a flat particle (a), and a sphere (b), for different values of h. For high values of h (which correspond to high porosity, low temperature and a large diffusion coefficient, the gas concentration varies only slightly from the exterior to the interior: the velocity of diffusion is high compared with the rate of chemical reaction, and the chemical regime predominates. Conversely, for a small value of h (low porosity, high temperature), the gas reacts in a thin layer and the velocity of diffusion is far less than that of the chemical reaction. The diffusion regime is then predominant.

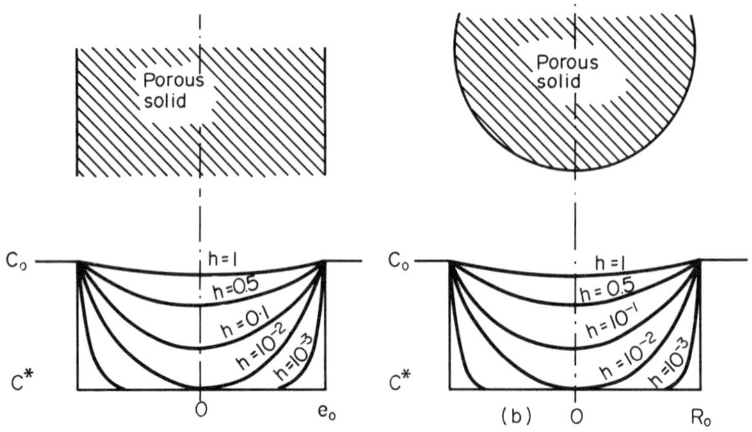

Fig. 36. Concentration of the gaseous reactant in a porous solid for different values of h

(a) Flat particle; (b) sphere

(C) RATE OF TRANSFORMATION OF REACTANT

The determination of the total fraction, F, of reactant transformed is complex. It is necessary to know the variation with time, dF/dt, at each point of the particle, and to integrate the function for the total volume from 0 to t.

This study is limited to two extreme conditions, which correspond, in one case, to uniform transformation throughout the particle, and in the other, to localised transformation in a thin layer.

Uniform transformation throughout the particle. When h is large compared with the dimensions of the particle, the composition of the gas is approximately equal to C_o throughout. The absence of a gradient indicates that the velocity of diffusion is much higher than that of chemical reaction, and the overall velocity of transformation is approximately equal to the maximum velocity of reaction (Fig. 28(b)):

$$r_g = r_r = BS_i(C_o - C^*) \tag{4.7}$$

As reaction occurs throughout the whole volume, the surface reaction is equal to $s_e V_o$. Taking into account (4.17) or (4.18) relating r_g and dF/dt, we obtain:

$$\frac{dF}{dt} = \frac{r_g}{qV_o} = \frac{B}{qV_o} s_e V_o (C_o - C^*) = \frac{B}{q} s_e (C_o - C^*) \tag{4.75}$$

and by integration between t = 0 and t, F = 0 and F:

$$F = \frac{Bs_e(C_o - C^*)}{q} t = \frac{t}{\tau_{chem.}} \tag{4.76}$$

The total time of transformation $\tau_{chem.}$ is then equal to:

$$\tau_{chem.} = \frac{q}{Bs_e(C_o - C^*)} \tag{4.77}$$

The fraction transformed does not depend on the volume nor the shape of the particle, since the transformation progresses at the same speed everywhere.

Localised transformation in a thin layer. When h is small, transformation takes place in a thin layer of solid (Fig. 36), and after a certain time, three different zones will exist; one where the solid has not yet been transformed, one in the course of transformation where there are simultaneous diffusion and reaction and which moves toward the centre, and finally, a completely transformed zone through which the gas diffuses.

The velocity of diffusion will vary in each zone, depending on whether the porosity of the solid product is equal to, or different from, that of the reactant. When the porosity of the product is of the same as that of the reactant, the velocity of transformation is controlled by diffusion in both zones. The gas concentration gradient does not change on passing from the transformation zone to the transformed zone (Fig. 37(a)) since the flux is conservative. As the zone undergoing transformation is thin, this case is very similar to that of a dense solid in a diffusion regime or the mixed regime, previously described, and the relative penetration, f, can be used to define the progress of transformation.

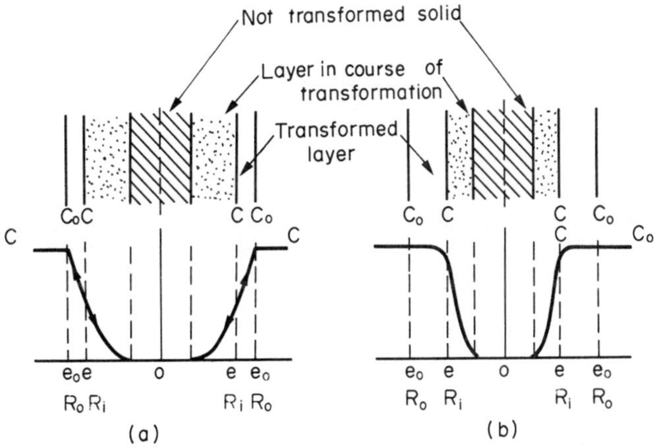

Fig. 37. Localised transformation in a narrow front

(a) The porosities of solid reactant and solid product are the same

(b) The porosity of the solid product is much greater than that of the solid reactant

A more interesting case occurs when the transformed layer is much more porous than the original reactant (Fig. 37(b)), and does not offer any resistance to the passage of gases. The velocity of transformation is limited only by the rate of diffusion through the solid in course of transformation, where the porosity is assumed to be constant. This velocity may be determined by the flow that crosses the transformed zone (where the flow is conservative) and, in particular, through the surface that separates the zone in course of transformation from the transformed zone.

Flat particle. The transformation velocity is equal to:

$$r_g = -DS\left[\frac{dC}{dx}\right]_{x=e} \tag{4.78}$$

D is the diffusion coefficient in the zone in course of transformation. The concentration gradient at the surface that separates the two zones is obtained by differentiation of (4.72). Substituting this value in (4.78), we obtain:

$$r_g = DS\,\frac{C_o - C^*}{h}\tanh(e/h) \tag{4.79}$$

The hyperbolic tangent of e/h gives a value near to 1, when e/h is above 2 ($\tanh 2 = 0.965$). If h is small, as has been assumed, and the thickness of the particle is large in comparison with h, the velocity of transformation is:

$$r_g = DS\,\frac{C_o - C^*}{h} \tag{4.80}$$

The thin transformation front advances at a constant velocity, and the law of advance deduced from (4.17) may be expressed by the relative penetration, f, or the fraction of reactant transformed, F:

$$f = F = \frac{t}{\tau_{diff.}} \tag{4.81}$$

with

$$\tau_{diff.} = \frac{qe_o h}{D(C_o - C^*)} \tag{4.82}$$

Sphere. The calculation for a spherical particle gives a result analogous to that for a flat particle, provided that h is small compared with R_i. The overall velocity can be written:

$$r_g = -DS_i\left[\frac{dC}{dR}\right]_{R=R_i} \tag{4.83}$$

The term dC/dR is determined by differentiating (4.74):

$$r_g = \frac{DS_i(C_o - C^*)}{h}\left[\frac{1}{\tanh(R_i/h)} - \frac{h}{R_i}\right] \tag{4.84}$$

When h is small compared with R_i, the term $\tanh(R_i/h)$ is close to 1, and h/R_i can be neglected. The expression for the velocity of transformation then takes on a simpler form:

$$r_g = \frac{DS_i(C_o - C^*)}{h} \tag{4.85}$$

As the layer in which the transformation takes place is thin, the relative penetration, f, may be utilised to define the degree of transformation. The relationship between r_g and df/dt provided by (4.19) gives:

$$\frac{df}{dt} = \frac{r_g}{qR_oS_i} = \frac{D(C_o - C^*)}{qR_oh} \tag{4.86}$$

By integrating (4.86) between t = 0 and t, f = 0 and f, we find:

$$f = \frac{D(C_o - C^*)}{qR_oh}\, t = \frac{t}{\tau_{diff.}} \tag{4.87}$$

in which:

$$\tau_{diff.} = \frac{qR_oh}{D(C_o - C^*)} \tag{4.88}$$

An analogy exists between (4.81) and (4.87) (diffusion regime in porous particles) and (4.26) and (4.44) (chemical regime in dense particles). This comes from the fact that the transformation in both cases is highly concentrated. Nevertheless, it is interesting to compare the manner in which the reaction rates vary with temperature.

Transformation of a dense particle in a chemical regime. In a chemical regime, when the concentration at the reaction interface of a dense solid is equal to C_o, the overall velocity is:

$$r_g = r_r = kRT\left(1 + \frac{1}{K}\right)S_i(C_o - C^*) \tag{4.7}$$

k and K vary with temperature, as indicated in (4.9a) and (4.9b), and the apparent activation energy can be deduced directly from them.

Transformation of a porous particle in a diffusion regime. Replacing h by its value in (4.85), we obtain:

$$r_g = \frac{DS_i(C_o - C^*)}{h} = \sqrt{kRT(1 + 1/K)Ds_e}\;\; S_i(C_o - C^*) \tag{4.89}$$

The specific surface, s_e, does not change and the diffusion coefficient D varies little with temperature. The fact that k and K, the values of which are greatly affected by temperature, are present in a square root expression in (4.89), indicates that the apparent activation energy should be half of that for a dense solid.

An apparent activation energy of about 40 kcal mole^{-1} has been found experimentally for the reduction of a dense particle of FeO, and about 20 kcal mole^{-1} for a porous particle.

4.2. KINETICS OF ISOTHERMAL TRANSFORMATION OF A PARTICULATE BED BY A GAS

A bed of particulate ore, concentrate or pellets, can be considered, as a first approximation, as a collection of spheres. The gas circulates between the particles by convection (generally forced), and through them by diffusion.

4.2.1. PHENOMENA WHICH GOVERN TRANSFORMATION OF A BED OF SOLID PARTICLES

In the case of a single spherical particle, it can be assumed that the gas composition is virtually unchanged on passing over it, but this hypothesis cannot be applied to a bed of particles, and the variations of the composition of solid and gas relative to the depth of the bed as well as to the elapsed time must be considered. Under these conditions, it is necessary to relate the extent of transformation of solid and gas by mass balances in addition to the kinetic data.

(A) APPARENT ORDER OF REACTION WITH RESPECT TO THE SOLID

The proportionality constant $1/\tau$ given by (4.45) and (4.82) is related to $(C_o - C^*)$, the difference between the concentration in the gas surrounding the sphere, and that at the equilibrium. Thus, the relative penetration, f, can be written:

$$f = b(C_o - C^*)t \tag{4.90}$$

In a study of the transformation of a bed, it can be assumed that the transformation of each particle follows a similar law. Nevertheless, as the gas concentration, C, varies with the time, t, and the height, z, of the column, the previous equation will take the form:

$$f = b[C(z, t) - C^*]t \tag{4.90a}$$

By comparing the value of df/dt in (4.19) with that derived from (4.90), we obtain:

$$\frac{df}{dt} = b(C - C^*) = \frac{r_g}{4\pi q R_o R_i^2} \tag{4.91}$$

On the other hand, a relationship is obtained between R_i and F from (4.15):

$$\frac{R_i}{R_o} = (1 - F)^{1/3} \tag{4.15}$$

Inserting this value in (4.91), r_g can be related to F and C:

$$r_g = 4\pi q b R_o^3 (1 - F)^{2/3}(C - C^*) = 3qbV_o(1 - F)^{2/3}(C - C^*) \tag{4.92}$$

Thus the transformation velocity proportional to the reactant gas concentration according to the earlier hypothesis, has an apparent order of 2/3 with respect to the solid. In many researches, this fractional order has been found for a single particle as well as for a thin bed. For this reason, the kinetic relation (4.92) will be maintained, and in a deep bed the velocity of transformation in a volume Sdz of thickness dz and of area S will be given by:

$$r_g = 3bq(1 - F)^{2/3}(C - C^*)Sdz \tag{4.93}$$

In this equation, q is the number of moles of solid product per unit volume of bed.

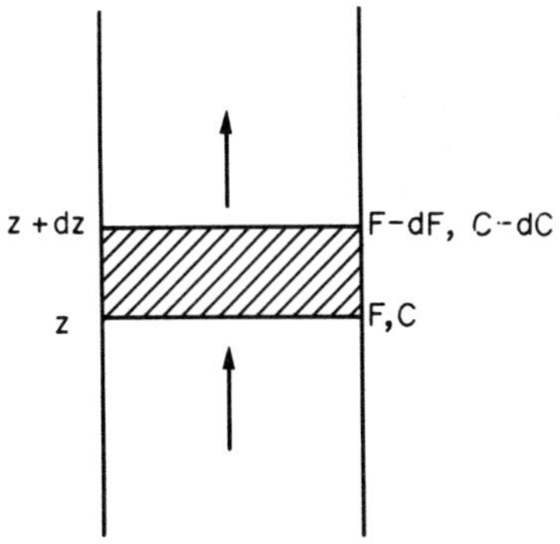

Fig. 38. Mass balance in a volume Sdz

(B) MASS BALANCE OF REACTANTS AND PRODUCTS

It is possible to deduce additional relationships between the transformation velocity in a volume Sdz, the fraction of transformed solid and the concentration of gaseous reactant, by means of mass balance.

Mass balance for the solid. The number of moles of solid transformed in the volume Sdz during the time dt is equal to the total number of moles of solid qSdz multiplied by the rate of production of transformed material dF/dt:

$$r_g(z) = qS \frac{dF}{dt} dz \tag{4.94}$$

Mass balance for the reactant gas. The quantity of gas transformed in the volume Sdz during the time dt is the difference between the quantities entering and leaving. The number of moles, n, of reactant gas entering is equal to the gas flux $n_g S$ multiplied by the fraction of gaseous reactant $C/\Sigma C_i$ and the time dt:

$$n = n_g S \frac{C}{\Sigma C_i} dt \tag{4.95}$$

ΣC_i is the sum of the concentrations of reactant, product, and inert gas. For an ideal gas, $C_i = P/RT$. This converts (4.95) to:

$$n = n_g S \frac{RT}{P} Cdt \tag{4.96}$$

In the same time, (n - dn) moles of reactant leave the volume Sdz. Assuming that the temperature and pressure are constant all through the bed, and that the flow of gas n_g does not change during passage, i.e. the number of moles of product is equal to the number of moles of reactant, then:

$$n - dn = n_g S \frac{RT}{P} (C - dC) dt \tag{4.97}$$

The transformation velocity in the volume Sdz may be expressed as a function of the gas concentration, in the form:

$$r_g(z) = -\frac{dn}{dt} = -n_g S \frac{RT}{P} dC \tag{4.98}$$

(C) GENERAL EQUATIONS

By eliminating $r_g(z)$ from (4.93), (4.94) and (4.98), two relationships are obtained between F and C:

$$q \frac{dF}{dt} = -\frac{RTn_g}{P} \frac{dC}{dz} \tag{4.99}$$

and

$$\frac{dF}{dt} = 3b(1 - F)^{2/3}(C - C^*) \tag{4.100}$$

These are not enough to describe the transformation of a bed completely, i.e. to determine C(t), C(z), F(t) and F(z) for the general case. Another relationship between F and F will be determined in a permanent regime.

4.2.2. TRANSFORMATION IN A STATIC BED

(A) QUALITATIVE DESCRIPTION

When transformation starts, the bed consists only of reactant solid. When gas is introduced, it reacts with the solid, and its concentration decreases. If the bed is deep enough, it will decrease from C_o to C^*, as can be seen in Fig. 39 for different times of reaction. During the same times, the fraction of solid transformed, F, increases in the manner indicated in Fig. 40. When transformation of the first layer is complete, curves C(z) and F(z) take on definite shapes that move upwards with a constant velocity.

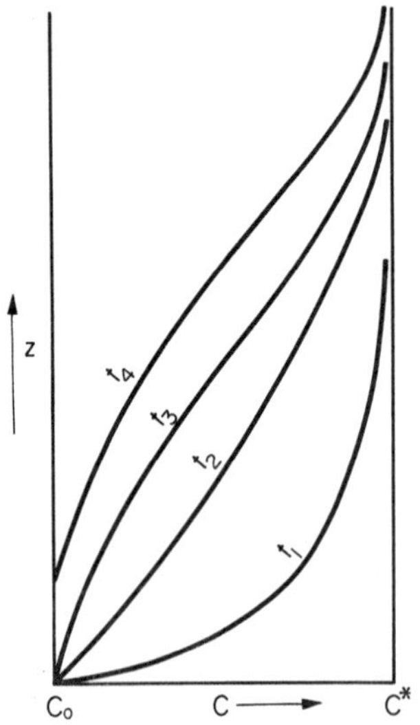

Fig. 39. Profile of concentration of the gaseous reactant in a static bed for different times

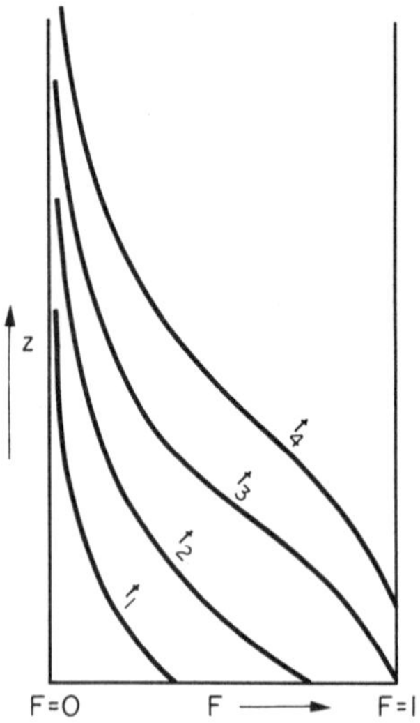

Fig. 40. Fraction of transformed solid in a static bed for different times

Until transformation of the first layer is complete, the regime is transient, since the shapes of $F(z)$ and $C(z)$ change with time. This is the initial transient period. Starting from the time when $C(z)$ and $F(z)$ take on definite shapes which are independent of time, the regime will be permanent. When the bed is almost completely transformed, the gas concentration at the exit approaches that at the entry, becoming identical with it when transformation is complete. This is the final transient regime. The relative importance of each regime depends on the height of the bed. If the bed is thick, the beginning and end of the transformation are of little importance, but in a shallow bed or in one of medium height, the permanent regime may never be established.

(B) PERMANENT REGIME

In a permanent regime, the transformation zone of height, H, is limited by a lower front in which the gas has a concentration, C_o, and where the solid reactant has been completely transformed, $(F = 1)$, and by an upper front where the gas reaches the equilibrium composition C^* and the solid reactant has not yet started to transform, $(F = 0)$. The lower front is well defined, since transformation of a layer, in the same way as for each sphere in it, is effected in a limited time. To prove this, it is enough to integrate (4,100) with $C = C_o$, between $F = 0$ and 1, $t = 0$ and τ. On the contrary, definition of the upper front is more difficult because the gas concentration approaches its equilibrium value, C^*, asymptotically. Comparing (4.99) and (4.100), the gas concentration gradient, dC/dz, is proportional to $(C - C^*)$ when F approaches zero, indicating an exponential variation of the function $C(z)$. Nevertheless, to fix a practical limit to the height, H, of the transformation zone, the upper front is defined as the plane where the concentration of the gaseous reactant approaches the equilibrium value C^* (i.e. $(C - C^*)/(C_o - C^*) = \varepsilon(\varepsilon \simeq 0)$).

Mass balance in a permanent regime. As has been illustrated (Fig. 39), the curve $C(z)$, i.e. the zone of transformation, is displaced upwards with a constant velocity dz/dt in a permanent regime. (4.99) can then be written:

$$q \, dF \frac{dz}{dt} = -\frac{RT}{P} n_g \, dC \tag{4.101}$$

The term dz/dt is constant, and $q\, dz/dt$ represents the velocity of advance of the zone, in moles of solid reactant transformed per unit time and section of the bed, or, in the case of a continuous process, the number of moles of solid reactant, n_s, charged and transformed per unit time and section. In (4.101), $q\, dz/dt$ is replaced by $- n_s$:

$$n_s \, dF = \frac{RT}{P} n_g \, dC \tag{4.102}$$

The change of sign is due to the fact that the gas and solid move in opposite directions. Integration of this equation between the boundary conditions, $F = 1$ and F, $C = C_o$ and C on the one side, and between $F = 1$ and 0, $C = C_o$ and C^* on the other, brings out the simple relationship between F and C in a permanent regime:

$$1 - F = \frac{C_o - C}{C_o - C^*} \qquad \text{or} \qquad F = \frac{C - C^*}{C_o - C^*} \tag{4.103}$$

(4.102) and (4.103) show that, in a permanent regime, F and C vary linearly and in the same direction (Fig. 41). The straight line EB is called the "operating line" and its slope, according to (4.102), is equal to:

$$\frac{dF}{dC} = \frac{RTn_g}{Pn_s}$$

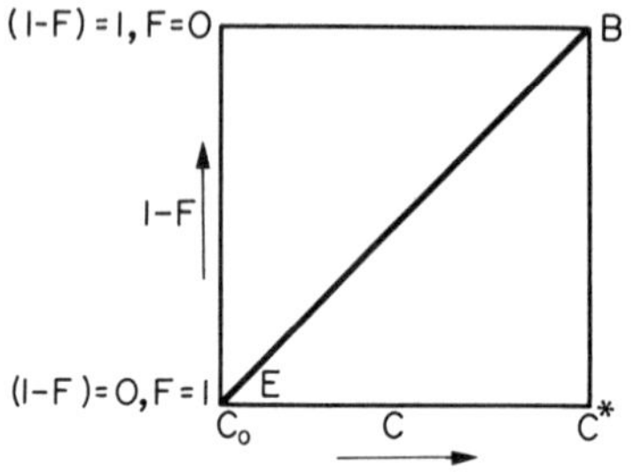

Fig. 41. Operating line in a permanent regime (ideal case)

This line represents graphically the mass balance in a permanent regime in the ideal case of a deep bed where the solid reactant comes out completely transformed at the bottom, and the gas is at the equilibrium composition at some level below the top.

Variation of F and C with time at a point in a bed in a permanent regime. In order to obtain the relationship between the fraction of solid transformed and the elapsed time at a given point in a deep static bed, (C - C*) in (4.100) is replaced by its value as a function of F from (4.103):

$$\frac{dF}{dt} = 3b(C_o - C^*)F(1 - F)^{2/3} \qquad (4.104)$$

Integrating between F = F and 1, t = t and t_p, where t_p is the practical transformation time of a zone of the bed, gives:

$$6b(C_o - C^*)(t_p - t) = -\ln\left[\frac{\left(1 - (1 - F)^{1/3}\right)^2}{(1 - F)^{2/3} + (1 - F)^{1/3} + 1}\right]$$

$$+ 2\sqrt{3}\arctan\left[\frac{2(1 - F)^{1/3} + 1}{\sqrt{3}}\right] - \frac{\pi\sqrt{3}}{3} \qquad (4.105$$

By giving F a value ε near to zero, which defines the upper limit of reactioı in the bed, t_p can be found from:

$$6b(C_o - C^*)t_p = \frac{\pi\sqrt{3}}{3} - \ln\left(\frac{\varepsilon^2}{27}\right) \qquad (4.10$$

According to (4.106), t_p depends on the value, ε, of F selected to define the upper limit of the transformation zone. In Fig. 42, (1 - F) is plotted as a function of time at any point of the bed.

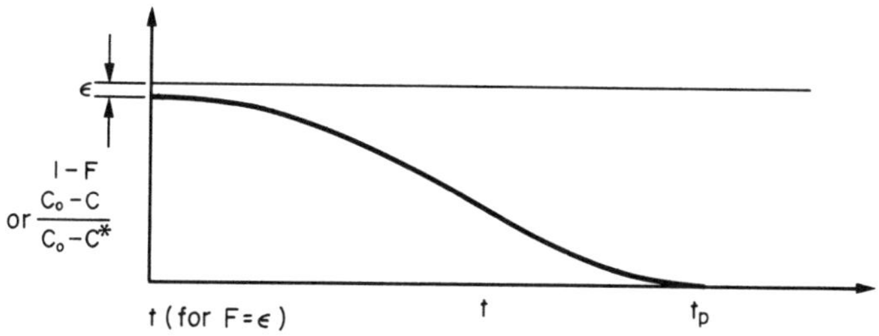

Fig. 42. Variation of F and C with time

The change of gas concentration with time at a point in the bed in a permanent regime can be deduced from the change in F(t) given by (4.105), replacing (1 - F) by $(C_o - C)/(C_o - C^*)$.

C(z) and F(z) profiles. A relationship can be obtained between dC/dz and C by comparing (4.99), (4.100) and (4.103):

$$\frac{dC}{dz} = -\frac{3bPq}{n_g RT}(C_o - C^*)\left[1 - \frac{C - C^*}{C_o - C^*}\right]^{2/3} \cdot \left[\frac{C - C^*}{C_o - C^*}\right] \tag{4.107}$$

This equation has the same form as (4.104). Integrating between $z = z_0$ and z, $C = C_o$ and C, gives:

$$\frac{3bPq}{RTn_g}(C_o - C^*)(z - z_o) = \frac{1}{2}\ln\left[\frac{\left(\frac{C_o - C}{C_o - C^*}\right)^{2/3} + \left(\frac{C_o - C}{C_o - C^*}\right)^{1/3} + 1}{\left\{1 - \left(\frac{C_o - C}{C_o - C^*}\right)^{1/3}\right\}^2}\right]$$

$$+ \sqrt{3}\arctan\left[\frac{2\left(\frac{C_o - C}{C_o - C^*}\right)^{1/3} + 1}{\sqrt{3}}\right] - \frac{\pi\sqrt{3}}{6} \tag{4.108}$$

The practical height H of the transformation zone is obtained by inserting a value of C close to C*, i.e.

$$(C - C^*)/(C_o - C^*) = \varepsilon \qquad \text{or} \qquad (C_o - C)/(C_o - C^*) = 1 - \varepsilon$$

$$\frac{3bPq}{n_g RT}(C_o - C^*)H = -\ln\frac{\varepsilon^2}{27} + \frac{\pi\sqrt{3}}{3} \tag{4.109}$$

(4.109) shows that the practical transformation height, H, is proportional to the gas flow, n_g. Figure 43 indicates the manner in which the gas concentration varies through the bed.

By replacing $(C_o - C)/(C_o - C^*)$ by $(1 - F)$ from (4.103), one obtains the F(z) profile which is similar to that for C(z).

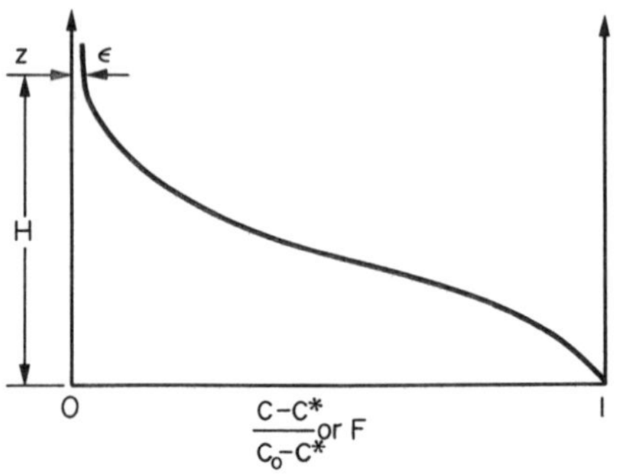

Fig. 43. C(z) and F(z) profile for a permanent regime

(C) OVERALL VELOCITY OF TRANSFORMATION AND FRACTION OF SOLID TRANSFORMED IN A BED

The overall velocity of transformation of a bed, r_{gl}, can be determined by the amount of gas transformed on passing through it. If C_o and C_s are the concentrations of reactive gas at the entrance and exit of the reactor, this velocity will be equal to the gas flow $n_g S$ multiplied by the fraction of gas transformed:

$$r_{gl} = n_g S \frac{RT}{P} (C_o - C_s) \qquad (4.110)$$

Deep bed. During the initial transient and permanent regimes, C_s remains practically equal to C*. During the final transient regime, C_s varies continuously, as is indicated by the curve C(t) in Fig. 44a. Figure 44a represents C_s as a function of time during the initial transient, permanent and final transient regimes. It can be shown that the shaded areas of Fig. 44a are equal, which gives an indication of the duration of the initial transient and permanent regimes.

The overall fraction of transformed solid can be calculated by integration of (4.18):

$$F_{gl} = \frac{1}{qV_o} \int_o^t r_{gl}\, dt = \frac{n_g S}{qV_o} \frac{RT}{P} \int_o^t (C_o - C_s)\, dt \qquad (4.111)$$

It is difficult to integrate (4.111) mathematically, but it may be carried out graphically, as in Fig. 44b.

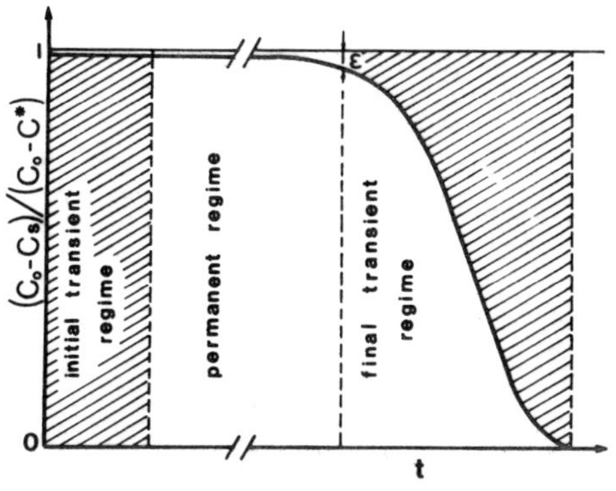

Fig. 44a. Concentration of reactant gas versus time at the exit of the reactor

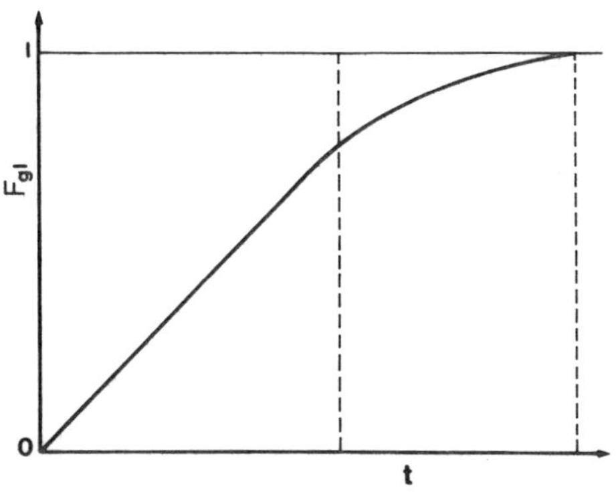

Fig. 44b. Proportion of solid transformed versus time

Beds of small and medium depth. A bed of medium depth is defined as one in which the depth is about equal to the height H of the practical zone of transformation. During the initial transient regime, the concentration C_s in the exit gas is near to that for equilibrium C*, but it decreases without reaching the permanent regime. The shape of the curve $C_s(t)$ is similar to that of Fig. 44a, but the part that corresponds to the permanent regime does not exist. In the HyL process for reduction of iron oxide ore, the bed can be considered to be of medium depth.

In a shallow bed, the kinetics of transformation are similar to those for a single sphere, and the concentration of the gaseous reactant varies little on passing through it. The composition of the exit gas always indicates less complete reaction than is required for equilibrium C*. A fluidised bed is technically shallow.

4.2.3. COUNTER CURRENT TRANSFORMATION

(A) THE PERMANENT REGIME IN A GAS-SOLID COUNTER-CURRENT BED

In a static bed, the permanent regime is established when the zone of transformation is displaced upward without changing, and at a velocity independent of time. A deep bed is required in order to establish a regime of this type. In a counter-current bed, it is not the zone of transformation that moves, but the solid, and if the gas flow and the feed rate of solid are constant, a permanent regime is established after a certain time. At a given point in the bed, the compositions of the solid and gas are independent of time. The composition profiles C(z) and F(z) may or may not be identical to those of a fixed bed in a permanent regime, depending on the operating conditions. If the profiles are those of a fixed bed in a permanent regime (i.e. if the exit gas is at the equilibrium composition at the upper front and the solid discharges completely transformed at the lower front), the regime is defined as ideal. If the profiles are different from these, the regime is not ideal.

The Ideal Regime

(1) When complete transformation of solid and maximum utilisation of gas are

desired (Fig. 45a(a)), it is possible to fix only one of the three following variables: (i) the gas flow rate, $n_g S$; (ii) the solid feed rate, $n_s S$; and (iii) the height of the transformation zone, H.

In fact, the height H of the transformation zone is a direct function of the gas flow $n_g S$, as indicated by (4.109). This flow, $n_g S$, is related to the feed rate of solid, $n_s S$ by the overall mass balance, obtained by integrating (4.102) between F = 0 and I, C = C* and C_o:

$$n_g \frac{RT}{P} (C_o - C^*) = n_s \qquad (4.112)$$

In an ideal permanent regime, the operating line (slope $dF/dC = RTn_g/Pn_s$) passes through points $E(F = 1, C = C_o)$ and $B(F = 0, C = C^*)$ (Fig. 45b(a)). Nevertheless, it should be noted that although the depth of the bed must not be less than the practical height of transformation for the ideal case, nothing prevents it from being greater. The composition of the gas and solid in the zone of the bed above the transformation zone does not change. This is the zone of chemical reserve.

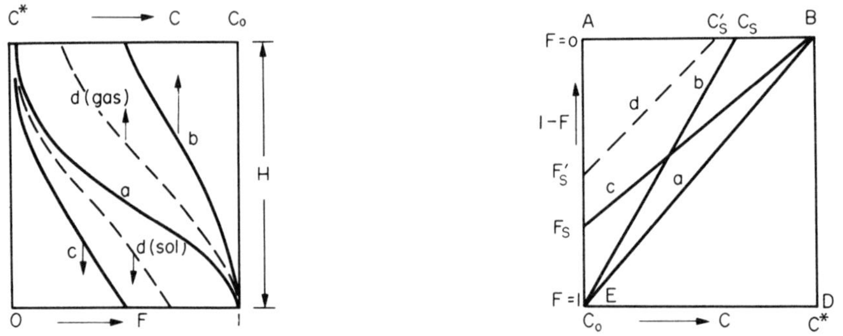

Fig. 45a. Profile C(z) and F(z) in a counter-current bed

Fig. 45b. Operating line in a counter-current bed

(a) Ideal regime: curves C(z) and F(z); (b) the gas flow rate is greater that ideal: curve C(z); (c) the solid feed rate is greater than ideal: curve F(z); (d) n_g and n_s are greater than ideal: two curves, C(z) and F(z)

Non-Ideal Regimes

(2) Assume that without changing the depth of the bed, the gas flow is increased to above that indicated by (4.112), and the solid feed rate is adjusted in such a way that the discharge is completely transformed. The gas utilisation will be less than the maximum, and it will emerge from the reactor at a concentration C_s, greater than C* (Fig. 45a(b)). The operating line will then pass through $E(F = 1, C = C_o)$ and $C_s(F = 0, C = C_s)$ and its slope $dF/dC = n_g RT/n_s P$ will be greater than that found in the previous case. The efficiency of utilisation of the gas η_g will be less than unity:

$$\eta_g = \frac{C_o - C_s}{C_o - C^*} < 1$$

(3) If, instead of increasing the supply of gas, the solid feed rate is increased, keeping the same depth of bed and the same gas flow rate, the fraction of solid transformed at the discharge will be less than unity (Fig. 45a(c)). The operating line, slope $dF/dC = n_g RT/n_s P$, less than that for the ideal case, will pass through $B(F = 0, C = C^*)$ and $F_s(F = F_s, C = C_o)$. The yield of the transformation will be less than the maximum (Fig. 45b(c)):

$$\eta_s = F_s < 1$$

(4) Finally, if the gas flow rate and solid feed rate are both higher than those which correspond to maximum utilisation of gas and complete transformation of solid, the operating line of slope $dF/dC = n_g RT/n_s P$, higher or lower than that of the ideal case, according to the respective values of n_g and n_s, will pass through $F_s'(F = F_s', C = C_o)$ and $C_s'(F = 0, C = C_s')$. The efficiency of utilisation of gas and the yield of solid transformation will both be less than 1 (Figs. 45a(c) and 45a(d)).

4.2.4. REACTIONS IN A FLUIDISED BED

(A) DESCRIPTION OF A FLUIDISED BED

A fluidised bed can be compared in operation to a continuous, back-mixed reactor of volume V, where gas and finely-divided solid form one nearly homogeneous phase, and the gas bubbles another. The flow of gas through the perforated gas distributor (n_g moles of gas per unit time, and cross-section of the bed), keeps the solid particles in circulation and suspension and reacts simultaneously with them. The gas enters with a composition C_o, and emerges with a composition C_s. The solid charged $(F = 0)$ at a feed rate n_s (moles per unit time and unit cross-sectional area) mixes with the partly transformed solid, reacts, and leaves the reactor with the composition F_s. If the gas-solid suspension is assumed to be a perfect mixture, the concentration of the gaseous reactant and the fraction of solid transformed inside the reactor are also equal, respectively to C_s and F_s (Fig. 46).

The following relationship is obtained between F_s and C_s by integration of the equation of mass balance (4.102) between $F = 0$ and F_s, $C = C_o$ and C_s:

$$n_s F_s = \frac{RT}{P} n_g (C_o - C_s) \qquad (4.113)$$

In a counter-current solid-gas bed (as the Wiberg process, for example), in a steady state, each particle remains in the reactor the same time, but the concentration of gaseous reactant which surrounds the particles varies with the height of the bed. In a back-mixed fluidised bed reactor, the gas around every particle has an approximately constant concentration, C_s, but the residence time of the particles in the bed is variable; some discharge immediately, while others remain inside the reactor for a longer period than the average. Under these circumstances, all of the particles will not have the same degree of transformation, and the transformed fraction, F_s, of the solid which discharges is an average.

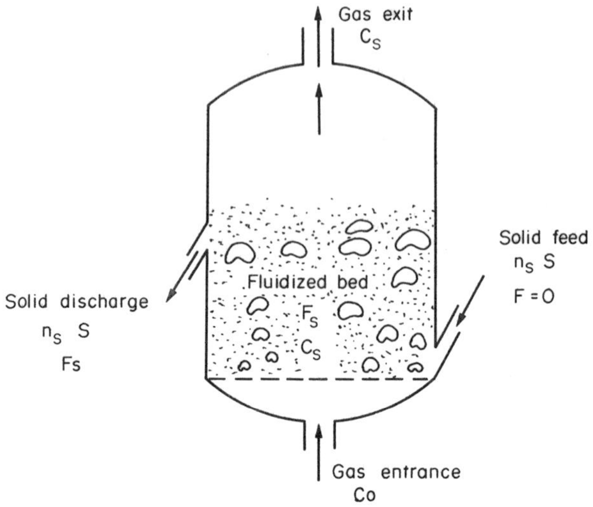

Fig. 46. A fluidised bed reactor (diagrammatic)

The average residence time of solid particles in the bed $\bar{t}$, is defined as the number of moles of solid qV which are inside the reactor of volume V, divided by the solid feed rate, n_sS, in moles per unit time:

$$\bar{t} = \frac{qV}{n_s S} = \frac{N\rho V}{n_s S} \tag{4.114}$$

q and N are the numbers of moles of solid and the number of particles per unit of bed volume, respectively, and ρ is the number of moles of solid per particle, assuming all to be of the same dimensions.

(B) RELATION BETWEEN $\bar{t}$ AND F_s OR C_s

The fraction transformed of each particle $F(t)$, as has been seen, depends on its residence time in the reactor. This fraction must be deduced from the kinetic equation; for example, equation (4.100), obtained for the particular case where, in the transformation of a spherical particle, the relative penetration is proportional to time. As the concentration of the gas phase surrounding each particle in a fluidised bed is assumed constant, and equal to C_s, the function $F(t)$ can be obtained by integration of this equation between $F = 0$ and F, $t = 0$ and t:

$$F(t) = 1 - [1 - b(C_s - C^*)t]^3 = 1 - (1 - \frac{t}{\tau})^3 \tag{4.115}$$

where τ, equal to $1/b(C_s - C^*)$ is the time for total transformation.
Let $I(t)$ and $D(t)$ be the distribution functions of the age of particles inside the reactor and of those at the discharge, respectively. These functions represent the fractions of particles with ages between t and $(t + dt)$ inside the reactor and at the exit. When the reactor is back-mixed, i.e. when the fraction transformed in the bed is the same as at the discharge, then:

$$I(t) \equiv D(t) \tag{4.116}$$

At the steady state, the number $NVI(t)dt$ of particles with ages between t and $(t + dt)$ is constant in the bed. This means that the number of particles reaching this age is equal to the number that leave it due to age: $NVI(t + dt)dt$, plus the number of this age leaving the reactor at the discharge, $(n_s S/\rho)dtD(t)dt$. The mass balance can then be written:

$$NVI(t) = NVI(t + dt) + \frac{n_s S}{\rho} D(t)dt \tag{4.117}$$

In this equation, NV represents the number of particles in the bed, and $(n_s S/\rho)dt$ the number of particles that leave the reactor during the time dt. As $D(t)$ is equal to $I(t)$, and $\bar{t}$ to $N\rho V/n_s S$, equation (4.117) may be simplified to:

$$D(t) = I(t) = -\bar{t}\,\frac{dI(t)}{dt} \tag{4.118}$$

Integration of (4.118) involves the introduction of a constant, K:

$$I(t) = Ke^{-t/\bar{t}} \tag{4.118a}$$

Determination of K is carried out on the basis that the sum of the fractions of particles of ages between 0 and ∞ is equal to 1; that is:

$$\int_0^\infty I(t)\,dt = \int_0^\infty Ke^{-t/\bar{t}} \tag{4.119}$$

Thus, the value of K is $1/\bar{t}$ and:

$$I(t) = \frac{1}{\bar{t}}\,e^{-t/\bar{t}} \tag{4.120}$$

Figure 47 represents the distribution functions $I(t)$ and $D(t)$ of the age of the particles against time.

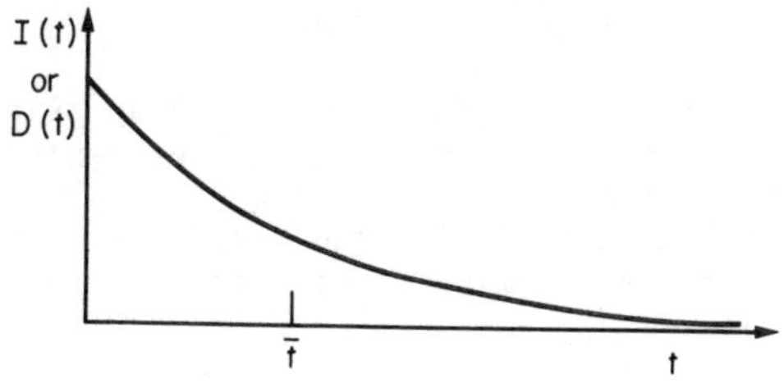

Fig. 47. Distribution functions of the age of particles against time

The average fraction of untransformed solid $(1 - F_s)$ at the discharge is the sum:

$$1 - F_s = \int_0^\infty (1 - F(t))D(t)\ dt = \frac{1}{\bar{t}} \int_0^\infty (1 - F(t))e^{-t/\bar{t}}\ dt \qquad (4.121a)$$

where $F(t)$ is given by (4.115) (chemical regime).

(4.121a) can be integrated as a development series*

$$1 - F_s = \frac{1}{4}\left(\frac{\tau}{\bar{t}}\right) - \frac{1}{20}\left(\frac{\tau}{\bar{t}}\right)^2 + \frac{1}{120}\left(\frac{\tau}{\bar{t}}\right)^3 \ldots \qquad (4.121b)$$

When $\tau/\bar{t}$ is less than 0.2, only the first term need be considered. (4.113), (4.114) and (4.121b) provide relationships which make it possible to calculate the production n_s, with a specified fraction of transformed solid F_s, from a reactor of volume V, using particles of a given dimension. (τ is proportional to the diameter of the particle according to (4.4.4)).

4.3. HETEROGENEOUS CATALYSIS

A catalyst is a material that, without participating directly in a reaction, accelerates it considerably. In heterogeneous catalysis, the catalyst is a solid capable of adsorbing and activating molecules of gaseous or liquid reactants and facilitating the reaction between them. Since adsorption is a surface phenomenon, the catalyst is generally a porous solid with a large specific surface.

A study of the kinetics of reaction with heterogeneous catalysis is similar to that of reaction in a porous solid. Nevertheless, two essential differences should be emphasised:

When a porous solid reacts, the reaction interface moves towards the interior of the particle. In heterogeneous catalysis, the catalyst does not participate directly in the reaction. It is the gaseous or liquid reactants which react on contact with it. As long as there are no foreign molecules strongly adsorbed on the surface of the catalyst in such a manner as to poison, the reaction interface does not change in the course of the reaction, and therefore the influence of the catalyst is independent of time.

The stages in the reaction of a solid with a fluid at high temperature are essentially: diffusion of the fluid reactant to the reaction interface, reaction at the interface and diffusion of the fluid product away from the interface. In heterogeneous catalysis, the stages of adsorption of reactants at the interface and desorption of the products should be added, and the sequence is as follows:

Diffusion of the reactants to the surface of the catalyst.

Adsorption of the reactants at the surface.

Chemical reaction between the reactants.

Desorption of the products of reaction.

Diffusion of the products away from the surface of the catalyst.

---

*D. Kunii and O. Levenspiel "Fluidization Engineering". J. Wiley & Sons Inc., New York 1969 p. 491.

4.3.1. LANGMUIR'S ISOTHERM

The three phenomena of adsorption, chemical reaction, and desorption, act together, and the resulting velocity can be expressed by a single equation using Langmuir's Theory. In this theory, it is assumed that the surface of the catalyst has a certain number of sites where the reactant molecules can be adsorbed. At a given pressure and temperature, a fraction θ of the sites is occupied, and the fraction $(1 - \theta)$ remains empty. The adsorption velocity, r_a, of molecules at the surface is proportional to the pressure and to the fraction $(1 - \theta)$ of the sites which are empty:

$$r_a = k_a P(1 - \theta) \tag{4.122}$$

On the other hand, the desorption velocity is proportional to θ:

$$r_d = k_d \theta \tag{4.123}$$

At a steady state, the two velocities are equal; therefore:

$$\theta = \frac{k_a P}{k_d + k_a P} = \frac{bP}{1 + bP} \tag{4.123a}$$

where $b = k_a/k_d$, k_a and k_d being the specific adsorption and desorption rate constants. Thus, θ is proportional to the pressure when it is low, $(bP \ll 1)$, and independent of it when it is high, $(bP \gg 1)$. The reaction velocity is proportional to the number of adsorbed molecules, i.e. to θ. Therefore, depending upon the value of b, the order of reaction will be zero or one.

It will be assumed, to simplify the calculations and to conserve the same form as before, that the resulting velocity of the three steps (adsorption, reaction, and desorption) is first order, and has a value similar to that already calculated for the transformation of a solid. Therefore:

$$r_r = BS_i(C - C^*) \tag{4.7}$$

The kinetics of reaction of a fluid crossing a bed of catalyst can then be treated in a similar way to the transformation of a porous solid, with the difference that the solid is not transformed during the reaction. As a first step, it is possible to calculate the fraction of the volume of the catalyst which is effective during the reaction; then, as before, the results obtained with a single sphere will be used to study the catalysis of a bed of spheres.

4.3.2. COEFFICIENT OF UTILISATION OF A CATALYST

As in the case of the transformation of a sphere of porous solid previously studied, the fluid penetrates the catalyst and reacts after being adsorbed. Since the catalytic reaction is assumed to be of the first order, the earlier calculations are still valid. In particular, the concentration of gaseous reactant varies with the distance R from the centre of the sphere, as shown by (4.74):

$$\frac{C - C^*}{C_o - C^*} = \frac{R_o \sinh(R/h)}{R \sinh(R_o/h)} \tag{4.74}$$

According to (4.67), $h^2 = D/Bs_e$. The overall velocity, r_g, of gas transformation inside the sphere is equal to the gas flux which crosses the external surface of the particle, i.e.:

$$r_g = DS_o \left[\frac{dC}{dR} \right]_{R=R_o} \tag{4.124}$$

Replacing in (4.124) the term $\left[\frac{dC}{dR} \right]_{R=R_o}$ by that resulting from differentiation of (4.74) gives:

$$r_g = \frac{DS_o(C_o - C^*)}{h} \left[\frac{1}{\tanh(R_o/h)} - \frac{h}{R_o} \right] \tag{4.125}$$

Depending on the value of h, reaction takes place throughout the particle (h is large), or only in a thin layer (h is small), as is seen in Fig. 36. It is interesting to compare the overall velocity given by (4.125) with that which would be obtained if all the catalyst was used effectively, i.e. if there was no delay due to diffusion. With a sphere of volume V_o ($V_o = \frac{1}{3} S_o R_o$) and specific surface, s_e, this velocity, r_g^o, would be equal to the maximum reaction velocity:

$$r_g^o = BV_o s_e (C_o - C^*) = \frac{1}{3} BS_o R_o s_e (C_o - C^*) \tag{4.126}$$

The coefficient of utilisation of the catalyst is obtained by comparing the two velocities:

$$\eta = \frac{r_g}{r_g^o} = \frac{3D}{Bs_e hR_o} \left[\frac{1}{\tanh(R_o/h)} - \frac{h}{R_o} \right] = 3\frac{h}{R_o} \left[\frac{1}{\tanh(R_o/h)} - \frac{h}{R_o} \right] \tag{4.127}$$

with $h^2 = D/Bs_e$, according to (4.67).

In practice, only a fraction η of the volume of the sphere is utilised in catalysis. The ratio, R_o/h, is called Thiele's criterion. With Thiele's criterion less than one, the utilisation coefficient is near to one. When this criterion is larger than 20, η is practically equal to $3h/R_o$. The chemical regime predominates in the first case, and diffusion in the second. The concentration of gaseous reactant C_o which surrounds the sphere has no influence upon the value of the utilisation coefficient, as can be deduced from (4.127).

When the catalyst is very effective, its utilisation coefficient is small, and only the external surfaces of the particles are involved. In consequence, when the price of the catalyst is high, as in the case of platinum, porcelain spheres or some other inert support, are coated with a thin layer of catalyst.

4.3.3. TRANSFORMATION OF A GASEOUS REACTANT THROUGH A BED OF CATALYST

A bed in the form of a column of spheres of catalyst can be divided into an infinite number of layers of thickness dz. Within each volume Sdz, the gas enters at a concentration C and leaves at a concentration (C - dC). The transformation velocity is that found in (4.98):

$$r_g(z) = - n_g S \frac{RT}{P} dC \tag{4.98}$$

This velocity can be related to the reactant concentration and to the utilisation coefficient by:

$$r_g(z) = r_g^o \eta = BS s_e \eta (C - C^*) \, dz \tag{4.128}$$

By equating (4.98) and (4.128), and integrating them, the concentration profile of the gaseous reactant can be calculated as a function of the height of the bed:

$$C - C^* = (C_o - C^*) \exp\left(- \frac{\beta z}{n_g} \right) \tag{4.129}$$

β equal to $Bs_e\eta P/RT$, depends on the reaction kinetics (coefficient B), the specific surface, s_e, the utilisation coefficient, η, the pressure, and the temperature. Figure 48 depicts the variation in gaseous reactant concentration as a function of the height of the bed.

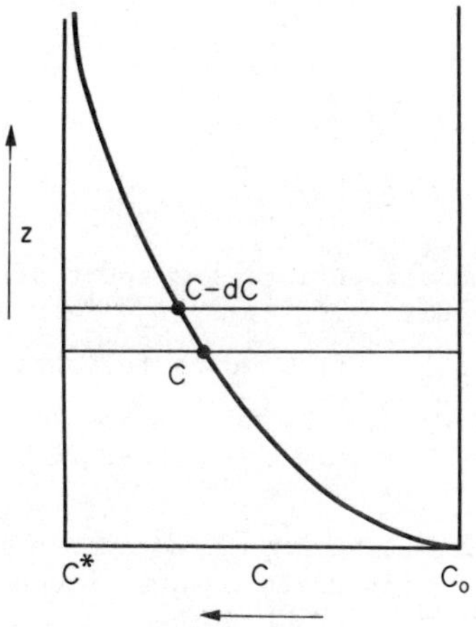

Fig. 48. Profile of the gaseous reactant concentration, C(z), for a first order catalytic reaction

The height, H, necessary to transform the gaseous reactant to a concentration near to equilibrium, i.e. $(C - C^*)/(C_o - C^*) = \varepsilon$, can be calculated in the same way as for a bed of solid transformed by a gas. Then we have:

$$H = - \frac{n_g}{\beta} \log \varepsilon \tag{4.130}$$

As is the case during transformation of a bed, the effective transformation height is proportional to the gas flow rate, and depends on the value of ε selected. Consequently, if the gas flow rate is fixed, the height of the reactor can be calculated, and vice versa.

4.4. MASS TRANSFER BETWEEN IMMISCIBLE LIQUID PHASES

When two immiscible phases, both containing a common transferable species, are brought into contact, the system reaches equilibrium by repetition of the following cycle:

A. The species is transferred across the interface, with or without reaction.

B. The species transfers from the bulk of Phase 1 to the interface, in order to restore the deficiency created by step A.

C. The species transfers from the interface to the bulk of Phase 2, in order to relieve the excess created by step A.

The immiscible phases can be:

1. Water and an organic liquid, or two liquid metals.
2. Liquid metal and slag or reactive gas, with reaction occurring at the interface to produce a compound of the metal.

The general laws governing these exchanges will be considered briefly in this section, together with the phenomena which can modify the kinetics of exchange, such as stirring, adsorption of a species at the interface and nucleation of a second phase in the bulk.

4.4.1. GENERAL LAWS GOVERNING EXCHANGES

TRANSFER IN A FLUID

In the absence of stirring and convection, transport of a species in a fluid is effected by diffusion and the diffusion rate r_d in the direction z is governed by Fick's first law, which can be expressed as follows:

$$r_d = dn/dt = - DS\ \partial C/\partial z \qquad (4.131)$$

When the fluid is well stirred, it becomes homogeneous in the bulk, but there is a layer of thickness δ, near the interface with a second phase, where transfer is still by diffusion. The transfer rate through this boundary layer can be written:

$$r_d = \frac{DS}{\delta}(C^b - C^i) = kS(C^b - C^i) \qquad (4.132)$$

where C^b and C^i are the concentrates in the bulk and at the interface, k is the mass transfer coefficient and has the dimensions of lt^{-1}. The distribution of solute species under these conditions is illustrated in Fig. 49.

TRANSPORT BETWEEN TWO STIRRED FLUIDS

When there is no reaction between the phases, or the reaction is fast compared with the rate of transport of solute, this takes place by movement from bulk to interface in Phase 1 and interface to bulk in Phase 2. Assuming that equilibrium

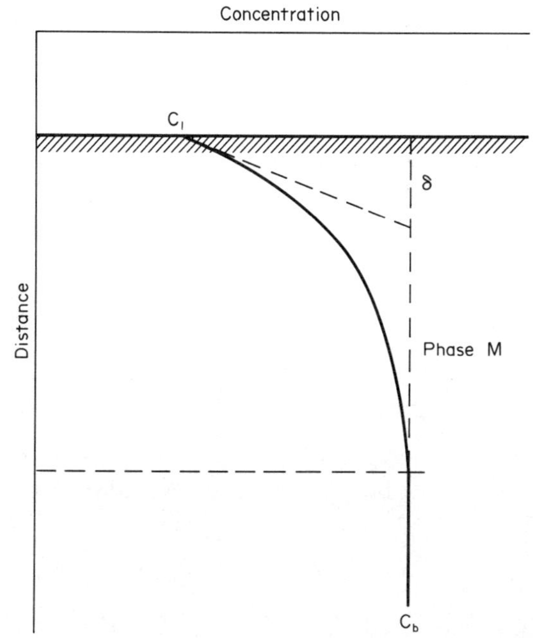

Fig. 49. Concentration vs distance in stirred phase M

With acknowledgement to F.D. Richardson, *Physical Chemistry of Melts in Metallurgy*, Academic Press, London (1974)

is reached at the interface, i.e.

$$C_2^i/C_1^i = C_2^*/C_1^* = K,$$

where C^* is the equilibrium concentration, K the partition coefficient and there is no accumulation of solute at the interface, the overall rate of transfer r_g is equal to:

$$r_g = r_{d_1} = r_{d_2} = k_1S(C_1 - C_1^i) = k_2S(C_2^i - C_2) \quad (4.133)$$

or

$$r_g/k_1S = C_1 - C_1^i \quad (4.133a)$$

and

$$r_g/k_2S = C_2^i - C_2 = KC_1^i - C_2 \quad (4.133b)$$

Multiplying (4.133a) by K and summing:

$$\frac{r_g}{S}\left(\frac{K}{k_1} + \frac{1}{k_2}\right) = KC_1 - C_2 \quad (4.134)$$

The term $KC_1 - C_2$ represents the driving force for the transfer of solute from Phase 1 to Phase 2 and the term in brackets represents the resistance to transfer compounded of one term for each phase.

FIRST ORDER REACTION AT THE INTERFACE

When metal is in contact with slag or gas, there is, in addition to the possibility of a common solute, the chance of a reaction. For example, in the reaction:

$$(MO)_1 = (\underline{M})_2 + (\underline{O})_2 \tag{4.135}$$

if the reaction kinetics are first order in both directions, the reaction rate r_r is equal to:

$$r_r = S\left(k_r^1 C_1^i - k_r^2 C_2^i\right) \tag{4.136}$$

where k_r^1 and k_r^2 are the velocity constants of reaction (4.135) in the two directions. At equilibrium, $C^i = C^*$ and r_r is zero.

$$k_r^1 C_1^* = k_r^2 C_2^* \qquad \text{or} \qquad \frac{k_r^1}{k_r^2} = \frac{C_2^*}{C_1^*} = K \tag{4.137}$$

K is the equilibrium constant for reaction (4.135)† and the reaction rate can be written:

$$r_r = k_r S\left(KC_1^i - C_2^i\right) \tag{4.138}$$

In the same way, when there is no accumulation of reagents or products at the interface, i.e. $r_g = r_{d_1} = r_{d_2} = r_r$, the transfer rate can be obtained by combining (4.138) with (4.133) to give:

$$\frac{r_g}{S}\left(\frac{K}{k_1} + \frac{1}{k_2} + \frac{1}{k_r}\right) = KC_1 - C_2 \tag{4.139}$$

---

†K is expressed in terms of concentration and is not constant unless the activity coefficients are constant.

In the equilibrium $(Fe)_m + (O)_m = (FeO)_s$, K can be calculated from the values of $C^*_{(O)_m}$ and $C^*_{(O)_s}$. At 1873 K, the solubility of $(O)_{Fe}$ in equilibrium with wustite is 0.23% by weight. The density of Fe is 7.2 g cm^{-3} and that of FeO is 4.5, thus:

$$C^*_{(O)_{Fe}} = \frac{0.23 \times 10^{-2} \times 7.2}{16},$$

$$C^*_{(O)_s} = \frac{1 \times 4.5}{72} \qquad \text{and} \qquad K = C_{(O)_{Fe}}/C_{(O)_s} = 0.017$$

(4.139) is of the same form as (4.134) with a supplementary resistance term arising from reaction (4.135).

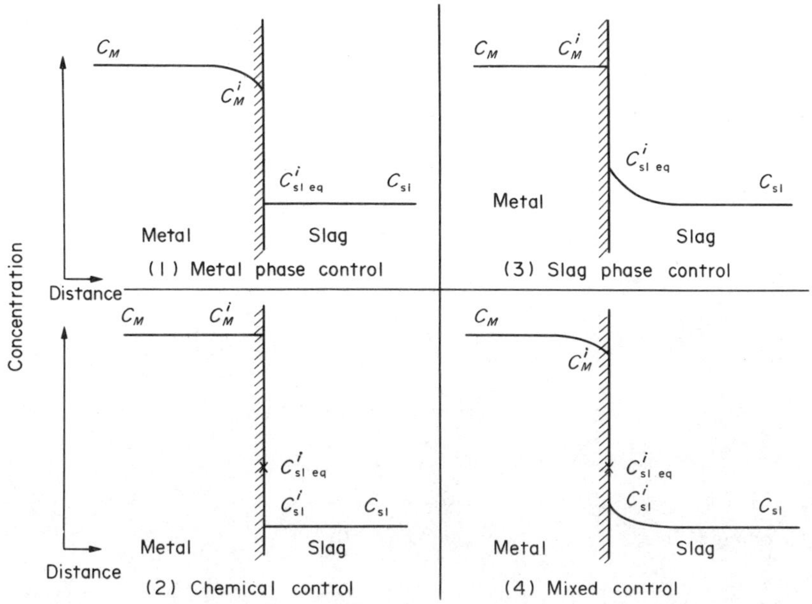

Fig. 50. Time-average concentration-distance curves across the phase boundary for a reaction involving a metal and slag under various conditions: (1) metal phase control, (2) chemical control, (3) slag phase control, (4) mixed control. Interfacial concentrations indicated by superscript i. $C^i_{sl,eq}$, is concentration in the slag which is or would be in equilibrium with C^i_M.

With acknowledgement to F.D. Richardson, *Physical Chemistry of Melts in Metallurgy*, Academic Press, London (1974).

RELATIONSHIP OF CONCENTRATION TO TIME

The variation of concentration with time can only be calculated with knowledge of the mass balance. Since the total number of moles of solute in the two phases is constant:

$$n_T = n_1 + n_2 = n_1^* + n_2^*$$

or

$$- dn_1/dt = dn_2/dt = r_g \tag{4.140}$$

Assuming that the changes in volumes V_1 and V_2 due to transfer of solute are too small to be considered, equations (4.139) and (4.140), make it possible to write:

$$dn_2/dt = \frac{C_2^*S}{R}\left\{\frac{C_1}{C_1^*} - \frac{C_2}{C_2^*}\right\} = \frac{-\,SC_2^*n_T(n_2 - n_2^*)}{Rn_1^*n_2^*} \tag{4.141}$$

where n_1^* and n_2^* are the numbers of moles in equilibrium in the two phases and R is the overall resistance term. By integrating, this becomes:

$$\frac{n_2 - n_2^o}{n_2^* - n_2^o} = \frac{C_2 - C_2^o}{C_2^* - C_2^o} = 1 - e^{-At} \tag{4.142}$$

with $A = \frac{S}{R}\left(\frac{1}{V_2} + \frac{K}{V_1}\right)$ and $R = \frac{K}{k_1} + \frac{1}{k_2} + \frac{1}{k_r}$.

n_2^o and C_2^o are the number of moles and concentration of solute in Phase 2 for $t = 0$.

Using (4.142) it is possible to calculate the time needed for the concentration in Phase 2 to change from C_2^o to a value near C_2^*, i.e.

$$\frac{C_2 - C_2^o}{C_2^* - C_2^o} = 1 - \varepsilon$$

The time for this can be calculated from $t^* = -\ln \varepsilon/A$.

SLAG - METAL - GAS REACTION

The reaction

$$(FeO)_s + (\underline{C})_m = (Fe)_m + [CO]_g \tag{4.143}$$

can be broken down into:

$$(FeO)_s = (Fe)_m + (\underline{O})_m \tag{4.143a}$$

and

$$(\underline{C})_m + (\underline{O})_m = [CO]_g \tag{4.143b}$$

Reaction (4.143a) takes place at the slag/metal interface, but (4.143b) can only proceed in the metal if bubbles of CO can form. This requires special circumstances, such as the presence of bubbles of another gas being passed through the melt, or already existing in cavities in the furnace refractories. The limiting step for (4.143b) is transfer in the metal to these bubbles, and when $(\underline{C})_m$ is much greater than $(\underline{O})_m$ the critical rate which must be considered as that of the transfer of $(\underline{O})_m$. Thus:

$$r_g = k_1'S_{mg}\left\{C_{(O)_m} - C^i_{(O)_m}\right\} \tag{4.144}$$

where S_{mg} is the area of the metal/gas interface, k'_1 the mass transfer coefficient at the surface and $C_{(O)_m}$ and $C^i_{(O)_m}$ the concentrations of (O) in the bulk and at the interface. When the concentration of $(\underline{C})_m$ is large relative to that of $(O)_m$, $C^i_{(O)_m}$ is proportional to P_{CO}:

$$C^i_{(O)_m} = KP_{CO}/C^i_{(C)_m} = K'P_{CO}$$

In this way, (4.144) becomes:

$$r_g/k'_1 S_{mg} = C_{(O)_m} - K'P_{CO} \tag{4.145}$$

With no accumulation of $(\underline{O})$ in the liquid metal, the flux of $(\underline{O})$ entering (4.143a) and the flux leaving the gas phase are equal.

Combining (4.139) and (4.145) we obtain:

$$\frac{r_g}{S_{ms}} \cdot \left(\frac{1}{k_2} + \frac{K}{k_1} + \frac{1}{k_r} + \frac{S_{ms}}{S_{mg}} \cdot \frac{1}{k'_1}\right) = KC_{(\underline{O})_s} - K'P_{CO} \tag{4.146}$$

The only factors in the driving force are $C_{(\underline{O})_s}$ and P_{CO}. The following example illustrates the influence of the supplementary resistance term $1/k'_1$ on the kinetics of reaction.

EXAMPLE 1. SILICON TRANSFER FROM SLAG TO METAL

The kinetics of the reaction:

$$(SiO_2)_s + 2(\underline{C})_m = (Si)_m + 2[CO]_g \tag{4.147}$$

have been studied experimentally[8] using carbon-saturated iron and slags containing CaO and SiO_2 (36% + 64%, with $a_{SiO_2} = 0.87$) as well as BaO, CaO and SiO_2 (50:15:35%, with $a_{SiO_2} = 0.07$). In some experiments, CO was injected at the slag/metal interface and in others it was not. Figure 51 shows that reaction is much more rapid when CO is injected and this can be explained in the following way: reaction (4.147) is the sum of:

$$(SiO_2)_s = (\underline{Si})_m + 2(\underline{O})_m \tag{4.147a}$$

and

$$2(\underline{C})_m + 2(\underline{O})_m = 2[CO]_g \tag{4.147b}$$

↖ nucleation dependent

When gas bubbles are provided at the interface, (4.147b) is very fast and the overall rate is limited only by the rate of (4.147a). The speed of this reaction increases with a_{SiO_2} in the slag. If no bubbles are provided, (4.147a) can only proceed at the rate at which oxygen is removed by (4.147b), i.e. very slowly. Homogeneous nucleation of CO bubbles in liquid Fe requires a very high degree of supersaturation and so (4.147b) takes place only where bubbles are available, i.e. the refractory surface, generally at some distance from the interface, and the overall reaction (4.147) is slow.

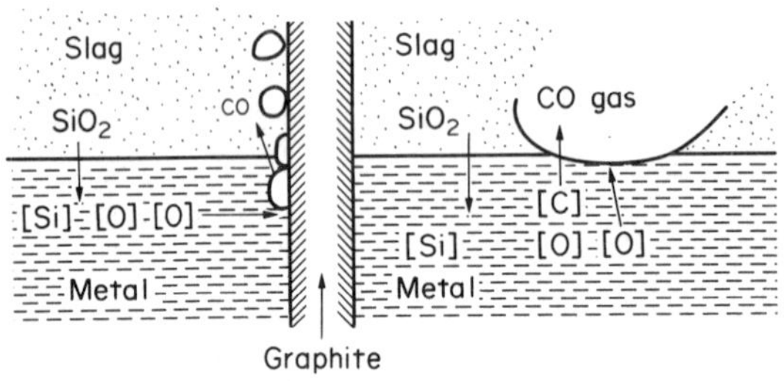

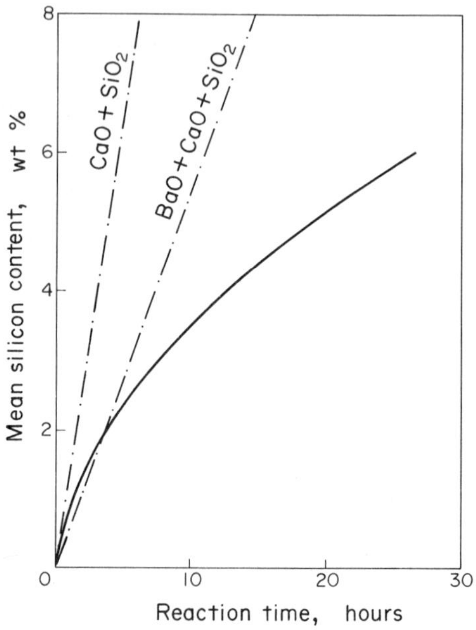

Fig. 51. Reduction of Si from slag into iron at 1873 K. With and without bubbles of CO. ———— CaO + SiO_2 without interface bubbles. • —— • —— CaO + SiO_2 and CaO + SiO_2 + BaO slags with interface bubbles.

With acknowledgement to F.D. Richardson, *Physical Chemistry of Melts in Metallurgy*, Academic Press, London (1974).

EXAMPLE 2. DESULPHURISATION OF IRON BY MAGNESIUM

This process is very effective because $K_{Mg-S} = (\%\ Mg)_m\ (\%\ S)_m = 1.8 \times 10^{-6}$ at 1535°C. Irons and Guthrie[9] injected Mg vapour into molten iron which had been strongly deoxidised by saturation with carbon. They examined the rate of removal of S, experimentally and theoretically, while the melt was being stirred vigorously by magnetic induction. The possible stages in the complete reaction are:

(a) Dissolution of [Mg] in the melt.

(a') Transfer of $(\underline{Mg})_m$ throughout the melt to MgS inclusions.

(b) Transfer of $(\underline{S})_m$ to the bubbles of Mg.

(b') Transfer of $(\underline{S})_m$ to inclusions of MgS.

(c) Reaction of $(\underline{S})_m$ and $(\underline{Mg})_m$ at the bubble surfaces.

(c') Reaction of $(\underline{S})_m$ and $(\underline{Mg})_m$ at MgS inclusions.

Since the flux of [Mg] to $(\underline{Mg})_m$ is much greater in step (a) than the flux of $(\underline{S})_m$ to the bubbles in step (b), most of the Mg dissolves in the iron, migrates and reacts at the small (2μ) MgS inclusions (step (c')) arising from particles formed on bubbles in step (c) and stripped as they rise. The enlarged MgS inclusions are swept quickly from the melt by the rising [Mg] bubbles, magnetic stirring and low density, giving a steady-state and roughly constant inclusion content. The low value found by the authors for the mass transport coefficient of Mg from the bubbles to the bulk (k_{Mg} = 0.046) may be due to the interfering effect of the layer of MgS on the [Mg] bubbles.

4.4.2. PHENOMENA WHICH CAN MODIFY KINETICS

Stirring, adsorption and nucleation are examined briefly for their influence on reaction rates.

STIRRING

This form of agitation can be carried out mechanically, or by convection, magnetic induction, or gas bubbling. Mechanical stirring is limited at high temperature by weakening of the materials of stirrer construction and by chemical attack. It can be particularly effective in aqueous solutions, especially when there is a high degree of shearing. Electromagnetic induction has been widely used in alloy steelmaking in electric arc furnaces, where the economic and technical advantages of rapid reaction to produce homogeneous melts are most obvious.

Gas bubbling can be used to combine stirring with chemical reaction, as in the high speed refining of steel and copper in converters. Thus, the high speed of removal of C from iron in pneumatic steelmaking can be attributed to the high level of $(O)_m$ and the efficient stirring by bubbles into which CO can diffuse. Experience with the Kaldo process showed that, if mechanical stirring by rotation of the converter is added to bubble stirring, the rate of removal of C is increased. Convection stirring, resulting from the difference in density between hot and cold metal, is demonstrated in submerged electric arc smelting of steel and copper matte. The fast upward current of hot liquid around the electrodes gives rise to a slower outward and downward flow to complete the circulation and mixing.

When an eddy arises in the bulk, due to any kind of stirring, the initial vertical motion is transformed into horizontal motion at a free surface or interface (Fig. 55) and the eddy disappears, exposing fresh surface and increasing the value of the transfer coefficient.

ADSORPTION AND INTERFACIAL TENSION

The adsorption of solute A at an interface is defined:[10]

$$\Gamma_A = n_A^i / S \tag{4.148}$$

where n_A^i is the difference between the total numbers of moles of A in the system and the numbers in the bulk of each phase.

$$n_A^i = n_A - \left(n_{A_1} + n_{A_2}\right).$$

S is the area of the interface and Γ_A is the excess of solute at the interface or surface. The value can be calculated, using the Gibbs adsorption isotherm:

$$\Gamma_A = -\frac{1}{RT}\left(\frac{\partial \gamma}{\partial \ln a_A}\right)_T \tag{4.149}$$

where γ is the surface tension. When the solute obeys Henry's Law, i.e. the activity coefficient is a constant, the equation can be written:

$$\Gamma_A = -\frac{N_A}{RT}\left(\frac{\partial \gamma}{\partial N_A}\right)_T \tag{4.150}$$

The greater the value of $\partial\gamma/\partial N_A$, the greater the adsorption of A. This can be illustrated by noting the influence of some elements on the surface tension of liquid iron[11] (Fig. 52a) and aluminium (Fig. 52b).[12]

The adsorption of a solute at an interface can modify the kinetics of interchange between the phases in two ways:[13]

(1) The adsorption of a non-reactive but surface-active solute can prevent a solute of lower surfactant capacity from reaching the surface and reacting.

(2) The decrease of interfacial tension γ_{ms} due to adsorption can be sufficient to increase the interfacial area. In a similar way, changes of surface tension γ_{sv} and γ_{mv} will modify the spreading of slag on metal and the area of the exchange surface (Fig. 54).

SURFACE TENSION OF METALS

Table 2 gives values for the surface tension of some liquid metals, γ_{mv} at temperatures very close to the melting points.

TABLE 2 Surface Tension of Some Liquid Metals

Metal	Pb	Zn	Al	Ag	Cu	Ni	Fe	H_2O
Melting point K	600	692	933	1234	1356	1727	1809	273
Surface tension (mjoules m^{-2} or ergs cm^{-2})	480	750	850	915	1280	1934	1790	75

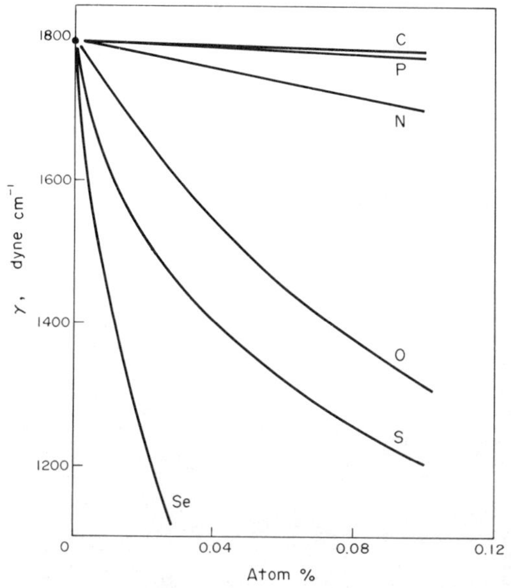

Fig. 52a. Surface tensions for solutions of C, P, N, O, S and Se in liquid iron at 1550°C. After Kozakevitch.

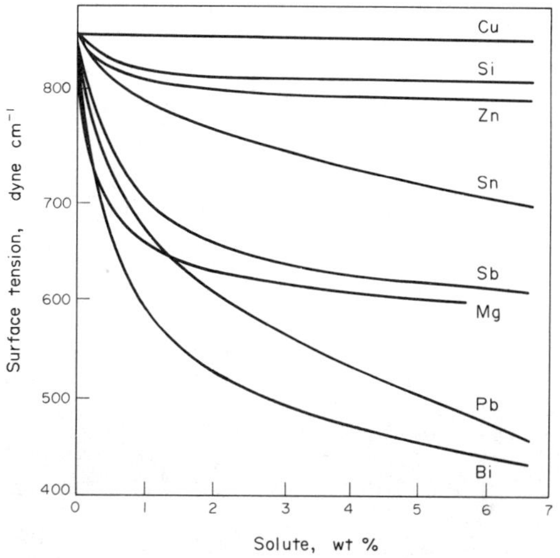

Fig. 52b. Surface tensions for aluminium containing various metals at 50 to 80°C above liquidus. After Korol'kov.

With acknowledgement to F.D. Richardson, *Physical Chemistry of Melts in Metallurgy*, Academic Press, London (1974)

The surface tension values of metals are very high compared with that of water and they increase with the melting points to the extent that $\gamma_{mv} \simeq T_f$. The temperature coefficient $\partial\gamma/\partial T$ is approximately -0.2 mjoules m^{-2} K^{-1} for all metals. The addition of a metallic solute affects the surface tension only when the value for the solute is lower. This is illustrated in Fig. 52b.

The effects of non-metallic solutes are shown in Fig. 52a and listed in Table 3.

TABLE 3 $\partial\gamma_{mv}/\partial$at% for Solutes in Liquid Fe[11]

Group in Periodic Table	3	4	5	6
Element	B ∿ 26	C ∿ 3	N ∿ 850	O 8600
	Al ∿ 38	Si ∿ 6	P ∿ 13	S 15400
		Sh ∿ 1600	As ∿ 200	Se 54600
			Sb ∿ 3900	Te > 54600

It can be seen that C, Si and P have very little surface activity, but O and S are very active and when the content is more than 0.02 - 0.03%, coverage of the surface is almost complete.

SURFACE TENSION OF SLAGS

The values for oxides, chlorides and sulphides are substantially less than those for metals.

TABLE 4 Surface Tension Values for Some Oxides, Chlorides and Sulphides

Compound	T_fK	γ(mjoules m^{-2})
NaCl	1273	98
FeO	1693	585
Al_2O_3	2323	690
SiO_2	2073	307
P_2O_5	373	60
Cu_2S	1473	400

As with metals, a solute only affects the surface tension of the solvent if its own value is lower (Fig. 53).[14]

INTERFACIAL TENSION BETWEEN SLAG AND METAL

Interfacial tension values, γ_{ms}, between metals and slags are between γ_{mv}, the values for metals, and γ_{sv}, the values for slags, and range between 600 and 1200 joules m^{-2}, but they can have very low values when a species is adsorbed at the interface, or when reaction takes place through it (dynamic tension). When the interfacial tension is very low, the energy required to allow spherical metal inclusions to pass into the slag is very small and it can be effected by only a small amount of stirring, with a significant improvement in the kinetics of exchange and reaction.

SPREADING OF SLAG ON METAL[15]

Reaction kinetics depend on the area of contact between metal and slag, as influenced by slag spread. This can be defined by the spreading coefficient F:

$$F = \gamma_{mv} - \gamma_{sv} - \gamma_{ms} \qquad (4.151)$$

where γ_{mv}, γ_{sv} and γ_{ms} are the interfacial tensions as shown in Fig. 54.

When F is positive, i.e. γ_{mv} is greater than $(\gamma_{sv} + \gamma_{ms})$, then the slag area increases, but when it is negative the slag tends to form circular lenses, especially when the amount is small. Adsorption of a species at the metal surface, or at the metal/slag interface, can modify the value of F.

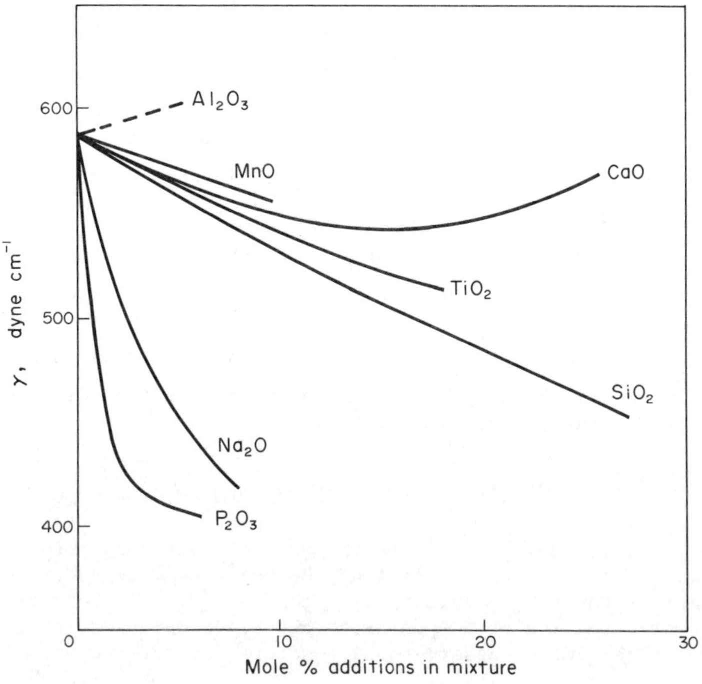

Fig. 53. Surface tension of mixtures with FeO at 1400 - 1420°C

With acknowledgement to F.D. Richardson, *Physical Chemistry of Melts in Metallurgy*, Academic Press, London (1974)

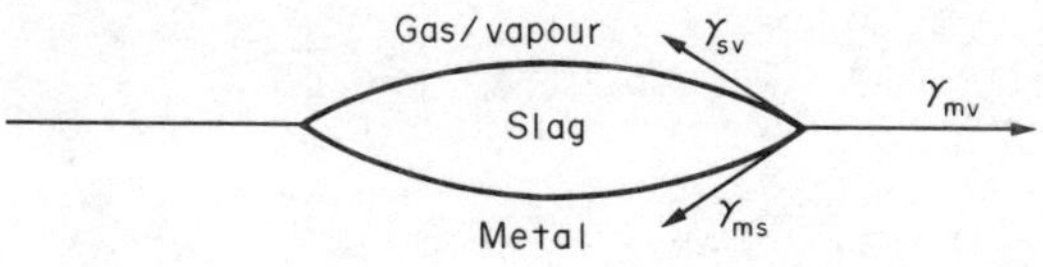

Fig. 54. Contact angle for slag drop on liquid metal.

With acknowledgement to F.D. Richardson, *Physical Chemistry of Melts in Metallurgy*, Academic Press, London (1974)

WETTING OF REFRACTORIES BY METAL AND SLAG

Refractories usually consist of sintered grains of oxide and are porous. The rate of decay of the refractory accelerates as the pores are filled, either by metal or slag. The capillary rise h of a liquid in a tube or pore of radius r is given by:

$$h = 2(\gamma_{rv} - \gamma_{rm})/\rho g r$$

or $$h = 2(\gamma_{rv} - \gamma_{rs})/\rho gr \qquad (4.152)$$

where γ_{rv}, γ_{rm} and γ_{rs} are the interfacial tensions between refractory and atmosphere, refractory and liquid metal and refractory and liquid slag respectively. The first is of the same order as that for liquid slag (γ_{rv} for solid Al_2O_3 at 2073 K is 905 mjoules m^{-2} (17) or less(18) and 690 mjoules m^{-2} for liquid Al_2O_3 at 2323 K). γ_{rm} is of the same order as γ_{sm}. Thus, when there is neither reaction between refractory and metal, nor adsorption of a solute at the interface, γ_{rm} is greater than γ_{rv} and the metal does not enter the pores. Therefore these perform a useful function in nucleating bubbles of CO and accelerating C removal. The superficial tension between liquid slag and refractory, γ_{rs}, is low compared with the surface tension of the refractory, γ_{rv}, and slag can enter the pores. This greatly increases the area of contact between slag and refractory and accelerates destruction of the latter by solution.

ADSORPTION IN A STIRRED SYSTEM(13)

As described earlier, eddies create fresh surfaces, but adsorption acts in the opposite way in concentrating material at the surface. The coverage of a surface by a surfactant is rapid and substantially complete. It is interesting to compare the mean time of retention of a species at a surface during stirring with the rate of adsorption. The time necessary to half-saturate a surface by a surfactant is very small. ((O) in Fe requires 10^{-4} sec.) For a well-stirred system, in which the mass transfer coefficient $k = 1$ cm sec^{-1}, the mean retention time is also about 10^{-4} sec, and for a lightly stirred system ($k = 10^{-2}$ cm sec^{-1}) the time is about 1 sec. Thus the coverage of a surface is virtually complete if there is only a small amount of stirring, but strong stirring can eliminate the effect of adsorption.

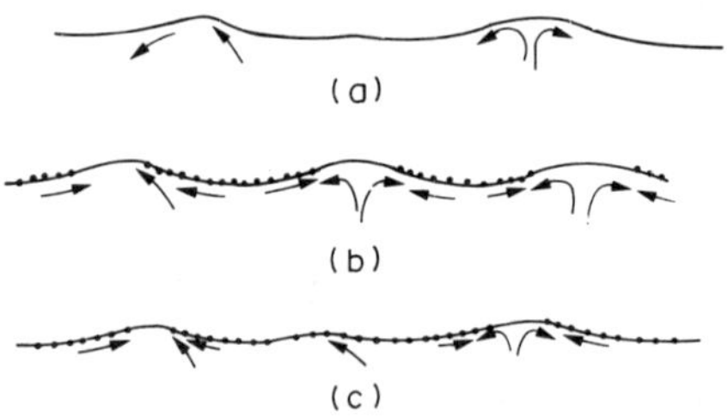

Fig. 55. Eddies at the interfaces.
(a) Without a surface-active additive.
(b) With a surface-active additive, substantial renewal of the surface
(c) With a surface-active additive, renewal of the surface

F.D. Richardson. Mem. Sci. Rev. Met. Nov. 1978, 630

It should be noted that substantial adsorption can damp down surface eddies, roughly in the same way that oil on the sea can diminish the amplitude of small waves. Conversely, a difference of surface tension between two areas of a surface, due to local adsorption of a solute, or a difference of temperature (the Marangoni effect), can provoke surface movement and stirring of a relatively thick surface layer.

EFFECT OF SURFACE-ACTIVE AGENT ON REMOVAL OF N_2 *FROM LIQUID Fe*

The effect of oxygen on the rate of removal of N_2 from inductively stirred iron containing 0.009% and 0.08% (O) is shown in Fig. 56.(16) Assuming that transfer into the gas phase is rapid, the rate r_g is equal to:

$$r_g = - dn/dt = - VdC/dt = kS(C - C^*) \qquad (4.153)$$

where C and C^* are the bulk and equilibrium concentrations of nitrogen in the presence of the gas phase and C^o is the concentration at time zero. Integrating (4.153) gives:

$$\ln (C - C^*)/(C^o - C^*) = \frac{kSt}{V} \qquad (4.154)$$

For a given oxygen content, the bulk concentration varies with the ratio S/V, the ratio of surface to volume, as shown in Fig. 56.

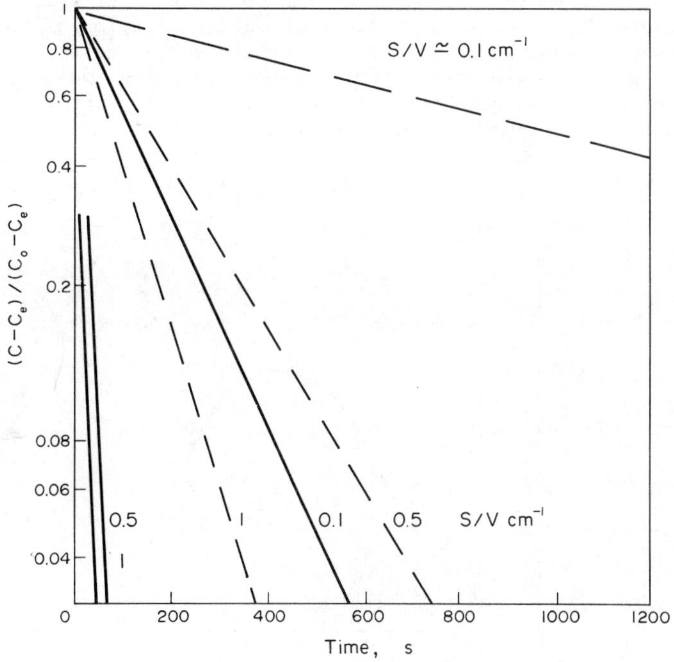

Fig. 56. Effects of surface-active agent on transfer of N_2 from Fe to gas. C_o, C and C_e are concentrations after time zero, time t and at equilibrium. S/V is the ratio of area to volume.

— — — 0.08% (O); ——— 0.009% (O)

With acknowledgement to F.D. Richardson, *Physical Chemistry of Melts in Metallurgy*, Academic Press, London (1974)

The mass transfer coefficient k can be assumed to be proportional to the fraction $1 - \theta$ of the surface not covered by adsorbed oxygen. According to Langmuir's Law:

$$\theta = \frac{bC_{(O)_{Fe}}}{1 + bC_{(O)_{Fe}}} \qquad \text{or} \qquad 1 - \theta = \frac{\theta}{bC_{(O)_{Fe}}}$$

where $C_{(O)_{Fe}}$ is the concentration of (O) in the liquid Fe and b is a constant. When θ is near to 1, i.e. when $C_{(O)_{Fe}} \gtrsim 0.02\%$, the mass transfer coefficient is proportional to $1/C_{(O)_{Fe}}$ and diminishes when $C_{(O)_{Fe}}$ increases. The same phenomenon occurs when the metal contains sulphur. Thus the kinetics of transfer of a species from one phase to another can be slowed down by the presence of a surface-active element, as shown in Fig. 56.

NUCLEATION OF BUBBLES

In the last example, nitrogen is removed from liquid iron by passage through the metal/gas surface. Bubbles of N_2 will nucleate in the liquid only if it is supersaturated with dissolved nitrogen and the conditions for this to occur can be explained by reference to the theory of homogeneous nucleation of bubbles in a liquid. Microbubbles ranging in size from a few molecules exist in this liquid and the free energy expression for their formation contains two terms, one a volume term corresponding to the formation of gas from the liquid and the other a surface tension term corresponding to the formation of a liquid/gas surface:

$$\Delta G = \frac{4}{3}\pi r^3 \Delta G_V + 4\pi r^2 \gamma \tag{4.155}$$

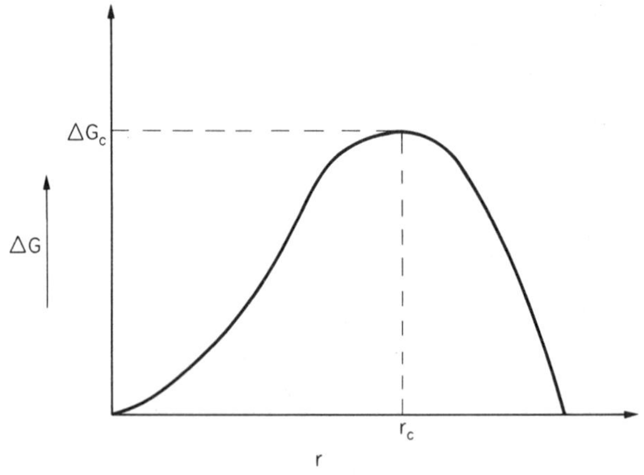

Fig. 57. Relationship of free energy of formation of bubble to radius

where r is the radius of the bubble and ΔG_V the free energy per unit volume of spontaneous bubble formation. ΔG reaches a maximum when r is a critical value r_C. For $r > r_C$, the increase corresponds to a decrease of ΔG and a change to a more stable state; so only bubbles with $r > r_C$ will grow. The value of the corresponding critical radius can be obtained by making $\partial\Delta G/\partial R = 0$, then:

$$r_C = -2\gamma/\Delta G_V \tag{4.156}$$

The radius of the bubble is related to the internal and external pressures P_i and P_a and the surface tension by the Laplace equation;

$$P_i = P_a + 2\gamma/r \tag{4.157}$$

When $r = r_C$, $P_i = P_C$ the critical pressure. From (4.155), (4.156) and (4.157) the free energy corresponding to r_C for bubbles in equilibrium with the supersaturated metal is:

$$\Delta G_C = 16\pi\gamma^3/3(P_C - P_a)^2 \tag{4.158}$$

The number of bubbles reaching the critical radius per cm^3 per second is given by:

$$J = K_B \exp\left(\frac{-\Delta G_C}{kT}\right) = K_B \exp\left(\frac{-16\pi\gamma^3}{3kT(P_C - P_a)^2}\right) \tag{4.159}$$

K_B has a value of about 10^{34} cm^{-3} sec^{-1} and k is the Boltzmann constant. (4.159) can be used to calculate the critical pressure inside a bubble in equilibrium with supersaturated metal. With $J = 1$ cm^3 sec^{-1} (a value at which nucleation is readily observable) and any gas, the critical pressure is extremely high, in that $P_C - P_a = 10^4$ atm. Thus the probability of homogeneous nucleation is effectively zero. In practice, nucleation does occur with $P_C - P_a \simeq 10^2$ atm. But it can be assumed that there is a contribution of heterogeneous nucleation from refractory micropores or pre-existing bubbles.

REFERENCES

1. A. Rist and N. Meysson, *Revue de Metallurgie*, 61, No. 2, 121-145 (1964).

2. P. Dutilloy, P.C. Ghosh and A. Rist, *Revue de Metallurgie*, 61, No. 12, 1044-1060 (1964).

3. P. Dutilloy, P.C. Ghosh and A. Rist, *Revue de Metallurgie*, 62, No. 1, 15-35 (1965).

4. A. Rist and G. Bonnivard, *Revue de Metallurgie*, 63, No. 3, 198-210 (1966).

5. A. Rist and G. Bonnivard, *Revue de Metallurgie*, 63, No. 4, 296-312 (1966).

6. P. Barrett and L. Bonnetain, *Bull. Soc. Chim. de France*, No. 3, 576-583 (1961).

7. F.D. Richardson, *Physical Chemistry of Melts in Metallurgy*, Vol. 2, Academic Press, London (1974).

8. E.T. Turkdogan, P. Grierson and J.P. Beisler, *Trans. Mat. Soc. of A.I.M.E.*, 12B, 755 (1981).

9. G.A. Irons and R.I.L. Guthrie, *A.S.M. and Met. Soc. of A.I.M.E.*, 12B, 755 (1981).

10. R. Defay and J. Prigogine, *Surface Tension and Adsorption*, Longmans, London (1966).

11. P. Kozakevitch, *Surface Phenomena of Metals*, Soc. Chem. Ind. Monograph 28, p. 223 (1968).

12. A.M. Karol'kov, *Casting Properties of Metals and Alloys*, p. 57, Consultants Bureau, New York (1957).

13. F.D. Richardson, *Mem. Sci. Rev. Met.*, II, 627 (1978).

14. P. Kozakevitch, *Revue de Metallurgie*, 46, 505 (1949).

15. P. Kozakevitch, G. Urbain and M. Sage, *Revue de Metallurgie*, VII, 2, 1 (1955).

16. T.R. Meadowcroft and J.F. Elliott, *Vacuum Met. Conf. 1963*, Am. Vac. Soc., Boston (1964).

17. W.D. Kingery, *Introduction to Ceramics*, p. 164, John Wiley & Sons Inc., New York (1960).

18. A. Bondi, "Spreading of Liquid Metals on Solid Surfaces. Surface Chemistry of High Energy Substances", *Chem. Rev.*, 52, 417 (1953).

19. J.P. Hirth and G.M. Pound, "Condensation and Evaporation", *Progress in Materials Science*, Pergamon Press, Oxford (1963).

GENERAL READING

Darken, L.S. and R.W. Gurry, *The Physical Chemistry of Metals*, McGraw-Hill, New York (1953).

Eyring, E.M., *Modern Chemical Kinetics*, Chapman & Hall, London (1965).

Hinshelwood, C.N., *Kinetics of Chemical Change*, Oxford Univ. Press (1942).

Kor, G.J.W. and F.D. Richardson, *Journal of the Iron and Steel Institute*, 206, 700 (1968).

Levenspiel, O., *Chemical Reaction Engineering*, J. Wiley & Sons Inc., New York (1962).

Parmetier, G. and P. Sochay, *Chimie Generale, Cinetique Chimique*, Masson et Cie, Paris (1964).

Petersen, E.E., *Chemical Reaction Analysis*, Prentice-Hall, New Jersey (1965).

Ward, R.G., *Journal of the Iron & Steel Institute*, 201, 11 (1963).

Yang, Ling and G. Derge, *Physical Chemistry of Process Metallurgy* (Ed. G.R. St. Pierre), Interscience, New York (1961).

This chapter is derived from the references cited, with material in Section 4.4, from F.D. Richardson.[7]

Chapter 5
EXTRACTION OF METALS

INTRODUCTION

Metals are found in the crust of the Earth as such and in a variety of chemical combinations. Because the atmosphere around the Earth contains oxygen, carbon dioxide and water vapour and because the initial concentrations of metal compounds were of sulphides and silicates in the cooling crust, most metals are found as oxides, sulphides or silicates. Sulphates and carbonates are obvious secondary products and there are those native metals where the binding energy to any of these elements or radicals has been too small to allow of formation or persistence under the ruling conditions. Gold and the platinum metals are among those found native, with copper, silver and mercury found as compounds, as well as in this form. Lead occurs as sulphide, sulphate and carbonate and iron as sulphide and oxide. More reactive metals can be divided into aluminium and silicon, which are found either as massive oxide deposits or in complex alumino-silicate clays and magnesium and calcium, occurring as mountain ranges of dolomite and limestone carbonates. The alkali metals and several others in the form of chlorides have formed soluble compounds and these have been transported into the sea, either to exist in solution, or to undergo hydrolysis and nucleation to form the nodules which are likely to constitute an important source of metals in the future.

With some exceptions, such as massive iron ore, limestone and quartz, the minerals are disseminated in the country rock or gangue and are mixed with metalliferous and other compounds containing values or impurities. Whether a given deposit is an ore in the technical sense depends on many factors, but commercial iron ore currently contains about 65% Fe, with low S and P, copper and nickel ores contain about 1% of the metal and gold ore is worked with a content of about 8 g per tonne or 8 ppm. The processes used for production of a given metal will depend on the form in which it is found as a mineral and on the chemical and physical properties of the metal and its compounds, in so far as they can be utilised to achieve maximum purity of the product, while consuming the minimum of energy, labour and reagents and generating the minimum of solid, liquid or gaseous pollutants. The maximum economic return must also be obtained from by-products and untreatable residues.

The first stage of treatment in most cases involves concentration of the valuable mineral and/or elimination of as much of the gangue and impurities as possible. This is carried out by such processes as flotation, gravity, or magnetic separation, after crushing and grinding to such a particle size as to "release" the mineral in substantially discrete lumps which are susceptible to concentration. This stage has a considerable effect on the nature and economics of subsequent processes and in some cases the attainment of satisfactory purity in the final product. Where the mineral, or an easily and cheaply prepared derivative is soluble in water or a cheap solvent, it may be most economical to take the suitably comminuted ore or processed concentrate and leach it to separate the metal. Copper from carbonate ore or roasted sulphide concentrate by sulphuric acid solution and gold from ore by sodium or potassium cyanide leaching are examples of this method. Concentrates or rich ores can be made to yield metal in one or several processes, either by direct reduction, (iron oxide by carbon or $CO + H_2$), or by conversion to reducible or soluble forms, followed by chemical or electroreduction. (Electrowinning of copper from leach solutions and cementation of gold from cyanide solution by zinc.) As the need for metals of very high purity increases, there is an extension of the methods whereby very pure compounds are prepared before metal production is attempted. This has been the case with tungsten and molybdenum and aluminium for the whole of this century and for zinc for as long as there has been electrolytic production, while beryllium and the nuclear energy metals provide other examples. The availability of energy, water and/or reductants, the liability to cause pollution and the effect on the recovery of valuable by-products can be important factory influencing the choice of the final process route.

5.1. STABILITY OF COMPOUNDS

For some time, the heat of formation of a compound was assumed to be the fundamental index of its stability, in that the greater the quantity of heat evolved on formation, the greater the amount of energy needed for dissociation. This concept ignored the influence of the entropy contribution to the available energy change involved in formation. When this was taken into account, the affinity, or Gibbs Free Energy of Formation was found to be the accurate index of stability, in that no exceptions have been found. In Chapter 3 it has been shown how these affinities are determined, how they are expressed as functions of temperature under standard conditions in the form of graphs and equations and the corrections to be applied when conditions are non-standard.

A study of the stability of compounds makes it possible to predict the state in which it is most probable that a metal will exist in nature, as well as the principles to be employed in the most efficient process for producing the pure metal.

5.1.1. AFFINITY OF ELEMENTS FOR OXYGEN

The Ellingham diagram for free energy of oxide formation (Fig. 23) has been studied in Chapter 3. The affinities of the metals for oxygen range from gold, stable under standard and atmospheric conditions, bismuth, mercury, and silver, the oxides of which decompose below 500°C, through the oxides of lead, copper, and iron, which are reducible by carbon, to those of aluminium, titanium, and calcium, requiring special processes for production of metal. The oxidation of carbon and the stability of the oxides is of great importance, because of the widespread use of carbon as a reducing agent for metal oxides. It can be seen from the plots of free energy changes involved in the oxidation of carbon, that carbon dioxide is more stable than carbon monoxide below about 710°C (983 K) and that the reverse is true above this temperature. Carbon monoxide formation is especially important because it takes place with a considerable increase of entropy. For this reason, the stability of CO increases with temperature, and it is theoretically possible to remove oxygen from any metal oxide by heating it to a sufficiently high temperature with carbon. The extractive metallurgy of metal oxides is essentially a study of their reduction by carbon, a gas, or another metal having a higher affinity for oxygen.

5.1.2. AFFINITY OF THE METALS FOR CARBON

In addition to being a reducing agent, carbon is also capable of combining with metals and so, when investigating the reduction of oxides by carbon, it is also necessary to determine the affinity of the metals for carbon. If this is high, the product will be a carbide and not a metal (Fig. 58).

The entropy change involved in the formation of carbides from the elements is small because the reactants and products are usually solid. Since $\Delta S^o = -d\Delta G^o/dT$ the Ellingham plots of free energy of carbide formation are near to horizontal. Methane and other hydrocarbons are exceptions in that the entropies of formation are strongly negative, because two or more moles of gas form one mole of hydrocarbon. The slopes of the standard free energy curves for these compounds are therefore large and positive, the compounds becoming less stable as the temperature rises. This property makes it possible to produce mixtures of CO and H_2 (used for the reduction of certain oxides) by cracking of CH_4 with steam.

Certain elements, such as zinc, copper, and the noble metals, have no affinity for carbon, while elements like Ti, Si, V, and Hf form very stable carbides. The transition elements, which dissolve carbon and carbides and form wide ranges of liquid and solid solution, can be distinguished from those elements which form refractory carbides such as Al_4C_3, CaC_2, and SiC, which are immiscible in the metal.

5.1.3. AFFINITY OF METALS FOR SULPHUR

The Ellingham diagram for sulphides (Fig. 58a) is similar in form to that for oxides in that the affinities of metals for sulphides decrease with temperature because, in most instances, there is a significant decrease of entropy when metal sulphides are formed. The high affinities of manganese, the alkali and alkaline earth metals, and such metals as cerium for sulphur, are of great importance relative to the removal of sulphur from iron and steel. However, except for the alkali metals, the affinities for sulphur are generally smaller than the affinities for oxygen. The relative affinities of some elements for O_2 and S_2 can be derived from Fig. 59, in which the standard free energies for reactions of the form:

$$2\langle MS\rangle + [O_2] = 2\langle MO\rangle + [S_2] \tag{5.1}$$

have been plotted against temperature for a number of metals, carbon, and hydrogen. The values for the free energies of reactions of the form given in (5.1) are the differences between those for the formation of oxides and sulphides:

$$2\langle M\rangle + [O_2] = 2\langle MO\rangle \tag{5.2}$$

$$2\langle M\rangle + [S_2] = 2\langle MS\rangle \tag{5.3}$$

The entropy change in (5.1) is small in most cases; thus, the relative affinity of metals for oxygen and sulphur does not change significantly with temperature except for Pb, Cu, and Ni.

In the case of complex equilibria between sulphides and oxides, knowledge of the relative affinities make it possible to predict the equilibrium composition. For example, since the standard free energy of oxidation of FeS is more negative than that for the oxidation of copper sulphide, equilibrium in the reaction:

$$(Cu_2O) + (FeS) \rightleftarrows (Cu_2S) + (FeO) \qquad (5.4)$$

will be displaced towards the formation of Cu_2S and FeO. The affinities of copper and iron for sulphur are similar in value, but the affinity of oxygen for iron is much greater than that for copper. Thus, in the oxidation of matte of Cu_2S and FeS, iron sulphide will be oxidised at first.

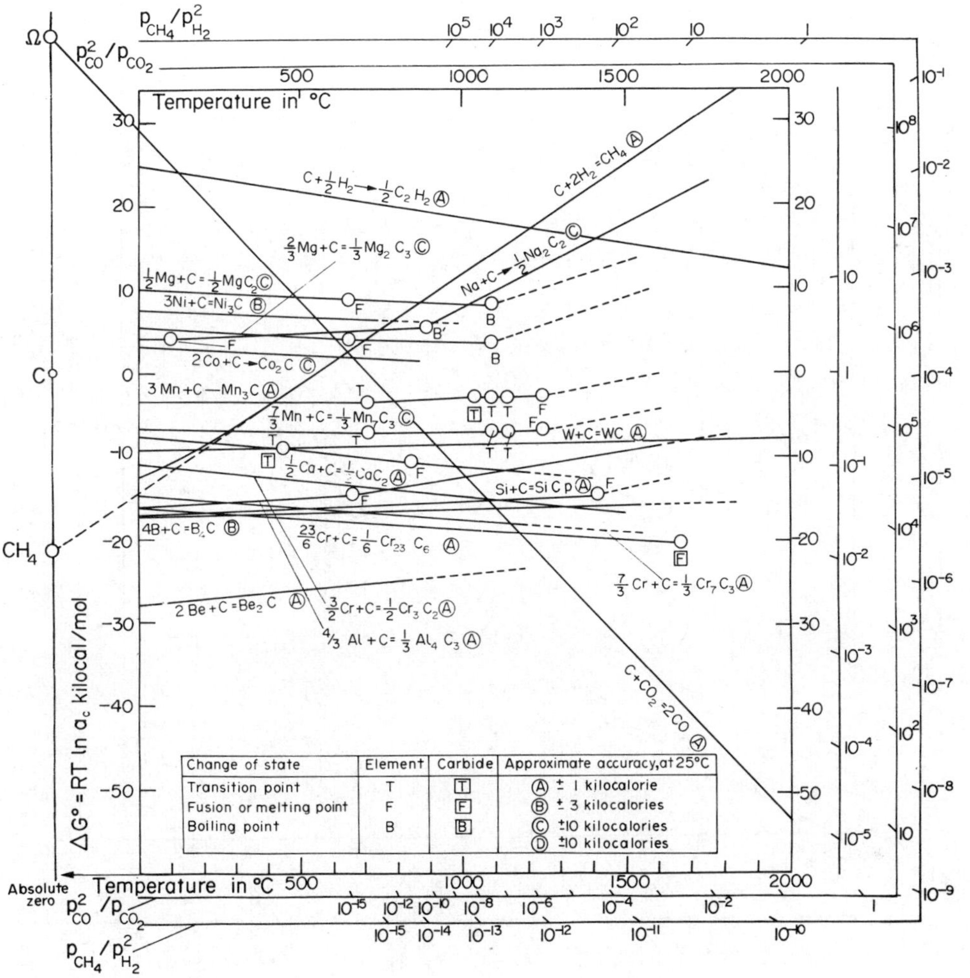

Fig. 58. Free energy of formation of carbides
(Courtesy of Mr. Olette and Mrs. Ancey-Moret)

The affinities of oxygen for sulphur in SO_2 and SO_3 have also been given in Fig. 59.

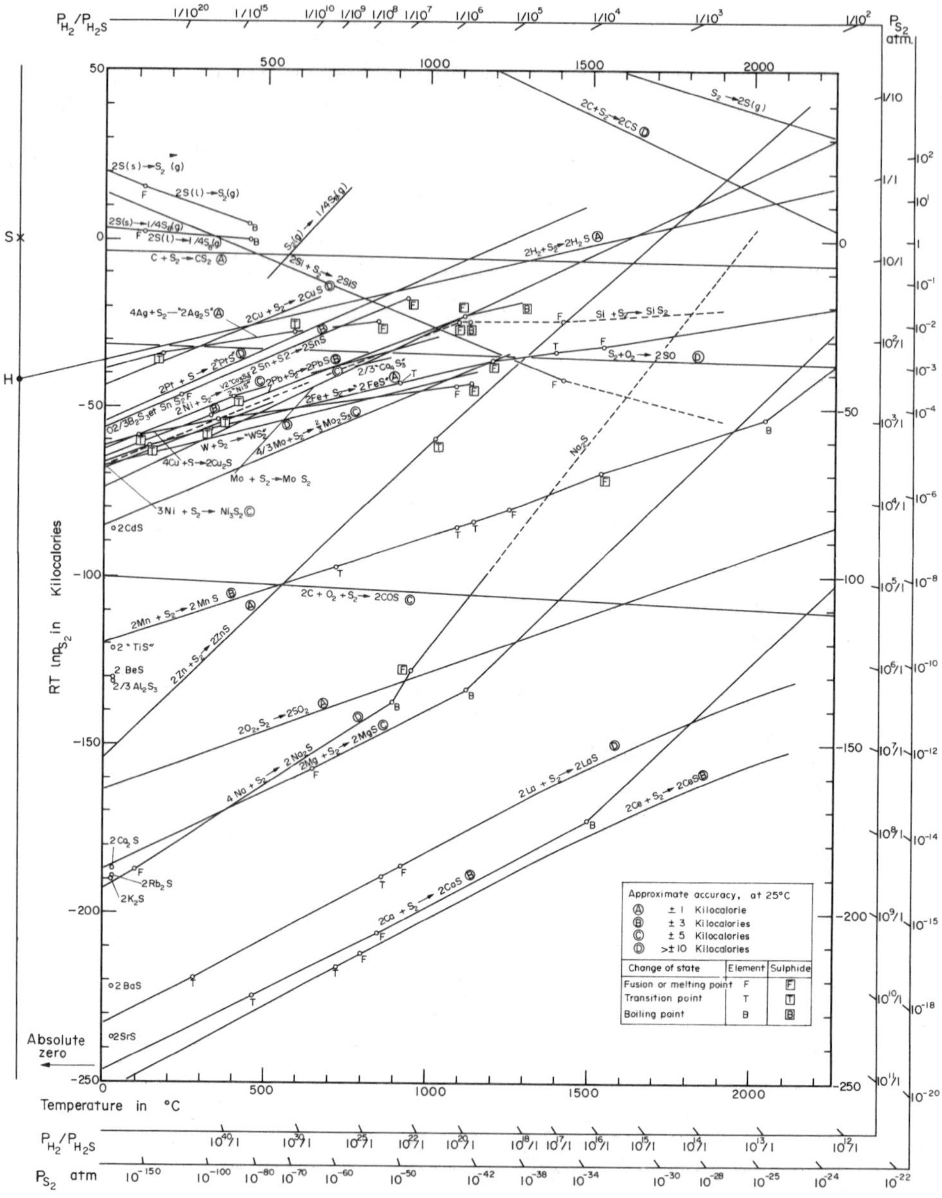

Fig. 58a. Free energy of formation of sulphides
(Courtesy of Mr. Olette and Mrs. Ancey-Moret)

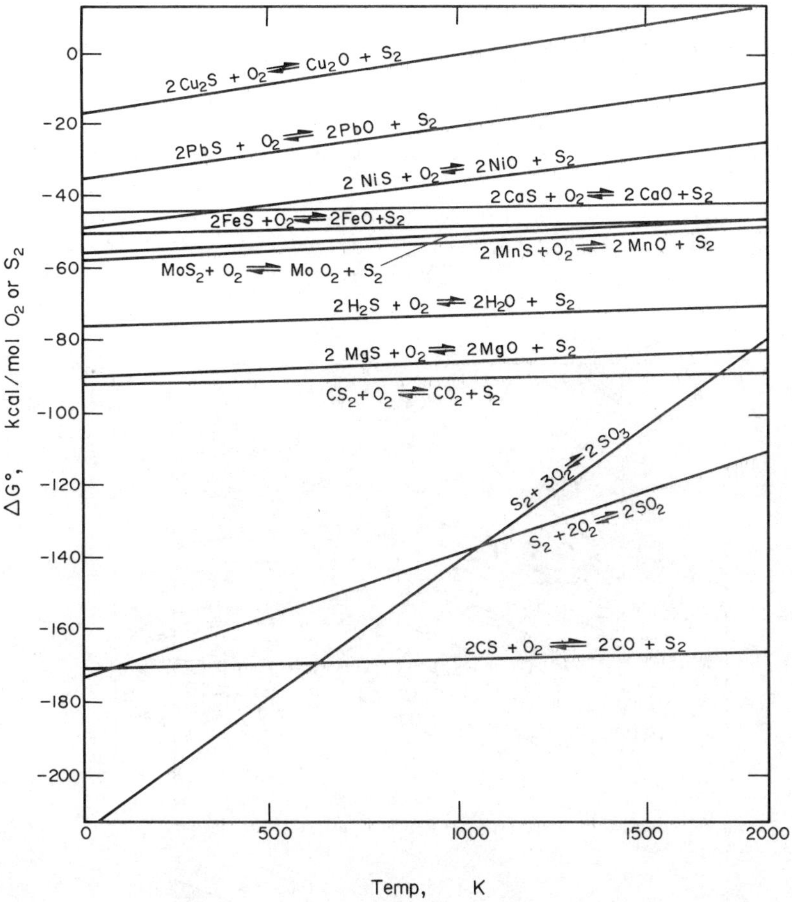

Fig. 59. Relative affinities of some elements for sulphur and oxygen

$$[S_2] + 2[O_2] \rightleftarrows 2[SO_2] \quad (5.5)$$

$$[S_2] + 3[O_2] \rightleftarrows 2[SO_3] \quad (5.6)$$

Below about 800°C (1073 K), SO_3 is more stable than SO_2, and above this temperature the stabilities are reversed. For oxidation of a sulphide according to:

$$2<MS> + 3[O_2] \rightleftarrows 2<MO> + [SO_2] \quad (5.7)$$

the free energy change can be calculated by adding the value for (5.1) to that for (5.5). Since ΔG^o for the oxidation of sulphur is always negative, ΔG^o for (5.7) is always more negative than for (5.1).

The affinities of hydrogen and carbon for sulphur are lower than those for oxygen. These are so low that sulphides cannot be reduced by either.

5.1.4. AFFINITY OF METALS FOR CHLORINE

While there are general similarities to the other free energy diagrams examined, the chloride diagram (Fig. 60a) indicates a number of significant differences from the oxides and sulphides. Carbon tetrachloride is thermodynamically unstable above about 750°C (1025 K), and C is only capable of reducing $AuCl_2$ and $AuCl_3$ to metal, while hydrogen is capable of reducing a limited range of chlorides, including those of Au, Cu, Hg, Ni, Fe, Cr, Mn, Ag, and Sn. Aluminium is especially interesting in that it forms two chlorides, $AlCl_3$ and the more volatile AlCl.

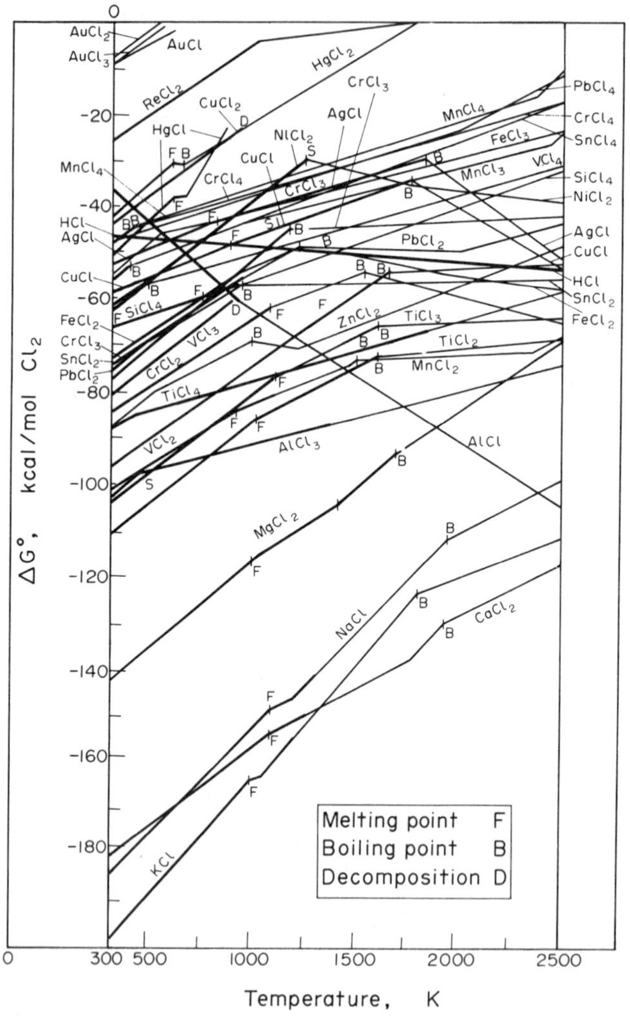

Fig. 60a. Free energy of formation of some chlorides

Since 2 moles of AlCl are formed from 1 mole of Cl_2, the entropy of formation is positive, and the stability of AlCl in contact with Al and Cl_2 increases with temperature. Several industrially important metals form chlorides which are volatile at temperatures between 400 and 1600 K: Cu, Ni, Be, Ti, V, Zr, and Cr, are included in this group. The availability of molten chlorides at temperatures between 800 and 1300 K has stimulated the utilisation of fused salt electrolysis as a method of metal extraction.

The differences in affinities of the elements for oxygen and chlorine are given in Fig. 60b. as a function of temperature. The values are those of the free energy of reactions of the type:

$$2\langle MO\rangle + 2[Cl_2] \rightleftarrows (MCl_2) + [O_2] \tag{5.8}$$

These indicate that certain oxides, Na_2O, CaO, ZnO, and PbO, can be converted directly to chlorides by reaction with chlorine, while others like Al, Ti, and Mg remain substantially as oxides under standard conditions.

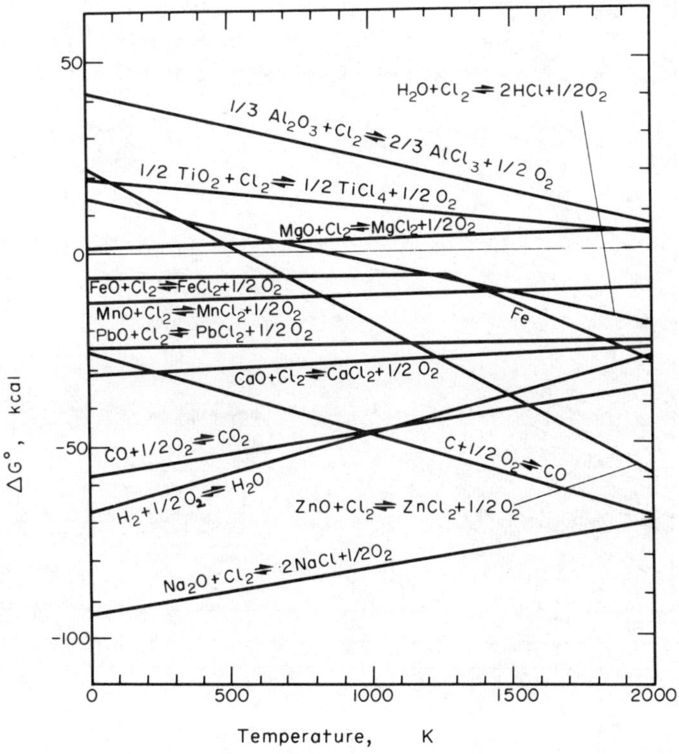

Fig. 60b. Relative affinities of some elements for chlorine and oxygen

5.1.5. AFFINITY OF METALS FOR NITROGEN

The data in Fig. 61 are principally of use in examining the possibility of nitrogen contamination when metals are heated in air, or air is blown through molten alloys. The instability of the nitrides of carbon, hydrogen, and iron is contrasted with the refractory properties of the nitrides of Si, B, Al, Ti, Zr, and Nb. The high affinities of Ti and Nb for N_2 are utilised in the production of non-ageing low carbon steel, since N_2 in solution in Fe is converted to Ti and Nb nitride precipitates. (*Note* that Ti and Nb are also included in "stabilised" stainless steel because of their high affinity for carbon.)

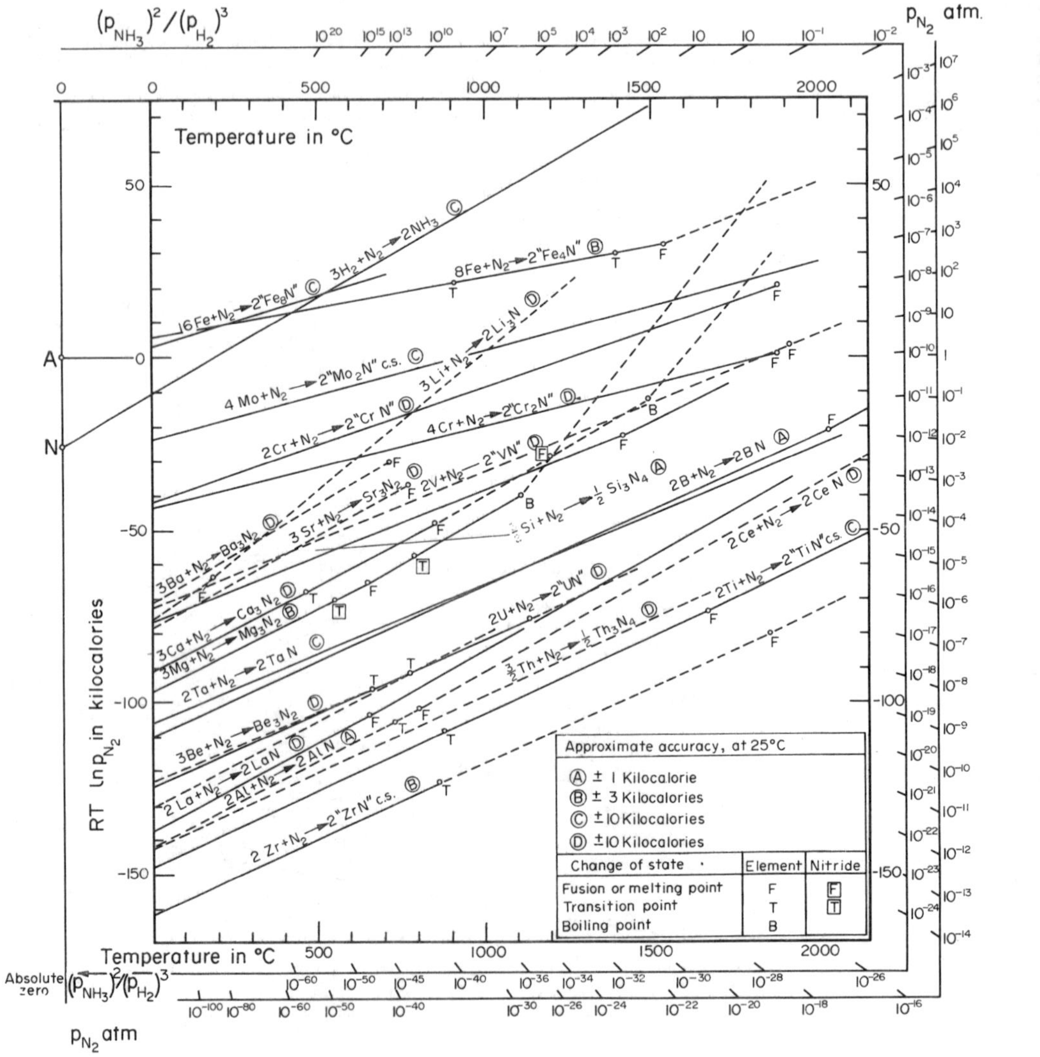

Fig. 61. Free energy of formation of nitrides
(Courtesy of Mr. Olette and Mrs. Ancey-Moret - IRSID)

5.1.6. COMPARISON OF AFFINITIES OF SOME METALS FOR S_2, O_2, Cl_2 AND F_2

Figure 62 gives values of the standard affinities of some metals for S, O, Cl_2 and F_2. It shows that they increase in virtually the same order and value as the Standard Potentials at 298 K from Ag to Ca. This can be expressed in the form:

$$A^{o} = nFE^{o} \tag{5.9}$$

With some exceptions, the fluorides are the most stable compounds, followed by the chlorides, oxides and sulphides, in that order. We can note that:

the affinities of Cu^{I} and Fe for S have similar values, but the affinity of Fe for O_2 is much greater than the affinity of Cu^{I} for O_2. This difference is important for its effect on the extractive metallurgy of copper.

C can be used only for the reduction of oxides because its affinity for S_2, Cl_2 and F_2, is lower than that of the metals.

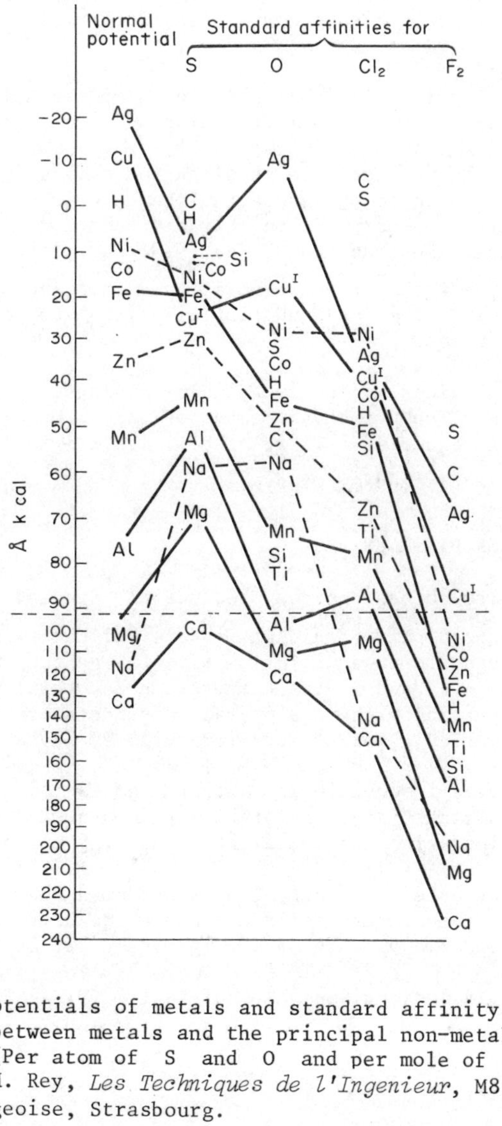

Fig. 62. Standard Potentials of metals and standard affinity values for compound formation between metals and the principal non-metals in k cal at 1000-1200°C.* (Per atom of S and O and per mole of Cl_2 and F_2).
Acknowledgement to M. Rey, *Les Techniques de l'Ingenieur*, M8, 2230, 1967. Imprimerie Strasbourgeoise, Strasbourg.

5.2. PROCESSING OF SULPHIDES

The extractive metallurgy of sulphides consists generally of a combination of oxidation and other processes, either to obtain the metal directly, or the oxide, which will be reduced, or the soluble sulphate, which will be subjected to leaching and electrowinning.

Predominance area diagrams can be constructed for systems containing O_2 and SO_2, using thermodynamic data for the reactions which are possible and the stable constituents under specified values of T, p_{O_2} and p_{SO_2} can be predicted.

5.2.1. CLASSIFICATION OF SULPHIDES

Metal sulphides can be classified according to their solubilities in the other phases which may be present during the extraction or refining process, principally metal and slag. According to this method of classification, it is possible to distinguish between:

Alkali and alkaline earth sulphides, soluble in molten salts and slags, but insoluble in metals. They ionise in solution and are ionic conductors.

Those sulphides which are insoluble in slags and only slightly soluble in metal, and which form single phases having the generic description of "mattes". These are semi-metallic in character and contain the sulphides of the transition metals and lead (Cu_2S, FeS, NiS, PbS, etc.). In general, the compounds are not perfectly stoichiometric, and their conductivity is more electronic than ionic. They are substantially less stable than the alkali and alkaline earth sulphides.

The intermediate sulphides, slightly soluble in slags, soluble in mattes, but insoluble in metals. The principal member of this group is MnS, with ZnS and Al_2S_3 as other examples. Their stabilities are intermediate between those of the first and second groups.

5.2.2. PREDOMINANCE AREA DIAGRAMS

During the roasting of sulphides, in the presence of O_2 and SO_2, several phases can be formed: the lower sulphide, sulphate, basic sulphate, oxide, or metal. When two condensed phases are in equilibrium with the gaseous phase, the variance of the system is equal to 2, and if the temperature and total pressure of the system are fixed, the partial pressures of the components of the gas phase are determined. If three variables are fixed, for example, T, p_{O_2}, and p_{SO_2}, the number of condensed phases in equilibrium with the gases is only one. Since the maximum value of the variance is 3, the stability domain of a phase can be represented in three dimensions with the coordinates p_{SO_2}, p_{O_2}, and T. This form of presentation is known as a Kellog diagram, or a predominance area diagram. In the simplified version, the temperature is fixed, and a two-dimension diagram with coordinates $\ln p_{SO_2}$ and $\ln p_{O_2}$ is used (Fig. 63a). It is also possible to plot $\ln p_{O_2}$ against T for a given value of p_{SO_2} (Fig. 63b).

A roasting reaction can be expressed in the form:

$$<MS> + 3/2[O_2] \rightleftarrows <MO> + [SO_2] \tag{5.10}$$

At equilibrium, the standard free energy change, ΔG^o_T, is equal to $-RT \ln K$, and since MO and MS are assumed to be present as pure solids, their activities are equal to 1; therefore:

$$\Delta G^o_T = RT \ln \frac{p_{O_2}^{3/2}}{p_{SO_2}} \tag{5.11a}$$

or

$$\ln p_{SO_2} = -\frac{\Delta G^o_T}{RT} + 3/2 \ln p_{O_2} \tag{5.11b}$$

The straight line plot representing this relationship on a p_{SO_2} vs p_{O_2} diagram divides it into two areas. In one, the oxide is stable, and in the other the sulphide, MO and MS coexist in the conditions represented by the line.

EXAMPLE

The Cu-S-O diagram has been constructed at 1000 K, a temperature which corresponds approximately to that of roasting in a multiple hearth furnace, or in a fluidised bed reactor. The compounds that can exist in contact with the atmosphere containing O_2 and SO_2 at a total pressure of 1 atm are: Cu, CuS, Cu_2O, CuO, Cu_2S, $CuSO_4$, and $CuO.CuSO_4$. If the free energy relationships of these seven compounds, taken in pairs, are used to determine values of p_{SO_2} and p_{O_2}, there will be 21 pairs of results. However, it is evident that only certain of these are possible. For example, the equilibrium Cu_2S/CuO is not possible because the Cu_2S must first be oxidised to Cu_2O, and CuO is only formed when p_{O_2} is too high for Cu_2S to exist. Tables 5-6 give the data and calculated values of $\log p_{O_2}$ and $\log p_{SO_2}$ for all the possible equilibria. The straight line plots of Fig. 63a have been drawn on the basis of the results given in Table 6. They define the domains of stability of each element or compound in the presence of SO_2 and O_2.

The line corresponding to the equilibrium:

$$[S_2] + 2[O_2] \rightleftarrows 2[SO_2] \tag{5.12}$$

has been inserted into this diagram and, at 1000 K, it divides the diagram into two regions: one where $p_{S_2} > 1$, and the other where $p_{S_2} < 1$ atm. From this it can be seen that at 1000 K, CuS decomposes to Cu_2S at a pressure of S_2 greater than 1 atm.

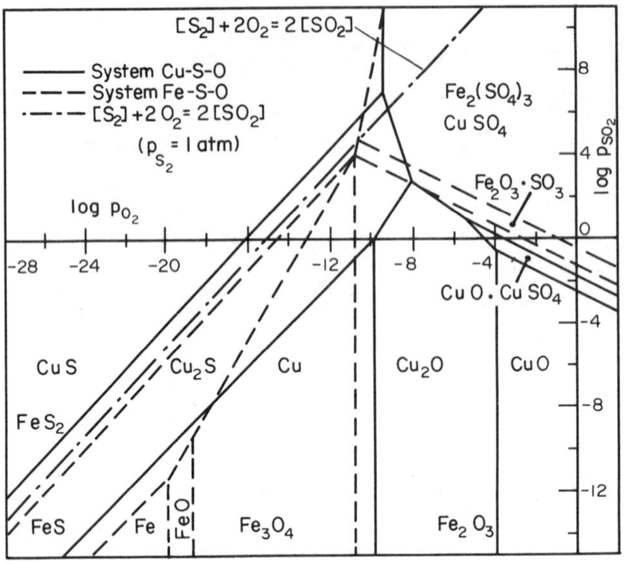

Fig. 63a. Predominance area diagram at 1000 K. ——— Cu-S-O system; — — — Fe-S-O system; ·—·—·—· $S_2 + 2O_2 = 2\ SO_2$

TABLE 5 Standard Free Energies for the Equilibrium Cu-O-S†

Reaction	ΔG^o = A + BT log T + CT (cal)
$2\langle Cu\rangle + \frac{1}{2}[O_2] = \langle Cu_2O\rangle$	− 40,500 − 3.92 T log T + 29.5 T
$\langle Cu_2O\rangle + \frac{1}{2}[O_2] = \langle 2\ CuO\rangle$	− 34,950 − 6.1 T log T + 44.3 T
$2\langle Cu\rangle + \frac{1}{2}[S_2] = \langle Cu_2S\rangle$	− 34,150 − 6.2 T log T + 28.7 T
$2\langle Cu_2S\rangle + [S_2] = 4\langle CuS\rangle$	− 45,200 + 54.0 T
$[S_2] + 2[O_2] = 2[SO_2]$	− 173,240 + 34.6 T
$[SO_2] + \frac{1}{2}[O_2] = [SO_3]$	− 22,600 + 21.36 T
$\langle Cu\rangle + 2[O_2] + \frac{1}{2}[S_2] = \langle CuSO_4\rangle$	− 183,000 + 88.4 T
$[SO_3] + 2\langle CuO\rangle = \langle CuO.CuSO_4\rangle$	− 49,910 − 3.32 T log T + 50.1 T

†Data from O. Kubaschewski, E.L. Evans and C.B. Alcock, *Metallurgical Thermochemistry*, 5th Ed., Pergamon Press, Oxford (1979).

TABLE 6 Relations between p_{O_2} and p_{SO_2} deduced from Table 5

Reaction	ΔG^o at 1000 K k cal	log K	$\log p_{O_2} = f(\log p_{SO_2})$
$2\langle Cu\rangle + \frac{1}{2}[O_2] = \langle Cu_2O\rangle$	-22·76	4·97	$\log p_{O_2} = -9{\cdot}95$
$\langle Cu_2O\rangle + \frac{1}{2}[O_2] = 2\langle CuO\rangle$	- 8·95	1·96	$\log p_{O_2} = -3{\cdot}92$
$2\langle CuS\rangle + [O_2] = \langle Cu_2S\rangle + [SO_2]$	-73·72	16·1	$\log p_{O_2} = -16{\cdot}1 + \log p_{SO_2}$
$\langle Cu_2S\rangle + [O_2] = 2\langle Cu\rangle + [SO_2]$	-45·27	9·90	$\log p_{O_2} = -9{\cdot}9 + \log p_{SO_2}$
$\langle Cu_2S\rangle + \frac{3}{2}[O_2] = \langle Cu_2O\rangle + [SO_2]$	-68·03	14·87	$\log p_{O_2} = -9{\cdot}91 + \frac{2}{3}\log p_{SO_2}$
$\langle CuS\rangle + 2[O_2] = \langle CuSO_4\rangle$	-84·775	18·53	$\log p_{O_2} = -9{\cdot}26$
$\langle Cu_2S\rangle + 3[O_2] + [SO_2] = 2\langle CuSO_4\rangle$	-95·83	20·95	$\log p_{O_2} = -6{\cdot}98 - \frac{1}{3}\log p_{SO_2}$
$\langle Cu_2O\rangle + \frac{3}{2}[O_2] + 2[SO_2] = 2\langle CuSO_4\rangle$	-27·80	6·08	$\log p_{O_2} = -4{\cdot}05 - \frac{4}{3}\log p_{SO_2}$
$\langle CuO{\cdot}CuSO_4\rangle + [SO_2] + \frac{1}{2}[O_2] = 2\langle CuSO_4\rangle$	- 7·84	1·71	$\log p_{O_2} = -3{\cdot}43 - 2\log p_{SO_2}$
$\langle Cu_2O\rangle + [SO_2] + \frac{1}{2}[O_2] = \langle CuO{\cdot}CuSO_4\rangle$	-19·96	4·36	$\log p_{O_2} = -4{\cdot}36 - \log p_{SO_2}$
$2\langle CuO\rangle + [SO_2] + \frac{1}{2}[O_2] = \langle CuO{\cdot}CuSO_4\rangle$	-11·01	2·41	$\log p_{O_2} = -4{\cdot}81 - 2\log p_{SO_2}$

The data indicate that, increasing the temperature decreases the value of the equilibrium constant; an effect which produces a displacement of the lines towards the right side in the predominance area diagram. Thus, for a given value of p_{O_2} and p_{SO_2}, a compound, stable at low temperature, becomes unstable at high temperature, and vice versa. At 1000 K, CuO is the most stable compound in the presence of air, provided that p_{SO_2} is low.

Sulphide ores are generally impure, as is the case of the copper sulphides which are associated with iron and molybdenum sulphides. By superimposing the predominance area diagrams of two or more metals M-O-S and M'-O-S, it is possible to determine the most stable compound of each metal under given conditions of temperature and partial pressures of SO_2 and O_2, and to specify the conditions for the separation of two or more compounds. Figure 63a is an example in which the systems Cu-O-S and Fe-O-S at 1000 K have been superimposed.* It can be seen that the iron oxides are stable over a wider range of conditions than for copper

*In this diagram, no account has been taken of reactions between constituents of the two systems, to give compounds such as $CuFeO_4$ and $CuFeS_2$, the formation of which decreases the activity of both constituents.

oxides. For this reason, the selective oxidation of pyrite is possible from a mineral or a complex concentrate.

Figure 63b was constructed by plotting $\log p_{O_2}$ (oxygen potential) against T for a given pressure of SO_2, in this case $p_{SO_2} = 0.2$ atm.

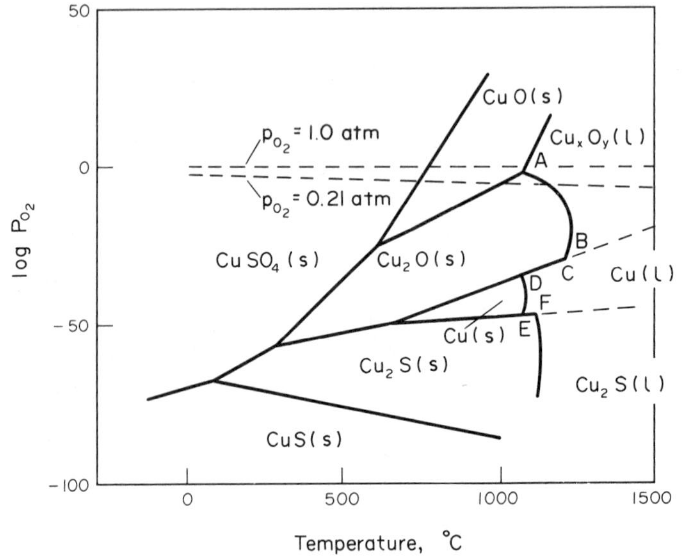

Fig. 63b. Predominance area diagram for the system Cu-O-S for $p_{SO_2} = 0.2$ atm

Acknowledgement to M. Rey, *Les Techniques de l'Ingenieur*, M8, 2230, 1967. Imprimeri Strasbourgeoise, Strasbourg.

This diagram can be used to establish the conditions of temperature and p_{O_2} necessary for the production of metallic copper or other compounds directly from a sulphide concentrate.

5.2.3. THERMAL DECOMPOSITION OF SULPHIDES

When a metal forms two sulphides, the one with the higher S content can be decomposed by heat, to evolve sulphur and form a lower sulphide. Iron pyrites, FeS_2, and Covellite, CuS, decompose endothermically according to the equations:

$$FeS_2 = FeS + 0.5\ S_2 \quad (T \sim 700°C) \tag{5.13}$$

$$2CuS = Cu_2S + 0.5\ S_2 \quad (T \sim 500°C) \tag{5.14}$$

In one variation of the flash smelting process, using specially designed burners, finely divided coal and pyrite are suspended in a stream of air and fuel. Reactior (5.13) is effected as a result of stoichiometric combustion of the fuel. The furnace gases contain S vapour and there is a molten bath of pyrrhotite and slag. The S is condensed to liquid and solidified as small spherical "prills". The pyrrhotite is tapped and granulated in a water jet before being roasted to oxide and SO_2 in a fluidised-bed furnace.

5.2.4. DIRECT PRODUCTION OF METAL FROM SULPHIDES

(a) While metallothermic reduction is possible in several instances, it is now rarely used. At one time, lead was made by reducing PbS with Fe:

$$\langle Fe \rangle + (PbS) = (Pb) + (FeS) \tag{5.15}$$

and although this method has been abandoned, the principle is still in use for the recovery of silver from the sulphide, using lead:

$$(Ag_2S) + (Pb) = (PbS) + 2(Ag) \tag{5.16}$$

Iron can also be added to lower the consumption of lead. The products are metal containing silver, other noble metals and excess lead and matte of lead sulphide and other metals. Excess lead is removed from the metal by oxidation.

(b) Copper and lead are obtained directly from the sulphides in the pyrometallurgical extraction of these metals. In copper metallurgy, the first step involves the physical separation of sulphides from gangue. The copper concentrate, or calcine, is charged into a reverberatory, flash or blast furnace with fluxes, and the charge is melted at 1500-1650 K. The copper and iron sulphides Cu_2S and FeS, which are stable in the absence of oxygen, are insoluble in the slag formed from the gangue and flux, and due to their greater specific gravity, form a separate lower liquid layer, called matte, which contains 25 to 70% Cu. In the second stage, liquid matte from the smelting furnace is fed into a converter, where air is blown through it by way of submerged tuyeres. Due to its larger affinity for oxygen, the iron is oxidised before the copper, as is seen in Fig. 59. The iron oxide formed, only slightly soluble in the matte, is stabilised in the form of fayalite, $2FeO \cdot SiO_2$, by silica in the slag. The reactions are as follows:

$$(FeS) + 3/2[O_2] \rightleftarrows (FeO) + [SO_2] \tag{5.17a}$$

$$2(FeO) + \langle SiO_2 \rangle \rightleftarrows (2FeO \cdot SiO_2)_S \tag{5.17b}$$

When all the iron has been oxidised, the copper sulphide is converted to copper metal and copper oxide, which is in turn reduced to metal by the remaining sulphide. The reactions that take place are:

$$(Cu_2S) + [O_2] \rightleftarrows 2(Cu) + [SO_2] \tag{5.18}$$

or

$$(Cu_2S) + 1.5[O_2] \rightleftarrows (Cu_2O) + [SO_2] \tag{5.19a}$$

combined with

$$2(Cu_2O) + (Cu_2S) \rightleftarrows 6(Cu) + [SO_2] \tag{5.19b}$$

From Fig. 63a it can be seen that above 1000 K, metallic copper is stable only when the oxygen partial pressure is below 10^{-11} atm. At pressures above this, the copper oxidises to Cu_2O, which is reduced again to metallic copper in the presence of Cu_2S. The copper produced, called "blister copper" is only slightly soluble in matte and slag. It forms, together with the noble metals, a separate phase in the bottom of the converter, below the tuyere level, due to its higher specific gravity. In this position, it is protected from further oxidation. Blister copper is refined by controlled oxidation of the impurities (As, Sb, Bi, etc.) with air, and is finally deoxidised with CO, H_2, hydrocarbons, or C.

The metallurgy of lead is related to that of copper, but the metal is generally produced by roasting the sulphide to oxide and then reducing the oxide to metal with carbon.

(c) In the same way, because of the low affinity of mercury for sulphur and oxygen at above 400°C, the metal can be produced by heating ore or concentrate in air to 600-700°C.

$$HgS + O_2 = Hg + SO_2 \qquad (5.20)$$

This can be carried out in a rotary kiln, using coal, oil, or gas as fuel, with excess air to oxidise the sulphur. The hot gases move counter-current to the solid charge and are cooled in external cast-iron condensers to recover liquid mercury.

(d) Molybdenite, MoS_2, a by-product of treatment of most porphyric copper ores, is the principal source of molybdenum. One method used to obtain the metal consists of roasting the MoS_2 to trioxide, MoO_3, and reducing the oxide with hydrogen or carbon. Part of the Mo is lost in the carbothermic reduction of the oxide in a low shaft electric furnace, because MoO_3 is volatile at above 620°C. To avoid these losses, the sulphide can be reduced directly to metal with carbon in the presence of calcium oxide. The overall reaction is:

$$<MoS_2> + 2<CaO> + 2<C> \rightleftarrows <Mo> + 2<CaS> + 2[CO] \qquad (5.21)$$

This can be broken down into two reactions which are thermodynamically possible. One relates to sulphur and oxygen interchange between molybdenum and calcium:

$$<MoS_2> + 2<CaO> \rightleftarrows <MoO_2> + 2<CaS> \qquad (5.21a)$$

and the other to the reduction with carbon of the oxide:

$$<MoO_2> + 2<C> \rightleftarrows <Mo> + 2[CO] \qquad (5.21b)$$

(5.21a) can easily be explained with the aid of Fig. 59, and (5.21b) by means of Fig. 23.

5.2.5. TRANSFORMATION OF SULPHIDES TO SULPHATES OR OXIDES

Metal sulphides can be converted to sulphates or oxides by roasting in air, before being reduced to metal. The roasting reactions can be written in the form:

$$\langle MS\rangle + 2[O_2] = \langle MSO_4\rangle \tag{5.22}$$

and

$$\langle MS\rangle + 1.5O_2 = \langle MO\rangle + [SO_2] \tag{5.23}$$

Whether the product obtained is oxide or sulphate depends on the conditions of temperature and pressure under which the roasting is carried out. At low temperature, the sulphates are formed. If the temperature is raised, the basic sulphate, and then the oxide, are obtained. At a certain temperature (that of inversion), sulphate and oxide will be in equilibrium:

$$2\langle MSO_4\rangle \rightleftarrows 2\langle MO\rangle + 2[SO_2] + [O_2] \quad (\text{or } [SO_3] \tag{5.24}$$

The standard free energy of this endothermic reaction (Fig. 64) becomes negative above the inversion temperature. The only stable compound in the presence of SO_2 and O_2 at atmospheric pressure is then the oxide. Decrease of pressure facilitates the formation of oxide.

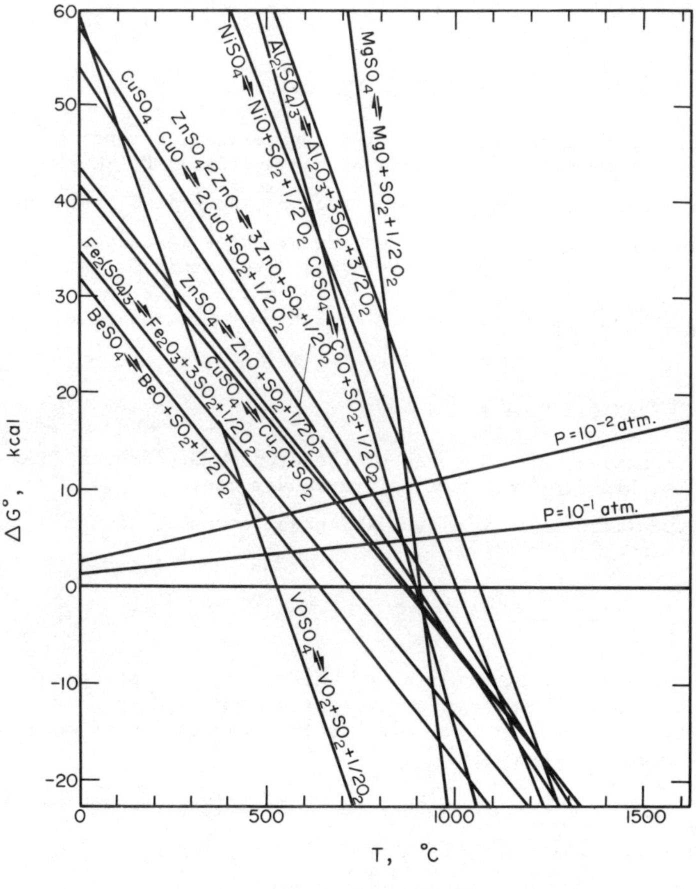

Fig. 64. Free energy of oxide formation from sulphate

The Roast-Leach-Electrowin (R.L.E.) process depends on the conversion of copper sulphides to soluble sulphates. In the first stage, roasting is carried out in a fluidised bed reactor at 680-690°C. By careful selection of the conditions (p_{O_2}, p_{SO_2}, and temperature), it is possible to form soluble copper sulphate and insoluble iron oxide, as was shown in Fig. 63a. The calcines are leached in dilute sulphuric acid, and the copper is recovered by electrowinning.

The roasting of sulphides to obtain oxides is extensively used; for example, in the metallurgy of lead, nickel, iron, cobalt, cadmium, zinc, molybdenum, rhenium, etc. The roasting is usually carried out in fluidised bed furnaces, multiple hearth furnaces, and rotary kilns, at a temperature higher than that of the inversion of reaction (5.24), to avoid forming the sulphate.

At the roasting temperature, the oxides are generally present in the solid state. Those of molybdenum and rhenium are sublimed at temperatures above 900 and 600 K, respectively, and the roasting of molybdenite, which sometimes contains rhenium sulphide, is effected between 850 and 870 K. Rhenium oxide, Re_2O_7, which is volatile at this temperature, is recovered from the roasted gases by condensation or solution in water.

5.3. OXIDE REDUCTION

Other oxidised compounds are also classified with oxides, including carbonates, MCO_3, and hydrates, $M(OH)_2$. The variance of the systems MCO_3-MO-CO_2 and $M(OH)_2$-MO-H_2O indicates that the decomposition pressure is a function of the temperature. The conversion of carbonates and hydrates to oxides is endothermic, as is that of sulphates, but it generally takes place at lower temperatures. Decomposition normally precedes reduction of the oxide. The reduction of oxides is effected by hydrogen, carbon monoxide, carbon, or a metal, or by electrolysis in a fused salt bath.

5.3.1. REDUCTION OF AN OXIDE BY A GAS

Hydrogen, carbon monoxide, and mixtures of these and hydrocarbons, are used to reduce oxides. The hydrocarbons must be previously converted to H_2 and CO by cracking, so as to eliminate carbon deposition and contamination of the reduced metal.

(A) PRODUCTION OF CO AND H_2

Hydrogen can be obtained by electrolysis of water, but this process is too expensive to be used extensively in metallurgical processes, and is only used in cases when a high purity is needed; for example, in the preparation of molybdenum and tungsten powders from their oxides.

Industrially, H_2 and CO are obtained from coal, oil, hydrocarbons, tar, or any other carbonaceous material.

(i) Reducing Gases Produced from Carbon: Boudouard's Curve

CO can be produced by incomplete combustion of carbon in oxygen, or by reaction between carbon and CO_2. The first reaction is exothermic (ΔH^o = - 26,400 cal $mole^{-1}$ of CO), but the second is endothermic (ΔH^o = + 41,000 cal $mole^{-1}$ of CO). The standard free energy of the reaction

$$CO_2 + C = 2CO \tag{5.25}$$

can be calculated from the free energies of formation of CO and CO_2 directly from the elements (Fig. 23). It is

$$\Delta G^o = + 40,800 - 41.7\ T \text{ cal mole}^{-1} \tag{5.26}$$

Since $\Delta G^o = - RT \ln K$ at equilibrium, the partial pressures of CO and CO_2 in equilibrium with carbon for a given temperature and total pressure can be calculated (the variance of the system is two). Boudouard's curve (Fig. 65) indicates the partial pressure of CO as a function of the temperature, for a total pressure of CO + CO_2 equal to 1 atm, in the presence of solid C. The value is close to zero at about 675 K and practically 1 atm above 1175 K.

If a mixture of CO and CO_2, in which the partial pressure of CO is different from that indicated by Boudouard's curve for the temperature, is passed over carbon, the system will react until this value is reached. Mixtures lying below the curve and containing an excess of CO_2 will change to consume CO_2 and C. Mixtures lying above the curve will react so as to generate CO_2 and deposit C until they reach equilibrium.

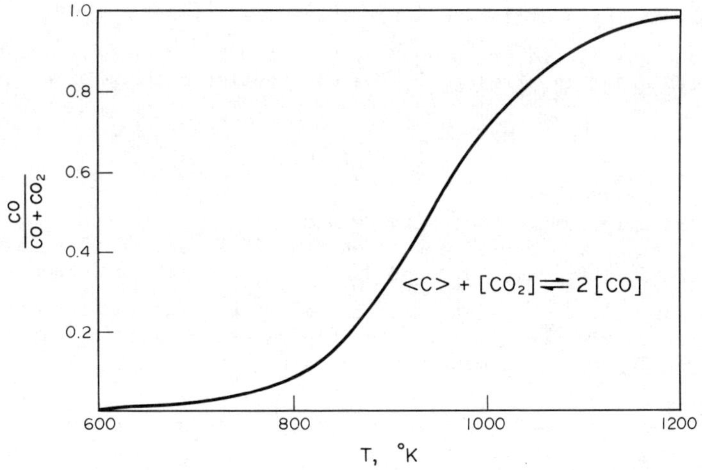

Fig. 65. Equilibrium C - O (1 atm) (Boudouard)

If carbon dioxide is replaced by steam, the following endothermic reactions take place:

$$[H_2O] + <C> \rightleftarrows [CO] + [H_2] \tag{5.27}$$

and

$$2[H_2O] + <C> \rightleftarrows [CO_2] + 2[H_2] \tag{5.28}$$

In the presence of carbon, two simultaneous equilibria will exist: $C-CO-CO_2$ and $C-H_2-H_2O$, and the composition of the mixture can be determined starting from the standard free energies of (5.27) and (5.28). At low temperature, CO is a more powerful reducing agent than H_2, whereas above 1093 K the opposite is the case. To obtain pure H_2, the mixture $CO-H_2$ is made to react with an excess of steam at a temperature below 1093 K. In this case:

$$[H_2O] + [CO] \rightleftarrows [CO_2] + [H_2] \tag{5.29}$$

The CO_2 formed is dissolved in lime water or H_2O under pressure.

Reactions (5.25) and (5.27) are strongly endothermic and an external source of heat is needed to maintain the carbon at a high temperature. In the Wiberg process, the heat is produced by current passing between electrodes immersed in a bed of coke. In the production of "water gas", air and steam are passed alternately into the reactor. Combustion of coke with air raises the temperature of the bed, and the gases produced by this reaction (CO_2 and N_2) are blown to waste. Steam is then passed over the superheated coke, decomposing to CO and H_2, and producing "water gas" in an endothermic reaction.

(ii) Reducing Gases from Hydrocarbons

The most widely employed hydrocarbon for the preparation of $CO-H_2$ mixtures is natural gas, but any liquid or gaseous hydrocarbon can be used. The hydrocarbons are cracked with steam in a reactor. The cracking of methane with steam occurs according to:

$$[CH_4] + [H_2O] \rightleftarrows [CO] + 3[H_2] \tag{5.30}$$

This reaction is only complete at high temperature. In order not to damage the catalyst, the process is carried out at about 1125 K, and results in the production of a mixture of CO, CO_2, H_2, CH_4, and excess H_2O, which is condensed. The partial pressures of the gases at equilibrium can be calculated according to the method indicated in Problem No. 11. Some processes use air (or oxygen) instead of water, achieving incomplete combustion according to:

$$[CH_4] + \tfrac{1}{2}[O_2] + \left(2[N_2]\right) \rightleftarrows [CO] + 2[H_2] + \left(2[N_2]\right) \tag{5.31}$$

Reaction (5.31) is exothermic and proceeds without external heat, but the product cannot contain more than 60% of reducing gas, CO and H_2 when air is used.

(B) THERMODYNAMICS OF OXIDE REDUCTION WITH CO AND H_2

The $\Delta G^o(T)$ diagram for oxides (Fig. 23) shows that the standard free energies of formation of water and carbon dioxide, starting from H_2, CO, and O_2, are similar to that for the formation of iron oxide, FeO. Thus, CO and H_2 can readily reduce all those oxides with standard free energies of formation less than that of wustite. For the reduction of an oxide, according to:

$$<MO> + [H_2] \quad (\text{or } [CO]) \rightleftarrows <M> + [H_2O] \quad (\text{or } [CO_2]) \tag{5.32}$$

the variance of the system formed by a non-volatile metal, its oxide and the gases H_2 and H_2O (or CO and CO_2), is two. If the total pressure and temperature are fixed, the partial pressures of the gases are determined. It is then possible, in the same form as for the equilibrium of $CO + CO_2$ with carbon, to trace the representative curve of the partial pressure of the reducing gas in equilibrium with the two condensed phases for each reaction as a function of temperature, at a given pressure, by solving the two equations:

$$\Delta G^o = -RT \ln \frac{p_{H_2O}}{p_{H_2}} \text{ or } -RT \ln \frac{p_{CO_2}}{p_{CO}} \tag{5.33}$$

$$p_{H_2} + p_{H_2O} = P \quad \text{or} \quad p_{CO_2} + p_{CO} = P \tag{5.34}$$

For a volatile metal it is necessary to specify the value of another variable, the partial pressure, as has been done for the reduction of ZnO to Zn by CO/CO_2 mixtures. Some metals, such as iron, manganese, tungsten, and molybdenum, form several oxides and reduction occurs in two or three stages. For iron oxides, the stages are as follows:

At a temperature above 840 K:

$$3<Fe_2O_3> + [H_2] \quad (\text{or } [CO]) \rightleftarrows 2<Fe_3O_4> + [H_2O] \quad (\text{or } [CO_2]) \tag{5.35}$$

Hematite → Magnetite

$$2<Fe_3O_4> + 2[H_2] \quad (\text{or } [CO]) \rightleftarrows 6<FeO> + 2[H_2O] \quad (\text{or } [CO_2] \tag{5.36}$$

Magnetite → Wustite

$$6<FeO> + 6[H_2] \quad (\text{or } [CO]) \rightleftarrows 6<Fe> + 6[H_2O] \quad (\text{or } [CO_2]) \tag{5.37}$$

Wustite → Iron

At temperatures lower than 840 K, magnetite is transformed directly to iron, according to:

$$<Fe_3O_4> + 8[H_2] \quad (\text{or } 8[CO]) \rightleftarrows 6<Fe> + 8[H_2O] \quad (\text{or } 8CO_2) \tag{5.38}$$

For each reaction the same number of atoms of iron have been kept, to indicate the respective amounts of oxygen reacting with the gas in each stage of the reduction of one oxide to the next.

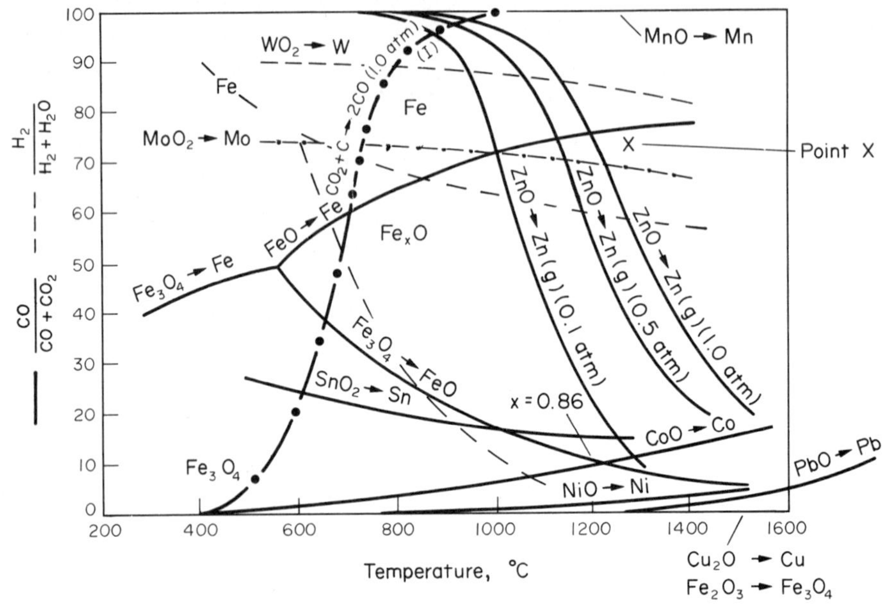

Fig. 66a. Equilibria of oxides with CO/CO_2 mixtures ——

and with H_2/H_2O mixtures — — —

Boudouard's curve —•—•—•— $CO_2 + C \rightleftarrows 2CO$

Chaudron's curves in Figs. 66a and 66b give the partial pressures of gases in equilibrium with iron and its oxides as a function of temperature. If, apart from the temperature and total pressure, one partial pressure is fixed, a domain is obtained with a single condensed phase (iron, wustite, magnetite, or hematite) delimited by the equilibrium lines.

It has already been seen that up to 820°C (1093 K), CO is a more energetic reducer than H_2, whereas at a higher temperature, the reverse occurs. Therefore, in order to reduce iron oxides, a partial pressure of H_2 greater than that of CO is required at temperatures lower than 820°C (1093 K), while a partial pressure of H_2 lower than that of CO is necessary at temperatures higher than 1093 K. This is why the equilibrium lines H_2O-H_2-condensed phases, and CO-CO_2-condensed phases intersect at 820°C (1093 K).

It should be noted that CO is not only a reducing agent but also a carburising agent. When the mixture CO-CO_2 has a composition which corresponds to the domain between Boudouard's curve and that of the equilibrium wustite-iron (Fig. 66b) part of the carbon will dissolve in iron, according to:

$$2[CO] \rightleftarrows (\underline{C}) + [CO_2] \qquad (5.39)$$

From the standard free energy of reaction:

$$\Delta G^o = -RT \ln \frac{a_{(\underline{C})} \cdot p_{CO_2}}{p_{CO}^2} \tag{5.40}$$

it is possible to determine the activity of carbon in iron in equilibrium with a gas of given composition.

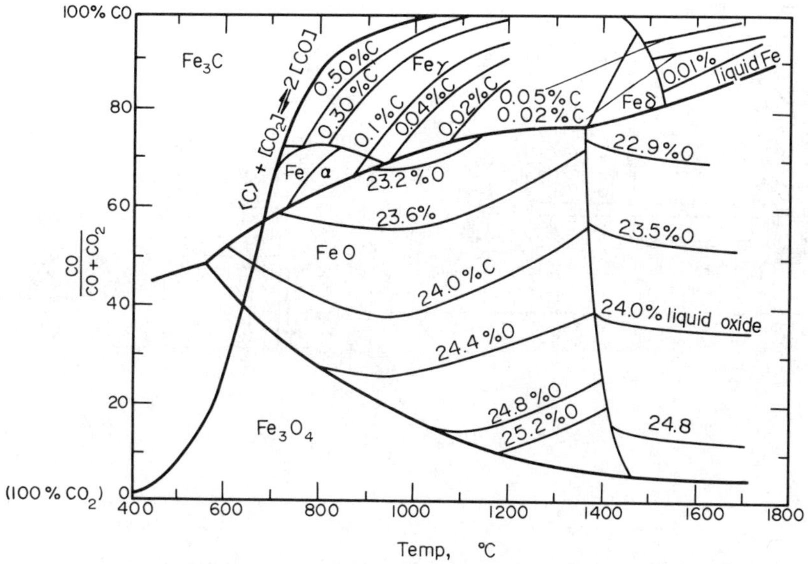

Fig. 66b. Equilibrium C-O-Fe with the degree of carburisation of Fe and oxidation of wustite in the presence of CO and CO_2

Wustite is not a stoichiometric compound, and its oxygen content varies with temperature and degree of reduction, as indicated by Fig. 66b. In the formula $FeO_{(1+x)}$, x varies from 0.05 to 0.12, depending on the conditions (see Fe-O diagram of Fig. 112).

(C) PROCESSES USING GAS AS A REDUCER

Figure 66a indicates the principal metal oxides which can be reduced by CO or H_2 and gives the conditions (T, P_{CO} or P_{H_2}) for metal production. In some cases the gas is produced by partial combustion of C added with the charge of oxide and flux, according to the reaction:

$$2C + O_2 = 2CO \tag{5.41}$$

Reduction to metal is then completed, either by CO when the oxide is solid, or by C when the rest of the charge has fused. Thus, in the production of Zn, Pb, Sn and Fe in the blast furnace, most of the reduction is effected by CO, but

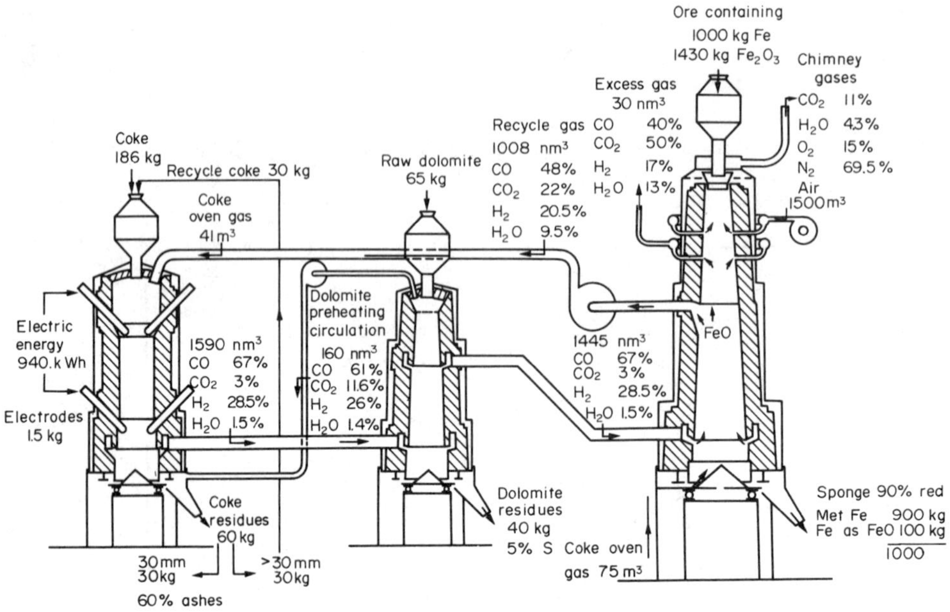

Fig. 67a. The Wiberg process

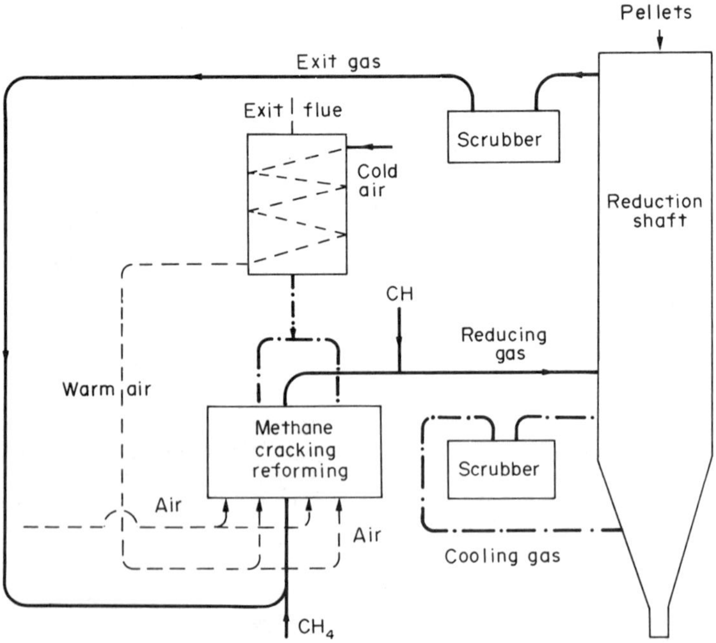

Fig. 67b. Simplified diagram of Midrex Process.
——— Principal Gas Flows ----- Air Flow
—•—•— Cooling Gas

solid C ensures a low metal content in the final slag. (See accounts of pyrometallurgy in Section 5 and Chapter 7.)

(i) Production of "Sponge Iron" by Direct Reduction (DR Processes)

Processes for this purpose have developed rapidly from a capacity of about 0.6 million tonne per year in 1960 to 2 million tonnes in 1970 and over 30 million tonnes in 1982. Unit sizes have increased from about 100,000 tonnes per year to 400,000 tonnes and then to over 1 million tonnes over the same period. The individual DR processes can be classified according to the type of furnace, the method of producing the reducing gas and the source of primary energy. About 85% of the total production is shared between the Midrex (57%) and H y L processes.

The Wiberg was the earliest large-scale unit using reducing gas. In this process (Fig. 67a), gas rich in CO enters the lower section of the shaft furnace at 1270 K and heats and reduces the charge as it passes up through it. Some is taken off at the level where it is in equilibrium with wustite and iron and the remainder passes on upward to reduce the higher oxides of iron to wustite. Some of this gas is burned with air in the top of the shaft to preheat the charge. The gas taken out at the equilibrium level is mixed with steam or coke oven gas and passed through a bed of electrically heated coke and then a bed of calcined dolomite, to remove the S picked up from the coke, before returning into the bottom of the reduction shaft. Coke and dolomite are fed into the regenerator and desulphuriser respectively. This process produces high purity iron sponge and consumes about 200 kg of coke and 1000 kWh of electricity per tonne of product.

The Hojalata y Lamina (H y L) process started as a discontinuous process developed in Monterey, Mexico. The reducing gas, obtained from natural gas by cracking with steam, contains, after condensing the excess water, 73% H_2, 16% CO, 4% CH_4 and 7% CO_2. This is heated to 1370 K and passed through a first reactor charged with iron ore already pre-reduced and pre-heated to 1170 K in a previous stage. The water formed is condensed at the exit of the reactor. The remainder of the gas is heated and passed to a second reactor where fresh ore is preheated and pre-reduced. The gas leaving this reactor is used to generate heat and energy in the plant. The operating cycle allows each reactor to be used for pre-reduction for 2 hours, and for metal production for the next 2 hours. The complete cycle lasts 5 hours, as 1 hour is needed to load and unload the reactors. The production of 1 tonne of metallic iron consumes from 450 m^3 to 500 m^3 of natural gas, depending upon the size of the reactors and the efficiency of heat recovery. A new version, H y L III, is continuous, using a shaft furnace and the consumption of gas is about 400 m^3 per tonne of iron.

In the Midrex process, the reducing gas is produced by reforming part of the exit gas at about 1000°C with natural gas, using the remainder of the exit gas as fuel, with preheated air. After any S is removed, the reducing gas, which contains about 50% H_2 and 35% CO (+ CO_2, CH_4 and N_2) is blown into the shaft furnace to heat and reduce the iron oxide pellets. The sponge iron at the bottom is cooled by circulating gas.

The fluidised-bed reactors used in several direct reduction processes (Novalfer, Nu-Iron, H-Iron and Flufer) incorporate several stages of fluidisation and overflow within one unit, so as to ensure thorough heating and exposure to reducing gas. The iron ore feed is coarser than 80 μm to avoid entrainment in the gas and it descends against an upward current of hydrogen under pressure (30 atm in the H-Iron process).

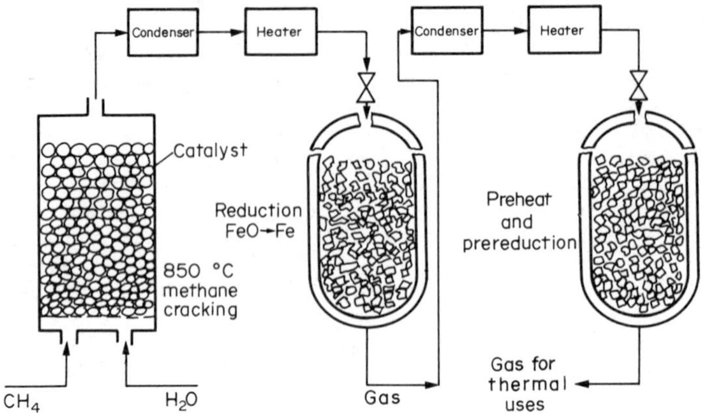

Fig. 67c. Simplified diagram of the H y L processes I and II

A mixture of ore, carbon and lime is fed into the rotary kilns in which the Krupp-Renn, RN, Lurgi and Accar processes are carried out. The carbon removes oxygen from the ore and the lime removes sulphur. Heat is supplied by burning pulverised coal or oil and CO is generated in the charge by reaction between carbon and the products of combustion. If the temperature is 1300°C, as in the Krupp-Renn, the charge becomes pasty and the metal agglomerates into globules or "luppen" which are recovered magnetically after ball milling the solidified product. Where the temperature is as low as 1000-1100°C, the reduced product is rapidly cooled and the powder or friable sponge is magnetically concentrated.

The sponge produced by direct reduction can be milled and screened for use in powder metallurgy if it is sufficiently pure. Otherwise the product, with 92-95% metallisation, can be charged into a blast furnace, a converter, or an electric furnace to produce pig iron or steel.

(ii) Reduction of Molybdenum and Tungsten Oxides by Hydrogen

Pure molybdenum and tungsten are obtained by reduction of their oxides with hydrogen. In the case of molybdenum, the oxide is produced by roasting molybdenite concentrates, MoS_2. The principal minerals of tungsten are scheelite, $CaWO_4$, and wolframite, $(Fe, Mn)WO_4$. In the most widely used extraction process, these are dissolved in an alkaline solution (NaOH or NH_4OH), forming the molybdate or tungstate. After filtering off the insoluble impurities, the oxides are precipitated by changing the pH of the solution. After drying, they are placed in trays and reduced in a tube furnace. Reduction is carried out with hydrogen in two stages. In the first stage, MoO_3 and WO_3 are reduced to MoO_2 and WO_2 at 870 K and 970 K respectively. In the second stage, these oxides are reduced to metal at 1270 K for molybdenum, and 1370 K for tungsten. The hydrogen circulates in counter-current, and the excess is burned at the exit of the furnace. A safety system prevents the entry of oxygen. The molybdenum or tungsten powder obtained is compacted into a bar by pressing, and is sintered in hydrogen at 2300 K by passing an electric current through it. Further swageing forms a compact and ductile cylinder suitable for mechanical working.

(D) MASS BALANCE IN THE REDUCTION OF A BED OF IRON OXIDE BY A GAS

Figures 66a and 66b give the molar fractions of reducing gas, CO or H_2, in equilibrium with two condensed phases (N_3^* for reaction (5.35), N_2^* for (5.36) and N_1^* for (5.37)). Taking a static bed of hematite traversed by a reducing gas having the composition represented by N_0 and assuming initially that the kinetics are instantaneous, it is possible, by means of a mass balance (Problem 13) to calculate the velocities of advance of the interfaces between hematite and magnetite, magnetite and wustite and wustite and iron, for a specified gas flow, $n_g S$. It can be seen that there are $(N_0 - N_1^*)n_g S$ moles of gas available for reducing wustite to iron (i.e. to remove about 2/3 of the oxygen originally bonded to iron as Fe_2O_3). There are $(N_1^* - N_2^*)n_g S$ moles available for reducing magnetite to wustite by removing 2/9 of the original oxygen and $(N_2^* - N_3^*)n_g S$ for removing the 1/9 necessary to convert hematite to magnetite. As Problem 13 indicates, the velocity of the wustite/iron interface is slower than those of the other two. When the kinetics are not instantaneous, the interfaces are replaced by zones of finite thickness, H_1, H_2 and H_3 (Fig. 68a) which move with the same velocities as those previously found for the interfaces, because they do not depend on the kinetics, but on the mass balance.

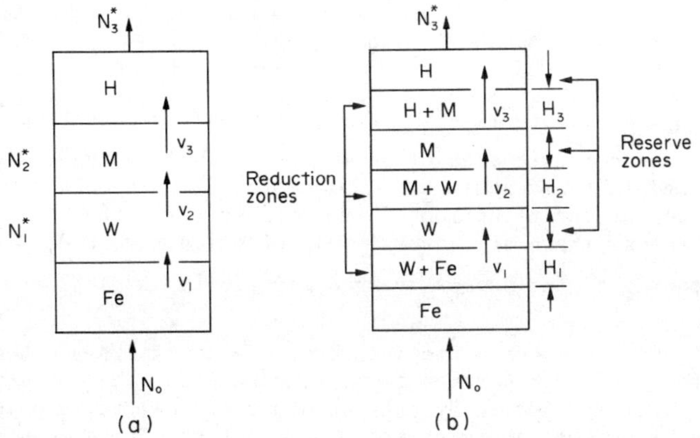

Fig. 68a. Velocity of advance of the reduction zones. (a) Instantaneous reactions; (b) Non-instantaneous reactions

In a counter-current system of ascending gas and descending iron ore, in order that the ore is completely reduced at the discharge and the gas utilisation is maximum, the ore must move downward with the same velocity v_1 as the upward movement of the wustite-iron interface in a static bed. The hematite-magnetite and magnetite-wustite reduction zones have a resultant velocity different from zero: $v_3 - v_1 > 0$ and $v_2 - v_1 > 0$, which means that they move upwards and stabilise at the top of the reactor after attainment of the permanent regime. The gas concentration profile in the reactor will then be that in Fig. 68(b): the exit gas does not reach equilibrium with magnetite and hematite because the reducing gas is too rich in CO and the gas flow too large in comparison with the quantity necessary to remove approximately one-third of the oxygen in hematite, which corresponds to the transformation of hematite to wustite $(FeO_{1.5} \rightarrow FeO_{1.05})$.

As indicated by Fig. 68b, the reactor may be divided into three zones: a lower zone where wustite is reduced to iron; a central zone, or chemical reserve, which contains pure wustite; and an upper zone where hematite and magnetite are reduced to wustite.

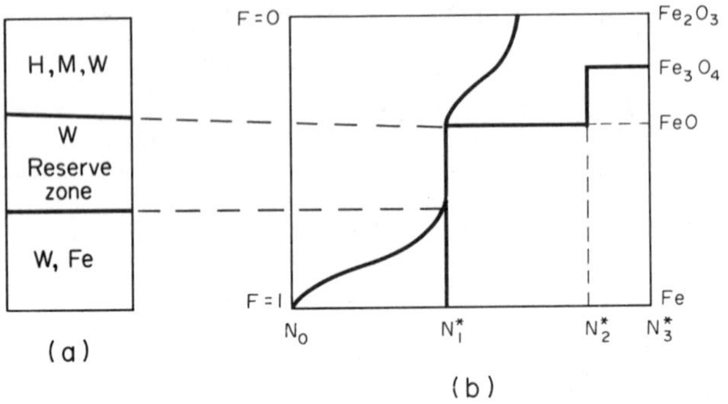

Fig. 68b. Gas concentration profile for the reduction of hematite to iron

The mass balance in a counter-current bed of iron oxide and gas can be depicted graphically with the aid of the operating line, seen in Chapter 4 (4.2.2). Equation (4.102) can be written:

$$dF/dN = n_g/n_s$$

where n_g and n_s are respectively the gas flow and the ore feed rate (in moles per unit of time and per unit area of bed), F and N are the fractions of solid and gas transformed in the reactions (5.35) to (5.37). The diagram representing the operating line for the reduction of iron oxides (Fig. 68c) is more complex than Fig. 45b, because there are three levels of oxidation, $FeO_{1.05}$, $FeO_{1.33}$ and $FeO_{1.5}$, corresponding to the successive gas equilibria N_1^*, N_2^* and N_3^*.

In ideal conditions, i.e. when a wustite reserve zone exists and the iron ore comes out from the discharge section completely reduced, the operating line passes through E and W (Fig. 68c). By extrapolation of EW, it is possible to obtain a concentration N_s of gas emerging from the reactor. If the gas supply is increased to above the value which corresponds to the disappearance of the wustite reserve zone, or the iron ore feed rate is increased, or both, the cases b, c, d, previously described, and depicted in Fig. 68c, are found.

The slope of the operating line dF/dN is equal to the ratio n_g/n_s. Thus, by adjusting the gas flow and the ore feed, it is possible to obtain the desired composition of solid, F_s and of gas, N_s, at the exit of the reactor.

In the Wiberg process, part of the gas is taken off at the level of the reserve zone of wustite, giving a smaller gas flow above this level. Therefore, above W the slope of the operating line changes suddenly and becomes proportional to the remaining gas flow which crosses the pre-reduction zone. The composition, N_s, of the gas leaving the pre-reduction zone can then be calculated and from this the amount of air for complete combustion of CO and H_2 to preheat the charge.

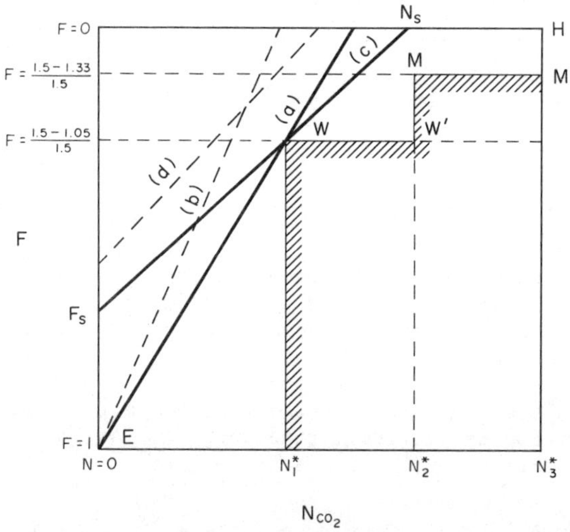

Fig. 68c. Operating lines for the reduction of hematite to iron by a gas. (a) Ideal case: a wustite reserve zone exists. (b) Large gas supply: the wustite reserve zone disappears. (c) The iron ore feed rate is too high to permit complete reduction of wustite. (d) Gas flow rate and iron ore feed rate are greater than ideal

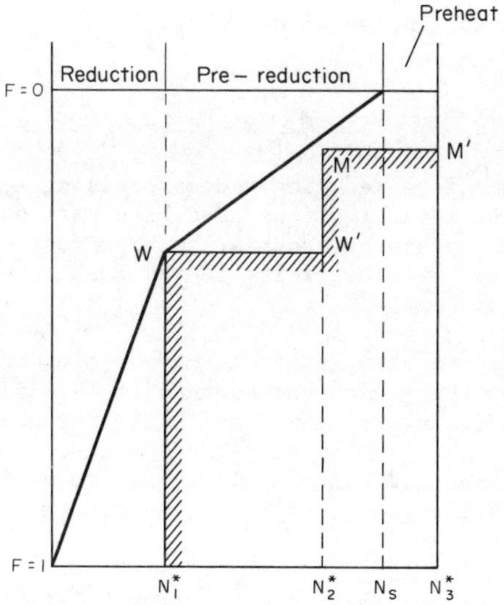

Fig. 68d. Operating line in Wiberg process. (The operating line for the iron blast furnace is studied in Chapter 7.)

5.3.2. REDUCTION OF METAL OXIDES BY CARBON (CARBOTHERMIC PROCESSES)

The reduction of an oxide by carbon can be expressed in the form:

$$(MO) + \langle C \rangle = (M) + [CO] \tag{5.42a}$$

The standard free energy of this reaction is the difference between the standard free energy of formation of carbon monoxide and that of the oxide $\Delta G^o_{CO} - \Delta G^o_{MO}$, each represented by a straight line in Fig. 23. From this figure, it can be seen that carbothermic reduction must be carried out at high temperature. However the enthalpy of (5.42a) is always positive and the reduction of an oxide by C involves the supply of heat, either by combustion of C or CO, or by electrical energy.

For easily reduced oxides, shaft or blast furnaces are the most efficient (see 5.5) as in the production of pig iron, zinc and tin. As illustrated above, much of the reduction is carried out by CO according to:

$$CO + MO \rightarrow CO_2 + M \tag{5.42b}$$

and when the metal is volatile, like Zn, there is the risk that (5.42b) will proceed in the opposite direction in the cooler part of the furnace and the metal will be oxidised.

The oxides of Si, Cr, Mn, Ti, W and Mo need high temperatures during reduction either because they are very stable, or because the metals are very refractory and they can be produced only in electric arc or resistance furnaces. The $\Delta G^o(T)$ diagrams for carbides show that most of the metals made in this way form more or less stable carbides. As the C in the system is at unit activity, either the carbide will be produced, or the metal will be saturated with C at the prevailing temperature.

(A) THE ELECTRIC REDUCTION FURNACE

In the blast furnace, coke is a reducing and carburising agent, as well as a heat source. For cost reasons, investigations have been carried out aimed at replacing the coke used for heating by another source, such as fuel oil, natural gas, or electrical energy. The use of electricity in the bosh in order to heat the charge has the following effects:

As it is not necessary to blow in air for combustion of coke, the amount of gas generated is only that which corresponds to (5.42a). This is substantially less than that produced in a conventional blast furnace.

Because this small amount of gas is only enough to heat the charge which is virtually in the reduction zone, it is not necessary to recover the sensible heat of the gases.

Electric reduction furnaces are therefore constructed with low shafts, but because the low temperature of the charge at any distance above the reduction zone does not permit the reduction of the oxides to metal by CO, this is effected almost 100% directly by carbon. The exit gas contains a high percentage of CO, and its heating and reduction powers (2500 cal m^{-3}) are much greater than those of blast furnace gas (900 cal m^{-3}).

Since the temperatures reached in the fusion zone of the electric reduction furnace are higher than those in conventional blast furnaces, it is possible to produce high grade ferroalloys or carbides, which are impossible to produce in the blast furnace. This type of furnace, sketched in Fig. 69a is used to produce pig iron, ferroalloys, and carbides, especially ferrosilicon, ferromanganese, ferrovanadium, and ferrochromium.

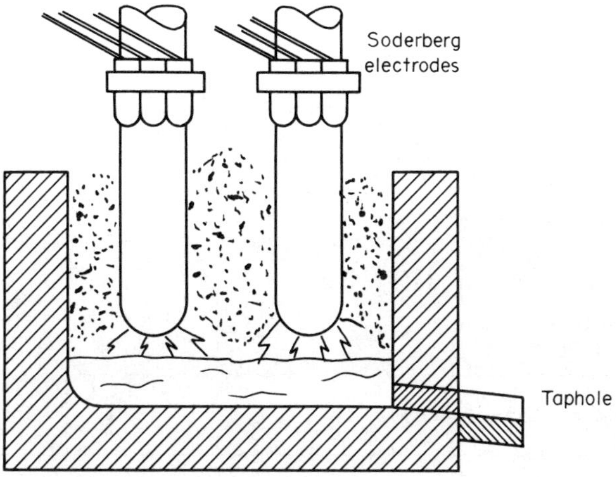

Fig. 69a. Electric reduction furnace

(i) Pig Iron Production in an Electric Reduction Furnace

The coke consumption may be readily calculated, because carbon serves only to reduce the charge and carburise the pig iron, and reduction by gas is nil. The coke may be of lower quality than that used in the blast furnace because degradation by compression in the low shaft is negligible.

The electrical energy consumption can be calculated from a thermal balance of the type given in Table 7.

As 1 kWh corresponds to 0.860 thermies, the consumption of electrical energy is 2400 kWh per tonne of pig iron.

The heating and reducing power of the gas (570 m^3, i.e. 1400 thermies per tonne of pig iron), can be used to preheat or preheat and pre-reduce the charge. The saving of electrical energy is 400 kWh in the first case, and 1200 kWh in the second. Several processes exist such as the Electrokemisk and the Strategic-Udy, which use the reducing power of CO in a rotary or shaft furnace.

(ii) Manufacture of Ferroalloys

Ferroalloys are very important raw materials for the manufacture of special and carbon steels. Some of them are used as deoxidisers in steelmaking (Fe-Mn, Fe-Si, Fe-Si-Mn), and others provide the necessary elements for special steels. Table 8 gives the ranges of composition of the addition elements that are most widely used. As can be seen in Table 8, the carbon content of stainless steels must be very low. If the ferroalloys used in the manufacture of stainless steel were produced by carbothermic reduction, they would be saturated with carbon and would introduce this element as an undesirable impurity. This was especially the case with ferrochromium, which was produced by silicon reduction. The new refining processes using oxygen, AOD (Argon/oxygen decarburization) and VOD, (vacuum/oxygen decarburization), make it possible to use high carbon ferroalloys (see Chapter 7).

TABLE 7 Consumption of Energy per Tonne of Pig Iron in an Electric Reduction Furnace (1 thermie = 4.186 mjoules)

Enthalpy of 1 tonne of pig iron at 1500°C	313 thermies
Enthalpy of 0·3 tonne of slag at 1500°C	132 thermies
Enthalpy of 0·72 tonne of CO at 1500°C	296 thermies
Reduction of 0·950 tonne of Fe starting from Fe_2O_3 and C	955 thermies
Reduction of 0·010 tonne of Si	53 thermies
Estimated losses (15%)	311 thermies
Total consumption of energy	2060 thermies

TABLE 8 Percentage of Addition Elements in Steels

C	Si	Mn	Cr	Ni	W	Mo	Others	Applications
0·3-0·7	0·9-1·2	0·9-1·8	-	-	-	-	-	Construction steels
0·5-0·6	1·5-2·0	0·6-0·9	1·6	-	-	-	-	Springs and Coils
0·05	2-3	0·3-0·7	-	-	-	-	-	Transformers
0·9-1·1	0·4-0·6	0·9-1·2	1·6	-	-	-	-	Ball Bearings
0·3-0·5	1·2-1·6	0·4	1·3-1·6	-	-	-	-	Tools
0·3-0·5	2-3	0·3-0·7	8-9·5	-	-	-	-	Resistance to high temperatures
0·04	-	-	18	8-12	-	2·5	-	Stainless steels
0·08	-	-	18	8-12	-	-	Ti=0·4	
0·7-0·9	0·2-0·5	0·2-0·4	3-5	-	1·2-2	7·5-8·5	V=0·3	High speed steels

Ferrosilicon. The reduction of silica (quartz) according to:

$$(SiO_2) + 2\langle C\rangle = (Si) + 2[CO] \tag{5.43}$$

requires temperatures higher than 1823 K, from thermodynamic data. At these temperatures, SiC would be formed in the presence of C:

$$(SiO_2) + \langle 2C \rangle = \langle SiC \rangle + 2[CO] \Delta G^o = 125,770 - 82T \text{ cal} \tag{5.44}$$

The production of Si instead of SiC in the electric furnace can be explained through the formation of the suboxide SiO and the temperature gradient which exists in the hot zone of the furnace. A study of the variance of the system Si-O-C shows that for a given temperature and total pressure, the maximum number of condensed phases (Si, SiC, SiO_2, C) which can coexist in the presence of the gaseous phase (SiO, CO) is two. From the standard free energies of the reactions:

$$SiO_2 + C = SiO + CO \tag{5.45a}$$

$$3SiO + CO = SiC + 2SiO_2 \tag{5.45b}$$

$$SiO_2 + Si = 2SiO \tag{5.45c}$$

$$SiC + SiO = 2Si + CO \tag{5.45d}$$

$$2C + SiO = SiC + CO \tag{5.45e}$$

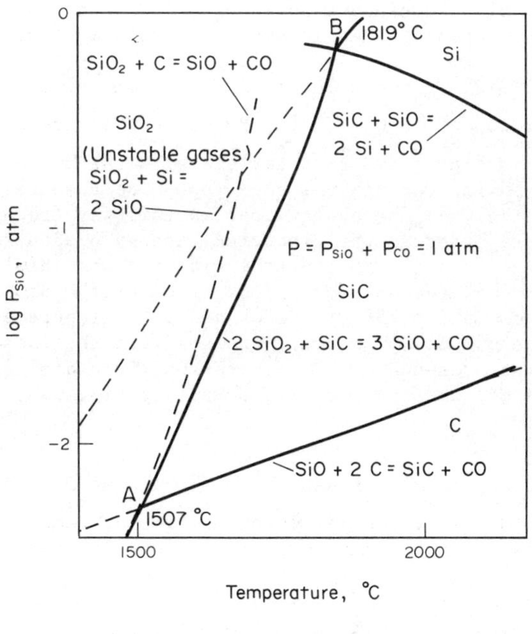

Fig. 69b. The predominance area diagram, log $P_{SiO} = f_{(T)}$, for the system Si-O-C when $P_{SiO} + P_{CO} = 1$ atm

(T. Rosenqvist, *Principles of Extractive Metallurgy*, McGraw-Hill, New York (1974) 378: A. Schei and K. Larsen, "A Stoichiometric Model of the Ferrosilicon Process", The Iron and Steel Society of AIME, 39th Electric Furnace Conference, Houston, 1981)

The reaction $C + SiO = Si + CO$ will not proceed at any temperature. Three condensed phases coexist only at 1507°C (SiO_2, SiC, C), Point A, and at 1819°C (Si, SiC, SiO_2), Point B. Above 1819°C, Si is stable only if $P_{SiO} > 0.67$ P. The reactions (c) and (d), which occur in the hottest zone of the furnace, near the electrodes, can be written in such a way that there are two reactions combined:

$$3SiO_2 + 3Si = 6SiO \qquad \text{and} \qquad 2SiC + 2SiO = 4Si + 2CO,$$

or in total

$$3SiO_2 + 2SiC = Si + 4SiO + 2CO \; (P_{SiO} \approx \frac{2}{3} P) \qquad (5.45f)$$

Liquid Si flows on the hearth and as the gas and vapour formed cool on leaving the hot zone, reactions (b) and (c) take place:

$$2C + SiO = CO + SiC \qquad \text{and then} \qquad 3SiO + CO = SiC + 2SiO_2$$

in total

$$2C + 4SiO = 2SiC + 2SiO_2 \qquad (5.45g)$$

The sum of (f) and (g) is equal to (5.43) and from the previous mechanism we can note that:

(a) When there is no excess of C, the whole amount is consumed in the "colder" zone, between 1510 and 1820°C and does not come into contact with Si, so preventing the formation of SiC.

(b) With excess C, including the C of the electrodes, there is enough to react with Si and cause a build-up of SiC on the hearth. The charge must be calculated accurately and the carbon must be reactive.

A great part of the SiO_2 reduced in (f) arises from (g) and recirculates in the furnace. The production of SiO_2 in the furnace at a more or less constant position tends to form a plug and to interfere with the descent of the charge. This is remedied by poking through the open top. The absence of a roof allows gas to burn above the charge and the energy lost in this way (recuperated in modern plants) is of the same order as the electrical energy needed for the production of Si. When reaction (g) is not complete, the gas contains SiO which burns to fumes of SiO_2. This may account for a loss of 6% of the SiO_2 when making 45% ferrosilicon, 12% when making 75% Si alloy and 20% for pure silicon. The power consumption per tonne of Si produced increases with the losses, reaching 10,700 kWh for pure Si. The presence of Fe in ferrosilicon diminishes the activity of Si (Figs. 109 and 138b) and enlarges its domain in Fig. 69b. This facilitates the reduction of SiO_2.

Ferromanganese. The higher manganese oxides (MnO_2, Mn_2O_3, and Mn_3O_4) are easily reducible to MnO (by thermal decomposition or reduction with a gas). The green-tinted manganese monoxide can only be reduced with carbon at high temperature, according to:

$$2(MnO) + 2\langle C\rangle \rightleftarrows 2(Mn) + 2[CO] \qquad \Delta G^o = 137{,}400 - 81.2\,T \text{ (cal)} \qquad (5.46)$$

Two grades of ferromanganese can be produced. In the blast furnace, "Spiegeleisen", containing about 20% Mn is made by smelting manganiferous iron ore, while ferromanganese containing 75-80% Mn is produced by smelting manganese ore in which iron is an impurity. In the latter case, 8-10% of the Mn passes to the slag, in spite of raising the CaO/SiO_2 ratio to nearly 2, and the coke consumption is considerably higher than for iron production. This high grade material is the normal product from an electric arc furnace. Manganese carbide will be produced preferentially by a reaction with a more negative energy balance than for (5.46):

$$2(MnO) + \frac{8}{3}\langle C\rangle \rightleftarrows \frac{2}{3}(Mn_3C) + 2[CO] \qquad \Delta G^o = 121{,}800 - 81.4\ T\ (cal) \qquad (5.47)$$

Ferroalloys made in this way will contain 7 to 8% carbon. This contamination by carbon is decreased by mixing liquid ferromanganese with molten manganese ore:

$$(Mn_3C) + (MnO) \rightleftarrows 4(Mn) + [CO] \qquad \Delta G^o = 91{,}700 - 40.2\ T\ (cal) \qquad (5.48)$$

Using this method, it is difficult to obtain a carbon content lower than about 3.5%. Pure manganese, or ferromanganese without carbon, is obtained by electrolysis or metallothermic reduction.

Ferrochrome. There are two reduction reactions for chromic oxide:

$$\frac{2}{3}(Cr_2O_3) + 2\langle C\rangle \rightleftarrows \frac{4}{3}(Cr) + 2[CO] \qquad \Delta G^o = 130{,}340 - 86.1\ T\ (cal) \qquad (5.49)$$

and

$$\frac{2}{3}(Cr_2O_3) + \frac{8}{7}\langle C\rangle \rightleftarrows \frac{4}{21}(Cr_7C_3) + 2[CO] \qquad \Delta G^o = 122{,}000 - 87.0\ T\ (cal) \qquad (5.50)$$

Reaction (5.50) has the more strongly negative free energy balance at the reduction temperature, and carbon-rich (70% Cr, 8% C) ferrochrome is obtained when chromite ($FeO \cdot Cr_2O_3$) is reduced by C in an electric furnace. A large part of the carbon can be eliminated by blowing air into the molten alloy at temperatures higher than 1973 K. In practice, carbon-free ferrochrome is produced by silico-thermic reduction (the Perrin process) (see Problem No. 26).

Other ferroalloys. It is also possible to produce the following ferroalloys by carbon reduction of the mixed oxides: Fe-Ti, Fe-V, Fe-Mo, Fe-W, Fe-Nb, etc. The first two contain from 25 to 30% Ti or V, and a high percentage of carbon. For the others (Fe-Mo, Fe-W, etc.), in which the alloy elements are less avid for carbon, the content of this element can be maintained at a low level.

(iii) Manufacture of Metallic Carbides

Metal carbides can be produced by carbon reduction of refractory oxides (CaC_2, Al_4C_3, SiC, etc.). Temperature of about 2300 K are needed for their formation, and these can only be reached in an electric furnace. The reactions that take place between coke and refractory oxides are:

$$\frac{2}{3}(Al_2O_3) + 3\langle C\rangle \rightleftarrows \frac{1}{3}\langle Al_4C_3\rangle + 2[CO] \qquad \Delta G^o = 192{,}200 - 76.8\ T\ (cal) \qquad (5.51)$$

$$(CaO) + 3<C> \rightleftarrows <CaC_2> + [CO] \qquad \Delta G^o = 103,620 - 45.9\,T\ (cal) \qquad (5.52)$$

$$(SiO_2) + 3<C> \rightleftarrows <SiC> + 2[CO] \qquad \Delta G^o = 125,770 - 82.2\,T\ (cal) \qquad (5.53)$$

Silicon carbide, or carborundum, is an electrically conducting refractory, and has a high hardness. The first two properties together are used in electrical resistance elements (instead of graphite), the second when the compound is made into refractory bricks for special uses, and when it is converted into SiC abrasive powders and shapes. Aluminium and calcium carbides are powerful reducing agents, both reacting strongly with water at ambient temperature to produce acetylene:

$$<CaC_2> + 2[H_2O] \rightleftarrows [C_2H_2] + <Ca(OH)_2> \qquad (5.54)$$

This reaction was used to produce acetylene when there were no other processes such as direct synthesis, starting from natural gas (electric arc cracking of CH_4). In metallurgy, the powerful reducing energy of carbides is used to produce alloys. For example, in the production of silicon-calcium, silica is reduced with carbon and calcium carbide according to:

$$2(SiO_2) + 2<C> + <CaC_2> \rightleftarrows (\underline{Ca}) + 2(\underline{Si}) + 4\,CO \qquad (5.55)$$

Calcium carbide is not stable in the presence of silicon. The desulphurising power of Ca is utilised by adding CaC_2 to molten cast iron:

$$<CaC_2> + (\underline{S})_{Fe} \rightleftarrows (CaS) + 2(C)_{Fe}$$

The residual $(\underline{S})_{Fe}$ is very low.

5.3.3. METALLOTHERMIC REDUCTION

From the study of Ellingham diagrams, it can be deduced that carbon reduction reactions of the type illustrated in (5.42) are endothermic and their standard free energy balance becomes more negative as the temperature rises. Thus, it is always necessary to supply external heat to the system (combustion of part of the carbon, electrical energy, etc.), and to use high temperatures to ensure that the reactions shall proceed in the desired direction. For metallothermic reduction of the type:

$$(MO) + (M') \rightleftarrows (M'O) + (M) \qquad (5.56)$$

there is very little change in the free energy of reaction if the temperature is altered. This is because the plots of the standard free energies of formation of oxides are almost parallel to each other. The practical value for ΔG^o of (5.56) will be close to the standard enthalpy, ΔH^o, except when the metal M is volatile. Thus, a reaction that is feasible because $\Delta G^o < 0$, will then be exothermic $(\Delta H^o < 0)$.

(A) EQUILIBRIUM BETWEEN METAL AND SLAG

At equilibrium, the standard free energy of reaction (5.56) is related to the activities of reactants and product by the expression given earlier:

$$\Delta G^o = - RT \ln \frac{a_{(M'O)} \cdot a_{(M)}}{a_{(MO)} \cdot a_{(M')}} \tag{5.57}$$

$$\frac{a_{M'}}{a_M} = \frac{\exp (\Delta G^o/RT)}{a_{MO}/a_{M'O}} \tag{5.58}$$

The extent of contamination of the reduced metal M by the reducer metal M' is a direct function of the ratio, $a_{M'} / a_M$. The conditions for reducing this contamination to a minimum are illustrated in Fig. 70, and are obtained when:

The standard free energy is the most negative possible. This is attained when the reducing agent M' has an affinity ($A^o = - \Delta G^o$) for oxygen which is much greater than that of the metal, M.

The temperature is as low as possible, subject to the metal and slag being sufficiently fluid to separate efficiently.

The ratio $a_{MO}/a_{M'O}$ is large; losses of oxide MO in the slag are allowable because they increase its activity and decrease the activity of M'O. In addition it is an advantage to add to the slag an oxide that forms stable compound with M'O. For example, to produce pure manganese by metallothermic reduction starting with Mn_3O_4, aluminium must be used instead of silicon, which would produce a solution of Si-Mn; it is necessary to operate at the lowest possible temperature, to allow for some losses of manganese oxide in the slag and to decrease the activity of the aluminium oxide formed by slagging with CaO. Thus, Al is at a very low residual level in the metal.

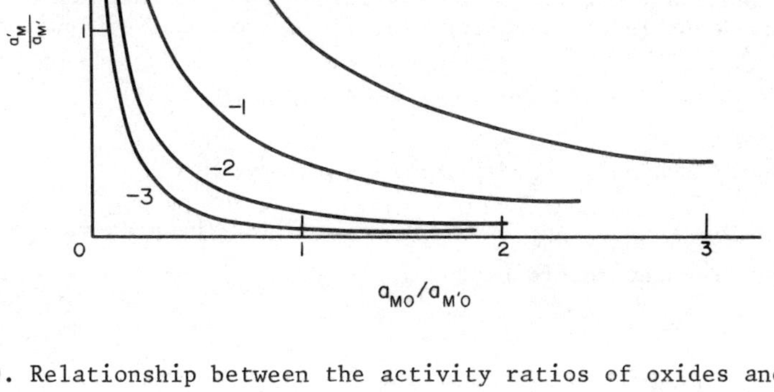

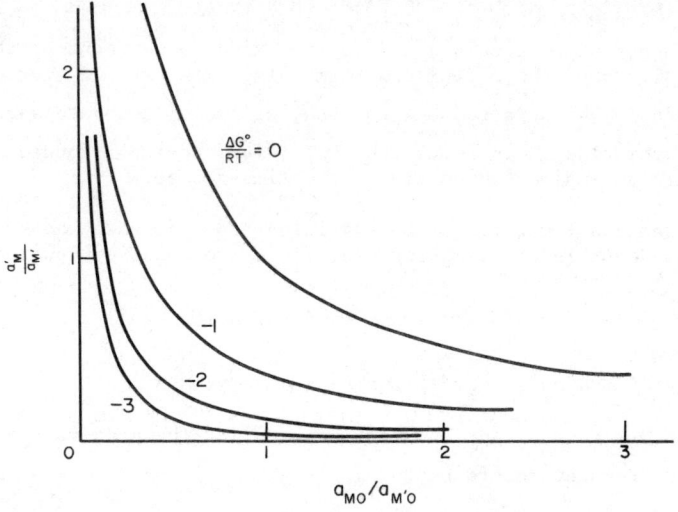

Fig. 70. Relationship between the activity ratios of oxides and metals

(B) HEAT OF REACTION

Knowledge of the enthalpy of the exothermic reaction (5.56) makes it possible to construct a thermal balance and to predict whether the heat generated will be enough to melt the metal and slag, and to heat them to such a temperature that the two phases will separate, due to their difference of density. The calculation should be carried out in a rigorous form (Problem No. 2). Nevertheless, in the particular case of aluminothermic reduction, a simplified calculation may be adequate: the ratio of the heat of reaction to the sum of the molecular weights of the products, $\Delta H^o/(M_M + M_{M'O})$ is calculated:

If the ratio is larger than 1100 cal g^{-1} (4500 joules g^{-1}), the reaction is violent, and may even be explosive.

Under 550 cal g^{-1} (2250 joules g^{-1}), the heat of reaction is insufficient to melt the products and heat them to the necessary separation temperature; an external supply of heat is necessary.

Between 550 and 1100 cal g^{-1} (2250 and 4500 joules g^{-1}), the reaction develops in a controlled manner without the need of an external supply of heat.

In order to control a violent reaction, an inert component may be added to the oxide to be reduced. It can be an oxide that does not participate in the reduction, but forms a fluid slag with the oxidised product (CaO for the aluminothermic process):

A reaction which is too slow, or for which the enthalpy of reaction is insufficient, may be accelerated in a number of ways:

The reductant and oxide are melted in separate furnaces, and mixed either in an arc furnace or in a ladle (silicothermic manufacture of ferrochrome).

A higher oxide of the same metal is added to the oxide to be reduced in such proportions that the reaction is effected under favourable conditions. For example, the aluminothermic reduction of manganese monoxide needs additional heat. This can be obtained by adding more strongly exothermic MnO_2 to the MnO which forms the bulk of the oxide component.

An extra supply of oxygen is added to the oxide to be reduced; this may be a peroxide which produces a flux and forms slag with Al_2O_3 (for example barium peroxide BaO_2 which is reduced to BaO, or sodium chlorate which can be reduced to chloride). In reduction by calcium, iodine is added as an oxidant to form calcium iodide, which fluxes the lime formed.

The amount of reductant metal must be sufficient to eliminate the oxygen from all of the compounds capable of being reduced, either to metal or lower oxide.

(C) EXAMPLES OF APPLICATION

(i) Manufacture of Manganese by Aluminothermic Reduction

Three oxides, MnO_2, Mn_3O_4, and MnO can be used for the preparation of manganese metal. They are reduced according to:

$$3\langle MnO_2\rangle + 4\langle Al\rangle \rightleftarrows 3(Mn) + 2\langle Al_2O_3\rangle \qquad \Delta H^o = -417,600 \text{ cal} \qquad (5.59)$$

$$3\langle Mn_3O_4\rangle + 8\langle Al\rangle \rightleftarrows 9(Mn) + 4\langle Al_2O_3\rangle \qquad \Delta H^o = -563,700 \text{ cal} \qquad (5.60)$$

$$3\langle MnO_2\rangle + 4\langle Al\rangle \rightleftarrows 3(Mn) + 2\langle Al_2O_3\rangle \qquad \Delta H^o = -114,000 \text{ cal} \qquad (5.61)$$

The enthalpies of these reactions per gram of product formed ($\Delta H/\Sigma M$) are equal, respectively, to 1123, 624, and 426 cal g^{-1} (4700, 2600 and 1780 joules g^{-1}). The first reaction would be too violent ($-\Delta H/\Sigma M > 1100$ cal g^{-1} or 4500 joules g^{-1}); the third could not be carried out without an external supply of energy ($\Delta H/\Sigma M <$ 550 cal g^{-1} or 2250 joules g^{-1}). Only the second can be carried out with relatively simple controls.

Manganese dioxide, or pyrolusite, is roasted for 3 hours at 1273 K to transform it to Mn_3O_4. Small amounts of manganese dioxide and granular aluminium are charged and ignited in an open magnesite crucible. Aluminium granules, Mn_3O_4 and powdered lime (to flux the alumina) are added separately and regularly to the incandescent charge in the calculated proportions, until the crucible is full. The amounts must be uniform and carefully proportioned in order to avoid local overheating and the overall temperature must be as low as possible, in order to limit the loss of Mn by volatilisation.

(ii) Manufacture of Ferrochrome by Silicothermic Reduction: The Perrin Process (See Problem 25)

The reduction of chromium oxide by silicon can proceed according to:

$$\frac{2}{3}(Cr_2O_3) + (Si) \rightleftarrows \frac{4}{3}(Cr) + (SiO_2) \qquad \Delta G^o = -31,730 + 3.7\,T \text{ (cal)} \qquad (5.62)$$

Most of the chromium used in the manufacture of low carbon special steels is obtained in the form of ferrochromium by reduction of chromite with silicon according to:

$$\frac{1}{2}(FeO\,.\,Cr_2O_3) + (Si) \rightleftarrows \underline{(Cr)} + \frac{1}{2}\underline{(Fe)} + \langle SiO_2\rangle \qquad \Delta G^o = -45,113 + 9.2\,T \text{ (cal)} \qquad (5.63)$$

The standard free energies for (5.62) and (5.63) are only slightly negative at the temperature of reaction (1900-2000 K), and reduction of the chromite, although facilitated by solution of the chromium in iron, is not complete due to the formation of chromium silicates with considerable losses of chromium (up to 28% Cr_2O_3 in the slag). The addition of lime for fluxing the silica decreases the percentage of chromium oxide in the slag to less than 8-10% Cr_2O_3, but since the amount of slag increases, the net loss of chromium is only slightly diminished.

Application of the Perrin Process considerably increases the recovery of chromium. It consists essentially of melting a slag of chromite and lime, containing 30% Cr_2O_3, in one arc furnace, and producing ferro-silicon-chrome (45% Si, 41% Cr) free of carbon in another by reacting a mixture of coke, chromite, and silica. The ferro-silicon-chrome is added to low grade slag (15% Cr_2O_3), originating from a previous treatment, in a ladle, to form two separate phases:

A low chromium lime silicate slag, which is discarded.

A ferro-silicon-chrome alloy with only 20-25% Si.

This alloy is poured, in turn, into a ladle containing the chromium-rich slag (30% Cr_2O_3) molten in the arc furnace to obtain:

A refined ferro-chromium, free of carbon and silicon (70% Cr).

A slag low in chromium (15% Cr_2O_3) which is treated in the next cycle (the first step of the process).

The advantages of such a process are: (1) A high reaction velocity, since the reaction interface between the reactants in the ladle is large due to the formation of drops in the molten mass. (ii) The losses of chromium are low, since the extraction is carried out in two (or more) steps.

The Ugine-Perrin process is also employed in the desulphurisation of steels. The steel is poured into a ladle containing a reducing and basic slag, prepared in a second furnace. The removal of sulphur is almost instantaneous.

(iii) Production of a Volatile Metal

High purity calcium, which is employed in the reduction of uranium fluoride of nuclear grade, is obtained as a vapour by aluminothermic reduction of lime (Problem 12).

5.4. METALLURGY OF HALIDES

The metal halides are soluble in water (with the exception of Ag^+, Hg_2^{++}, Cu^+ and Pb^{++} halides); which explains why the largest source of halides is sea water. Nevertheless, terrestrial ore deposits of alkaline and alkaline earth metals can be found, such as halite NaCl, sylvite KCl, sylvinite NaCl-KCl, carnallite $MgCl_2$. KCl, $6H_2O$, fluorite CaF_2 and cryolite Na_3AlF_6. Some natural bromides and iodides are also known, such as silver bromide and silver iodide, AgI and AgBr.

The alkaline and alkaline earth halides, the bonding of which is to a great extent ionic, differ from the halides of the transition metals, in which there is substantial covalent bonding, in that the former have high melting and vaporisation temperatures, although lower than those of the corresponding oxides, and are ionic conductors.

The chlorides are by far the most abundant of the natural halides, and most of the halides produced industrially are also chlorides. This accounts for the special importance of these compounds.

5.4.1. PRODUCTION OF CHLORIDES

The industrial production of chlorides is generally effected by treatment of oxides and relatively rarely from a metal. The reaction involved is basically the displacement of $O^=$ ions by Cl^- ions, sometimes accompanied by oxidation or reduction. The chloridising agents are Cl_2, HCl, and MCl_2, when necessary mixed with O_2 as an oxidising agent, or with C, CO, or H_2, as a reducing agent.

(A) DIRECT CHLORINATION

The behaviour of chlorine in the presence of a metal oxide depends on the relative affinities of chlorine and oxygen for the metal. The reaction:

$$[Cl_2] + <MO> \quad (MCl_2) + \tfrac{1}{2}[O_2] \qquad (5.64)$$

is thermodynamically possible only if the chloride is more stable than the oxide. The curves of Fig. 60b which represent the values of ΔG^o for reactions similar to (5.64) show that chlorination is possible for most metal oxides because, except for Mg, Al, Ti, Si and Ni, the chlorides are more stable than the corresponding oxides. Similar information is given in Fig. 62.

Chlorination of an oxide M'O by HCl or MCl_2 can be expressed in the form:

$$<M'O> + 2[HCl] \text{ (or } <MCl_2>) \rightleftarrows M'Cl_2 + [H_2O] \text{ (or } <MO>) \qquad (5.65)$$

This can be broken down into two reactions, the standard free energies of which are plotted in Fig. 60b.

$$<M'O> + [Cl_2] \rightleftarrows <M'Cl_2> + \tfrac{1}{2}[O_2] \qquad (5.65a)$$

and

$$[H_2O] \text{ (or } <MO>) + [Cl_2] \rightleftarrows 2[HCl] \text{ (or } <MCl>) + \tfrac{1}{2}[O_2] \qquad (5.65b)$$

The yield is good only when the value of ΔG^o for (5.65b) is less negative than for (5.65a). Fig. 60b shows that the chlorinating power of metal chlorides decreases from $AlCl_3$ to $TiCl_4$, $MgCl_2$, HCl and $FeCl_2$ to NaCl. Volatilisation of $M'Cl_2$ makes ΔG^o for (5.65) more negative and increases the concentration of this compound in the products. Conversely, the chlorinating capacity of MCl_2 in (5.65) is decreased if it is volatile.

The most widely used chlorinating agents are chlorine and the chlorides of hydrogen, magnesium, calcium, and iron. The chlorides of Al, Ti, and Mn are too expensive, and although NaCl is cheap, it is too stable to be effective. It is possible that NaCl could be used in the presence of silica, so that the sodium oxide would be converted to silicate and the equilibrium of (5.65) shifted to the right.

Ores which are mixtures of oxides can be chlorinated so as to effect extraction and purification at the same time. If ilmenite FeO . TiO_2 is attacked by Cl_2, the $TiCl_4$ and $FeCl_3$ which are formed can be separated by fractional distillation, because $TiCl_4$ boils at 136.5°C and $FeCl_3$ at 315°C. Chlorine losses, due to the formation of $FeCl_3$, which has little commercial value, raises the cost. An alternative is to chlorinate the iron selectively with HCl. The reaction:

$$<FeO> + 2[HCl] \rightleftarrows (FeCl_2) + [H_2O] \qquad (5.66)$$

is thermodynamically possible at any temperature, whereas:

$$\frac{1}{2}\ TiO_2 + 2\ HCl = \frac{1}{2}[TiCl_4] + H_2O \tag{5.67}$$

has a positive standard free energy. By use of this selective chlorination reaction, it is possible to produce valuable synthetic rutile from relatively cheap and abundant ilmenite.

(B) CHLORINATION WITH REDUCTION OR OXIDATION

When the metal exhibits two or more oxidation states, it is possible to produce a chloride in which the metal is in a different state of oxidation from that in which it was combined with oxygen. Chlorination is carried out simultaneously with oxidation or reduction.

Oxidising chlorination can be carried out with chlorine gas or a mixture of HCl and air or oxygen. When Cl_2 is reacted with a metal, the maximum degree of oxidation is obtained. This type of reaction is used industrially; for example, to recover tin from scrap tinplate and discarded tinplate containers. The tin coating is converted to $SnCl_4$ which is liquid at the temperature of operation, and drains to the bottom of the reaction vessel. There is slight attack on the iron base, with an increase in chlorine consumption.

When a metal can exist in two states of oxidation, the action of Cl_2 on the lower oxide is to raise the state of oxidation - as when FeO is converted to $FeCl_3$. This also occurs with the mixture of HCl and O_2:

$$2\langle FeO\rangle + \frac{1}{2}[O_2] + 6[HCl] \rightleftarrows 2(FeCl_3) + 3[H_2O] \tag{5.68}$$

The standard free energy of (5.68) per mole of Fe is similar in value to that for chlorination without oxidation (5.66), but the lower boiling point of $FeCl_3$ compared with that of $FeCl_2$ may be an advantage in special cases.

Reducing chlorination of an oxide is carried out by combining HCl or MCl_2 with C, CO, or H_2. There is an economy in the use of chlorinating agent because of the lower state of oxidation of the chloride product. In the extraction of tin from SnO_2, it is possible to produce either $SnCl_2$ or $SnCl_4$ as an intermediate product. $SnCl_2$ requires less chlorinating agent, but it is necessary to add a reducing agent in order to change the state of oxidation of the tin. The reaction is carried out in a fluidised bed reactor at 700-800°C using a mixture of HCl and CO:

$$\langle SnO_2\rangle + 2[HCl] + [CO] \rightleftarrows [SnCl_2] + [CO_2] + [H_2O] \tag{5.69}$$

The standard free energy of the reaction:

$$\Delta G^o = 6730 - 33.5\ T + 5.06\ T\log T + 2.22\times 10^{-3}\ T^2 \tag{5.70}$$

is slightly negative in the specified range of temperature, and since the temperature of reaction is above the boiling point of $SnCl_2$ (652°C), gaseous $SnCl_2$ is formed at equilibrium.

It is possible to combine Cl_2 and C, CO, or H_2 so as to maintain the same state of oxidation:

$$<MO> + [Cl_2] + <C> \rightleftarrows (MCl_2) + [CO] \tag{5.71}$$

or

$$<MO> + [Cl_2] + [CO] \text{ (or } [H_2]) \rightleftarrows (MCl_2) + [CO_2] \text{ (or } [H_2O]) \tag{5.72}$$

If the free energy of formation of CO, CO_2, or H_2O from O_2 and C, CO, and H_2 respectively is added to that for (5.64), the strongly negative values have a significant effect on the energy balance of the overall reaction and promote chlorination, not necessarily with coincident reduction in the state of oxidation. From (5.65) and (5.72) it can be seen that, stoichiometrically, the utilisation of Cl_2 and H_2 is equivalent to 2HCl. However, the free energy of formation of HCl:

$$[H_2] + [Cl_2] \rightleftarrows 2HCl \qquad \Delta G^o = -\,43{,}540 - 10.44\,T + 1.98\,T \log T \tag{5.73}$$

is strongly negative under the conditions of operation. This contribution to the overall energy balance is an advantage in facilitating chlorination, and it is preferable to use the H_2 and Cl_2 mixture.

Reaction (5.71) is utilised industrially in the production of $MgCl_2$ and $TiCl_4$. Briquettes of magnesite (or rutile), mixed with carbon and bonded with tar, are reacted with Cl_2 at 800°C. $MgCl_2$, which is liquid at this temperature, is separated by filtration through a porous hearth, while $TiCl_4$ is recovered as a vapour.

5.4.2. METAL PRODUCTION FROM CHLORIDES

As is the case with oxides, chlorides can be reduced to metals by thermal decomposition, by carbothermic or metallothermic reduction, by reduction with a gas, or by electrolysis. This last method is the only practicable means of producing magnesium and the alkaline metals, starting from their chlorides.

(A) REDUCTION OF CHLORIDES WITH CARBON

In the Mantos Blancos deposit near Antofagasta, Chile, the copper ore contains copper chloride, among other compounds. The ore is leached with sulphuric acid and CuCl is precipitated from the solution by treatment with SO_2. The high purity CuCl crystals are pelletised with lime and coke and, after drying, the pellets are charged into a batch rotary furnace, where the CuCl is reduced to metallic copper at 1200°C, according to the equation:

$$2(CuCl) + (CaO) + <C> \rightleftarrows 2(Cu) + (CaCl_2) + [CO] \tag{5.74}$$

This reaction can be divided into:

$$2(CuCl) + (CaO) \rightleftarrows (CaCl_2) + (Cu_2O) \qquad (5.74a)$$

and

$$(Cu_2O) + <C> \rightleftarrows 2(Cu) + [CO] \qquad (5.74b)$$

Calcium and copper have larger affinities for chlorine than for oxygen, and the free energy of (5.74a) is practically zero (Fig. 62), while the change for (5.74b) is strongly negative. This combination gives an overall negative value of ΔG^o for (5.74). The products are very pure copper, calcium chloride slag, and CO, which is burned at the exit of the furnace.

(B) REDUCTION OF CHLORIDES WITH GASES

Certain copper ores cannot be treated by conventional processes. Oxide ores with alkaline earth gangue require either ammonia, an expensive compound for leaching, or an excessive amount of sulphuric acid. Mixed copper sulphides, and oxides cannot be recovered completely either by leaching or flotation. The Copper Segregation Process* can be applied in such cases for complete extraction of copper. The process consists of oxidising any sulphide to oxide, followed by conversion of the oxide to chloride by means of a chlorinating agent, and reduction of the chloride so formed by gas.

If necessary, the ore is roasted at 700-800°C to convert sulphides to oxides in a suitable furnace (rotary kiln, multiple hearth, or fluidised bed). Otherwise, it may be heated in a more or less neutral atmosphere by the combustion of fuel and the hot products are charged continuously into a gas-tight reactor with small quantities (1 or 2% by weight) of NaCl as a chlorination agent and 2-3% of coke as a reductant. The overall change that takes place in the reactor is:

$$<Cu_2O> + 2<NaCl> + <SiO_2> + [H_2O] + <C> \rightleftarrows 2<Cu> + <Na_2SiO_3> \; 2[HCl] + [CO] \qquad (5.75)$$

This can be broken down into the following stages: Sodium chloride reacts with silica gangue and the water of hydration of the clays, to form hydrogen chloride and a silicate:

$$2<NaCl> + <SiO_2> + [H_2O] \rightleftarrows <Na_2SiO_3> + 2[HCl] \qquad (5.75a)$$

Hydrogen chloride reacts with copper oxide, forming volatile copper chloride Cu_2Cl_2:

$$<Cu_2O> + 2[HCl] \rightleftarrows [Cu_2Cl_2] + [H_2O] \qquad (5.75b)$$

The water reacts with coke to form CO and H_2:

*Another extraction process for mixed ores is the so-called LPF (leaching-precipitation-flotation) in which the copper oxides are leached in a first stage and precipitated with iron sponge in the second, floating the total (cement copper and sulphides) in a third stage, so obtaining a concentrate for melting in reverberatory furnace.

$$[H_2O] + <C> \rightleftarrows [CO] + [H_2] \qquad (5.75c)$$

The hydrogen reduces the copper chloride and regenerates hydrogen chloride:

$$[Cu_2Cl_2] + [H_2] \rightleftarrows 2<Cu> + 2[HCl] \qquad (5.75d)$$

The copper thus obtained is found in the form of small nodules and filaments over the coke, and is recovered by conventional flotation. Although the reducing agent is coke, direct reduction of Cu_2Cl_2 by carbon must involve direct contact. The growth of filaments at some distance from the surface of the coke can only be explained by the intervention of a reducing gas (CO or H_2).

(C) METALLOTHERMIC REDUCTION OF CHLORIDES (KROLL AND HUNTER PROCESSES)

Metallothermic processes are used on an industrial scale to produce metals of high value such as pure titanium, vanadium, or uranium, starting from their halides. The general reaction is as follows:

$$(MX_2) + (M') \rightleftarrows (M'X_2) + (M) \qquad (5.76)$$

The most widely employed reductants are magnesium, calcium, and sodium. $TiCl_4$ and VCl_4 are the raw materials from which titanium and vanadium are produced. The melting points are -23°C for $TiCl_4$ and -25°C for VCl_4, and the boiling points 135.8°C for $TiCl_4$ and 152°C for VCl_4. In the Kroll process, gaseous $TiCl_4$ or VCl_4 is fed into a steel reactor which contains molten magnesium at 800°C under an argon atmosphere to eliminate formation of oxides or nitrides. Titanium or vanadium sponge, impregnated with solid $MgCl_2$, is obtained on cooling. The metals are separated from the $MgCl_2$ by leaching, and the compacted sponge is melted, usually in a cold crucible arc furnace. The Hunter process is similar, but uses sodium as a reductant instead of magnesium.

Pure uranium (nuclear grade) is obtained by reduction of uranium tetrafluoride, UF_4, with calcium or magnesium in a calcium fluoride-lined crucible.

(D) THERMAL DECOMPOSITION OF HALIDES

The thermal decomposition of some halides is thermodynamically possible, especially for bromides and iodides, which are less stable than the chlorides or fluorides at high temperatures. Aluminium is of special interest because it forms two chlorides, $AlCl$ and $AlCl_3$. Their existence and their characteristics have been utilised in a refining process which is discussed later.

(i) The Van Arkel process consists of heating the impure metals (Ti, V, etc.) in the presence of iodine in an evacuated chamber fitted with a tungsten heating filament. The actual iodide formed depends on the temperature to which the impure metal is heated. In the case of titanium, the tetraiodide is formed at low temperatures (175°C) and the diiodide at higher values (525°C):

$$\langle Ti \rangle + 2[I_2] \xrightarrow{175^\circ C} [TiI_4] \xrightarrow{525^\circ C} [TiI_2] + [I_2] \qquad (5.77)$$

The vapours formed by these reactions migrate from the periphery of the chamber to the filament, which is at a temperature of about 1400°C, and decompose on the surface to give massive pure metal:

$$[TiI_4] \xrightarrow{1400^\circ C} \langle Ti \rangle + 2[I_2] \qquad (5.78)$$

The operating temperatures are selected so that iodides of other metals (Fe, etc.) which are impurities, are not volatile.

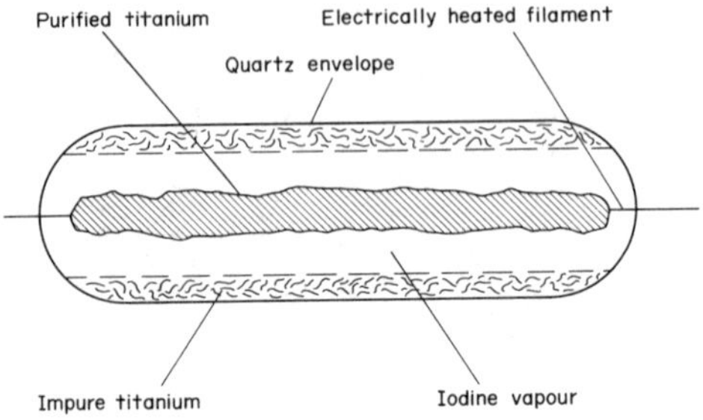

Fig. 71. The Van Arkel process

(ii) Aluminium, which is generally obtained by electrolysis of the oxide in molten fluoride, may also be produced or refined, starting from the subchloride AlCl. This, as indicated in Fig. 60a, is the most stable chloride at high temperatures, while the trichloride $AlCl_3$ is more stable at low temperatures. In the first step, alumina is reduced with carbon in the presence of $AlCl_3$, which acts as a chloridising agent:

$$\langle Al_2O_3 \rangle + 3\langle C \rangle + [AlCl_3] \xrightarrow[1\ atm]{1700^\circ C} 3[AlCl] + 3[CO] \qquad (5.79)$$

This reaction is of the same type as (5.71) and can be divided into two parts: reduction of alumina with carbon, and chlorination of aluminium with $AlCl_3$ to produce AlCl. The AlCl is then rapidly cooled to 700°C to avoid oxidation by CO. At this temperature it disproportionates to produce $AlCl_3$ vapour and liquid Al:

$$3[AlCl] \rightleftarrows [AlCl_3] + 2(Al) \qquad (5.80)$$

The $AlCl_3$ is recycled for use in reaction (5.79). Impurities in the alumina (iron, titanium, and silicon) could also be transformed to volatile chlorides ($TiCl_4$, $FeCl_2$, $SiCl_4$) which contaminate the aluminium subchloride. They are stable under the conditions of operation (700°C) and are recirculated with the $AlCl_3$ until the concentration is sufficient to require removal.

Figure 72 illustrates how commercial purity aluminium produced by electrolysis can be refined. At temperatures above 1000°C, the trichloride reacts with impure aluminium and forms the subchloride, which is disproportionate in contact with refined aluminium at 700°C (5.80).

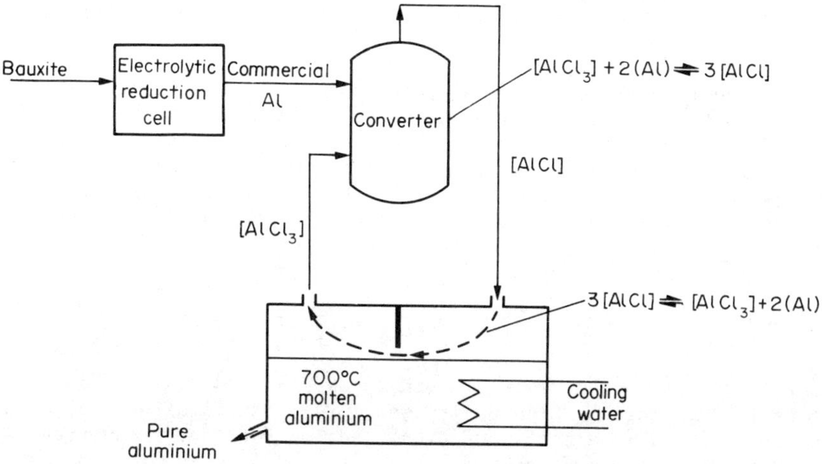

Fig. 72. The refining of aluminium

5.5. EXAMPLES OF EXTRACTIVE PROCESSES

The processes used to extract copper, nickel, lead, zinc and tin are described below in order to illustrate the way in which unit operations and chemical reactions, based on the special properties of the metals and their compounds, are combined to achieve efficient and economical production. The principal methods of processing ores which have been crushed so as to release (form separate particles) the mineral(s) from the gangue are described briefly in Section 5.5.6. Selections from these are used to prepare concentrates which are rich and/or pure enough to be treated by the more costly extraction processes in the most economical manner. The principal methods of producing metals from ores are summarised in Table 12. They illustrate the frequency with which chemical processing to prepare a pure compound of the metal is used, as well as the use of aqueous and fused salt electrolysis to produce high purity metal in the final stage. Vapo-metallurgy is familiar in the production of zinc, cadmium and nickel, but it is increasing, through the use of volatile halides, for several other metals.

5.5.1. EXTRACTIVE METALLURGY OF COPPER (at.wt. 63.5, density 8.96 g cm^{-3}, T_f 1083°C)

Copper is an excellent conductor of electricity and heat, is ductile enough to be drawn into fine wire and is resistant to atmospheric corrosion. Nearly half of the copper produced is used directly for electrical purposes and about one third is allowed and made into tubes, strip, extrusions and wire. About 10% is alloyed and made into castings. America, Canada, Chile, Eastern Europe, Peru, Zaire and Zambia are the principal producers, with world production totalling about 9 million tonnes in 1982 and increasing by about 4% per annum.

The principal minerals are sulphides and carbonates, the latter derived by oxidation of the former. The principal sulphides are chalcopyrite, $CuFeS_2$ and bornite, Cu_5FeS_4, found mixed with pyrite, FeS_2, and the sulphides of As, Sb, Bi,

Ni, Pb, Mo and Zn, as well as Ag and Au, in ores containing between 0.2 and 4%, with most between 0.5 and 1.5%. The other important primary sulphides are chalcocite, Cu_2S, and covellite, CuS. The oxidised minerals are mainly carbonates, malachite, $CuCO_3Cu(OH)_2$, and azurite, $2CuCO_3Cu(OH)_2$, with some important deposits of chrysocolla, $CuOSiO_22H_2O$. Mixed minerals are found where oxidation is incomplete.

The sulphide minerals are generally crushed, ground and concentrated by flotation, before pyrometallurgical treatment to produce anodes for electrolytic refining. The oxidised minerals are generally leached and the solutions concentrated and purified before copper is produced by electrowinning. Mixed ores may be treated either as oxides, after roasting, or by the LPF (Leach, Precipitation, Float) process, in which the crushed ore is leached to dissolve the oxidised mineral, the copper is precipitated by cementation with iron and the copper and copper sulphides are then recovered by flotation.

PYROMETALLURGY

Copper flotation concentrates contain 20-35% Cu, 20-30% Fe and 25-35% S, as well as 10-20% gangue. High purity metal is produced after separation of gangue, iron, sulphur and metallic impurities. Most of current production is from fossil fuel fired reverberatory furnaces, converters and refining furnaces, using separate operations for each, but there is increasing use of processes which combine two or more of these and which require much less fuel energy because they make use of that derived from the oxidation of iron and sulphur. The separate stages in the older process will be described, since each can be seen to contribute to the whole.

(i) Low grade concentrate may be roasted at 700-800°C, in order to convert part of the sulphides to oxides and SO_2. From Fig. 63a it will be seen that iron sulphides will oxidise more readily than copper sulphides, but some copper can be oxidised:

$$FeS_2 = FeS + \frac{1}{2} S_2 \tag{5.81a}$$

$$\frac{1}{2} S_2 + O_2 = SO_2 \tag{5.81b}$$

$$5CuFeS_2 + 2O_2 = Cu_5FeS_4 + 4FeS + 2SO_2 \tag{5.82}$$

$$3FeS + 5O_2 = Fe_3O_4 + 3SO_2 \tag{5.83}$$

$$2Cu_2S + 3O_2 = 2Cu_2O + 2SO_2 \tag{5.84}$$

(ii) High grade concentrate, or roasted low grade material, is mixed with flux and charged into a reverberatory furnace. Copper-rich slag from the converters is also poured in through the roof and heat is provided by the combustion of pulverised coal, oil or gas. The solids melt and react to form a slag containing less than 1% Cu, mainly iron silicate, of density about 3 g cm^{-3} and matte, containing 35-45% Cu of density 5.3 g cm^{-3}. These are immiscible and the slag at about 1200°C floats on the matte, which is at about 1150°C. The slag is discarded and the matte is tapped into ladles for transfer to the converters. Fe_3O_4 in the charge and in the converter slag is reduced to FeO by reaction with FeS:

$$3Fe_3O_4 + FeS = 10FeO + SO_2 \tag{5.85}$$

and this forms Fayalite, 2FeO . SiO_2.

(iii) Molten matte is poured into the converters and air is blown through it at 1200-1300°C in two stages. In the first, FeS, PbS and ZnS, are completely oxidised and other impurities are removed to varying degrees. The aim is to oxidise the FeS to FeO, leaving only Cu_2S and to convert the FeO to Fayalite (m.p. 1200°C) by reaction with added SiO_2. This is never complete and the slag at the end of this stage is saturated with Fe_3O_4 (m.p. 1697°C) and there may be 27% Fe_3O_4 and 10-15% Cu present. The slag is removed and further quantities of matte are poured in, blown and slagged until the converter is full of molten Cu_2S at about 1300°C. In the second stage, the aim is to produce metal low in iron and sulphur as a result of the reaction:

$$Cu_2S + O_2 = 2Cu + SO_2 \tag{5.86}$$

The first stage of this autogenous process is strongly exothermic and scrap anodes can be melted when added to cool the bath. The heat generated in the second stage is barely sufficient to compensate for the losses by radiation and in the exit gas.

Blowing is stopped at the first appearance of Cu_2O and the metal at this stage will contain about 0.6% O_2 and 0.01-0.03% S. This is "blister copper", so described because of the blisters which form on the upper surface of solidifying slabs, due to the formation of SO_2 gas by reaction between dissolved (S) and dissolved (O) under the thin solidified crust and its subsequent deformation.

(iv) This metal is still too impure to cast into anodes or slabs for working, and the oxygen, sulphur and metallic impurities must be further reduced. This is effected by further oxidation and removal of oxides, followed by controlled deoxidation to give the required structure. The blister copper is fed into cylindrical furnaces capable of holding up to 300 tonnes and air is blown through until there is a separate layer of Cu_2O, with other oxides in solution. At this stage, the metal will contain about 0.001% S and 1% O. The oxide layer is thickened and separated and the metal is deoxidised by blowing a reducing gas, such as propane or natural gas which has been desulphurised, through it. The progress of oxygen removal can be followed by an oxygen probe, or by casting small samples and noting the development of a flat upper surface. The product can be cast into anodes, or, if the precious metal content is low and certain impurities are absent, into slabs for rolling and other working processes. The final oxygen content is 0.05-0.15%.

In the past, when lump ores were readily available, the first stage was frequently carried out in a blast furnace, but by now only a very small number of blast furnaces operate with sintered concentrate. The efficiency and economy of the reverberatory were greatly improved by the use of submerged arc electric heating. The heat is developed by the passage of current through the slag and matte, but mainly through the high-resistance slag. The roof temperature is down to about 450°C from 1300°C and the slag is of lower viscosity and Cu content because of the removal of Fe_3O_4 and Cu_2O by immersed carbon electrodes in an agitated liquid.

The Outokumpu and INCO flash smelters were the earliest of those making use of the heat of oxidation. The Outokumpu furnace incorporates two vertical shafts, one at each end, on top of the lower section which acts as a settling chamber. Dried ground concentrate and air/oxygen mixtures are passed through a burner in the combustion shaft at about 1300°C and the molten products and SO_2 pass down into the settler. If the oxygen content of the air/O_2 mixture is 40%, this settler does not require supplementary heating. At lower levels of oxygen, fuel is burned for this purpose. Slag (1% Cu) and matte (45-50% Cu) separate in this chamber and flow out through separate tapholes. The gases, containing 10-15% SO_2, leave by way of the shaft at the other end and are used for acid manufacture. The slag is too rich to discard and must be treated, either by slow cooling and flotation, or electric furnace smelting with coke or pyrite to produce concentrate or matte and a discard slag containing 0.5% Cu. The use of electric heating in the settling chamber, as at Tamano has made possible the production of discardable slag containing about 0.6% Cu when the matte contains 60%.

The INCO furnace operates with 95% oxygen in the blast and a horizontal flame trajectory. The concentrate/blast mixture produces discard slag with 0.6% Cu with matte containing about 50%. The dust carry-over is very low and the requirement for external energy is less than for any other process. Both the Outokumpu and INCO processes have been used to produce blister copper, but for reasons of concentrate availability and slag losses, both are now used to produce matte which is oxidised to blister in a converter.

The Noranda process was initiated to produce blister copper, but for reasons of impurity control it is now producing 75% Cu matte which is processed in a Pierce-Smith converter. The charge of pelletised concentrate and flux is not dried. It is thrown by a slinger into the furnace, falling on to the slag surface. If necessary, low grade coal is added with the charge and burned with it on the top of the slag which floats on a layer of "white metal" (98-99% Cu_2S and 1-2% FeS) and this, in part, on a layer of blister copper containing about 0.4% Fe, 0.9-1.7% S and 0.3% O. Oxidation is effected by an air/oxygen blast blown through tuyeres immersed in the white metal layer. Concentrate is converted into white metal and a fayalite slag containing 15% Fe_3O_4 and 8-13% Cu. The amount of oxygen blown is related to the product required and the amount of fuel to be burned. Smelting is autogenous when the blast contains about 33% oxygen. The exit gases will contain about 4% SO_2 with an air blast and 13% when smelting is autogenous. The blister copper can be made into anode metal in a conventional refining furnace.

The KIVCET cyclone smelting furnace uses flash smelting, combined with electric heating of the settling zone. It is in use for treating complex concentrates containing copper, lead and zinc, using 92-95% oxygen blast. The feed must be dry (< 2% H_2O) and finely ground (< 1 mm). Concentrates contain 2-26% Cu, 19-30% Fe, and 15-42% S, as well as 0-23% Zn and 0-43% Pb, with about 12% gangue. Mattes range from 40-50% Cu, with up to 3.5% Zn and 5% Pb. That part of the furnace containing the burner and the gas offtake is separated from the electrically heated portion by a partition dipping into the slag. This allows of control of the atmosphere in the heated portion by adding crushed carbon and at the prevailing temperature of about 1250°C, zinc and lead are reduced from the oxides and volatilised. At the same time, Fe_3O_4 is eliminated and the effluent slag contains 0.35% Cu with 50% Cu matte, which is suitable for converting.

The Mitsubishi process was designed to produce blister copper directly from concentrates, using a combination of three stationary furnaces, a smelter, a converter and a slag cleaner, connected by troughs through which matte and slag flow continuously by gravity. The feed is a mixture of concentrates with 20% Cu, converter dust and fluxes, blown down vertically through stainless steel tubes using a blast containing 33% O_2. Crushed converter slag is also added. The smelter products, matte (65% Cu), slag (15% Cu) and gas (13-15% SO_2) pass respectively to the converter, the electric arc slag cleaning furnace and the acid plant. Quartz is added to the converter and the blast, with 26% O_2, makes blister copper with 0.1-0.2% S, slag with 15% Cu and gas with 13-15% SO_2. The blister flows to the refining furnace and the slag is cast and crushed. This slag is recycled to the smelter. The smelter slag flows to a forehearth and from there to the slag cleaner, where coke is added and copper is separated as matte, to leave a discard slag containing 0.5-0.6% Cu. The reclaimed matte passes to the converter. The operation has been developed to such a degree that it has been put under computer control. There are no conventional converters and the efficiency of conversion of S in the charge to SO_2 is high.

In those processes where the primary product is matte, conversion to blister copper will involve the use of a converter, in most cases of Pierce-Smith design (horizontal cyclindrical). This has the disadvantage of causing considerable loss of sulphur and pollution by fugitive emissions of gas during charging and slagging, when turned out of the blowing hood. Pollution can be reduced by large auxiliary hoods, but the capital cost is substantial. The Hoboken Siphon Converter is made in similar sizes and gas emissions are prevented by the device being under controlled suction, as well as there being a cover which can be used to close off the feed opening. The campaign length in both cases is governed by wear of the lining at the tuyere line and an attempt has been made to eliminate this problem by use of the Top Blown Rotary Converter, TBRC, which is modelled on the Basic Oxygen Steel converter and particularly the Kaldo version, in which the vessel is inclined and rotated while blowing is in progress. A number have been installed and used for short periods, but there has been no general changeover. Pure oxygen can be used because it is blown down on to the liquid charge from a water-cooled copper lance without damage to the lining, thus, conversion is accelerated and campaigns are much longer. The Mitsubishi converter is a stationary top-blown unit.

The production of anodes from blister is essentially as outlined in the section on Pyrometallurgy.

TABLE 9 Process Energy Requirements for Different Smelting Processes. (Kellogg and Henderson, A.I.M.E. Copper Symposium, Las Vegas, 1976, p. 373)

Primary Smelting Unit	BTU × 10^6/ST anodes	GJ/MT anodes
Wet Charge Reverberatory	18.465	21.455
Hot Calcine Charge Reverberatory	15.586	18.110
Electric Smelting Dry Calcine	24.188	28.104
Outokumpu Flash: no oxygenation	15.531	18.046
Outokumpu Flash: oxygenation	12.254	14.238
INCO Oxygen Flash	9.944	11.554
Mitsubishi: no oxygenation	13.954	16.213
Noranda: no oxygenation	21.152	24.577
Noranda: oxygenated air	12.279	14.267

In addition to being flat, the grain structure of the anodes must be such that they dissolve uniformly, i.e. the copper grains pass into solution, rather than falling out of the corroded matrix and contributing to the "slimes" which accumulate at the bottom of the electrolysis tank. There should be no large holes or the detachment of large pieces and anodes should be capable of use until only about 15% of the original is left.

HYDROMETALLURGY

Oxidised ores and roasted sulphides can be leached in sulphuric acid, or the roasting can be controlled to produce copper sulphate from the copper sulphides, at the same time as forming Fe_2O_3 from the iron sulphides. The copper can be selectively leached and used to prepare electrolyte. Copper sulphides can be leached directly with solutions of ferric salts, but these and other leach liquors containing iron must be purified before use. Ores with basic gangue are leached in ammoniacal solutions and reaction in any circumstances is accelerated by operation under pressure. Leaching may be carried out by percolation *in situ*, in dumps and in tanks capable of holding thousands of tonnes. Agitation leaching can also be carried out on more finely crushed material. The aim is to produce solutions containing 50 g l^{-1} Cu and low in Fe, to which about 150 g l^{-1} of H_2SO_4 is added to form the electrolyte (but see below).

For many years the recovery of copper from the dilute solutions, which were all that were available from some sources, was entirely by cementation with scrap iron, and cement copper is still a substantial form of raw copper. More recently, solvent extraction has been employed to produce strong solutions from dilute (Fig. 73), and the Nchanga plant in Zambia is an example of such an operation on a large scale.

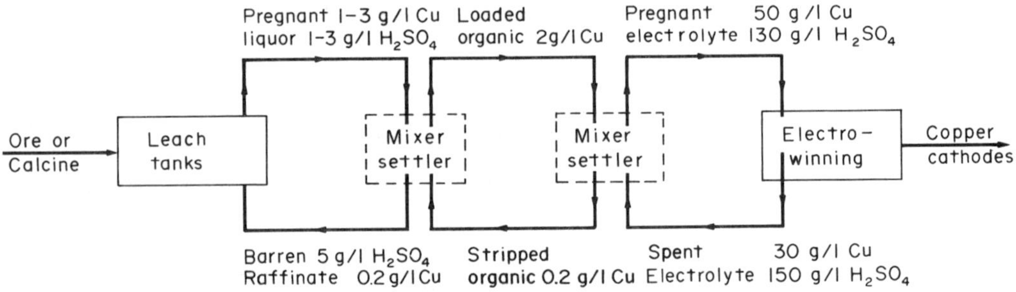

Fig. 73. Leaching, solvent extraction and electrowinning of copper

Solvent extraction effects the concentration of copper ions by first extracting them selectively from dilute, low-acid aqueous solution into an immiscible organic solvent and then returning them to a strongly acid aqueous solution, in the presence of certain chelating reagents. The principal examples of these are the LIX (hydroxy-oxime) and KELEX (hydroxy-quinoline) series of compounds. They are used in solution in kerosene or similar solvents in 10-20% v/v solutions and the transfer can be represented by the equation:

$$Cu^{2+}_{aqueous} + 2RH_{organic} = R_2Cu_{organic} + 2H^+_{aqueous} \qquad (5.87)$$

which indicates the importance of pH in influencing transfer in a given direction. From Fig. 73 it can be seen that liquors containing about 2 g l^{-1} can be processed to give liquor with 50 g l^{-1}.

After solvent extraction, electrowinning involves the electrolysis of solutions, using insoluble anodes of lead containing about 10% Sb and 0.5% Ag. The cathode reaction is the normal conversion of Cu ions to metal:

$$Cu^{2+} + 2e^{-} = Cu^{o} \qquad (5.88)$$

with a standard potential of +0.34 V. The anodic reaction in an aqueous solution is essentially:

$$H_2O = 2H^{+} + \frac{1}{2} O_2 + 2e^{-} \qquad (E^{o} = + 1.23\ V) \qquad (5.89)$$

This gives a theoretical deposition voltage of 0.89 V, but the actual value may be 2.5 V when the anode overvoltage and electrolyte potential drop are included.

This raises the power requirement to 2500 kWh per tonne Cu, which is ten times that for the production of cathode copper from soluble anodes.

Cathodes produced by electrowinning are too impure for electrical use except in some cases of solvent extraction, and final purification is effected by making into anodes and electrorefining with the product of pyrometallurgy.

Electrorefining is carried out by the electrochemical solution of anodes weighing about 350 kg, 1 m square and 5 cm thick, hanging from connecting lugs in a circulating acid copper sulphate solution, containing about 50 kg m^{-3} of Cu and 150 kg m^{-3} of H_2SO_4 at about 55°C and deposition on to cathodes of pure copper 1 m square and 2-5 mm thick, weighing 5-15 kg when new. The behaviour of the impurities in the anodes depends on their solubility in the electrolyte and the solubility of some of their compounds as well as on their position in the Electrochemical Series (Table 10).

TABLE 10

Element	Au	Ag	Se	Cu	As	Bi	Sb	H	Sn	Pb	Ni	Co	Fe
SEP (volts)	1.4	0.8	0.74	0.34	0.25	0.16	0.15	0.00	-0.13	-0.13	-0.23	-0.28	-0.41

Au and the Pt metals are electropositive to Cu and would deposit preferentially if they were in solution, but they are insoluble and fall to the slime. Ag is also electropositive, but any that enters the electrolyte is precipitated as AgCl by reaction with NaCl dissolved for that purpose. S, Se and Te enter the system as Cu_2S, Cu_2Se, Cu_2Te, Ag_2S, Ag_2Se and Ag_2Te inclusions in the anodes; all are insoluble and also pass to the slime under normal conditions. Should any Se dissolve, it is so electropositive that it is cemented out on to the anode surface. Pb forms insoluble $PbSO_4$ and Sn forms insoluble $Sn(OH)_2SO_4$. Both metals are electronegative to Cu. As and Sb are electronegative and form basic sulphates which tend to pass as solid "float slime" to the cathode. Ni, Co and Fe dissolve as soluble compounds, but they are sufficiently electronegative that they do not deposit. Dissolved Fe reduces the electrochemical efficiency by consuming electrons during the cycling of Fe^{2+} ions to the anode, to be oxidised to Fe^{3+} and back to the cathode to be reduced back to this state.

Copper and the soluble impurities Ni, Co, Fe and part of the As and Sb are prevented from accumulating beyond prescribed limits in solution by the bleeding off of 3-5% of the circulating electrolyte to a purification circuit and removal of all of the metals to leave concentrated H_2SO_4, "black acid", which is returned.

The anodes remain in the cells for up to 28 days and cathodes are removed and replaced by new ones after half and full time. At the end of the period, the scrap anodes are removed and the slimes taken out to be treated for recovery of precious and valuable elements. The cathodes are washed free of electrolyte and melted increasingly in an ASARCO furnace with no oxidation and cast into slabs for rolling or extrusion. There is also continuing development of equipment for continuous casting and working, which can produce rod or strip directly from molten metal.

Electrorefined copper is the purest form available and the statement of 99.99% purity is normally supplemented by details of the individual impurity content in parts per million (ppm). A series of examples gave the following figures:

Ag 4-10,	As 0.4-4,	Sb 0.5-3.5,	Bi 0-0.4,	Fe 0.5-2.5
Ni 0.7-1,	Pb 0.6-1.3,	Te 0.1-0.4,	Zn 0.1-0.3,	S 4-9 ppm.

5.5.2. METALLURGY OF NICKEL (at.wt. 58.7, density 8.90 g cm^{-3}, T_f 1453°C)

The principal uses of nickel are as a constituent of alloys and they range from the 9% Ni steel used for cryogenic purposes to refractory alloys for furnace parts exposed to temperatures up to 1200°C. In between, there are the high-strength, nonmagnetic and stainless steels, the nickel-base "superalloys", the unique zero-expansion alloy INVAR containing 35% Ni, the magnet alloys with Al and Co and many other alloy systems with Cu, Al, Mn and Zn. Ni powder is an efficient hydrogenation catalyst and the metal has an important field of use in chemical and pharmaceutical plant, as well as providing an important constituent of decorative and protective electroplating systems.

The principal minerals are the hydrated silicate, of mixed composition NiO, MgO, SiO_2, H_2O, characterised by Garnierite and representative of the lateritic ores which contain 75% of the world reserves of nickel and the sulphide $(Ni, Fe)_8S_9$, Pentlandite, which has provided a high proportion of world output in recent years. The oxide/silicate is associated with compounds of Fe, Cr and Co and the sulphide with minerals of Fe, Cu, Co and the Pt metals. The sulphides are found in Canada and the USSR, as well as S. Africa, W. Australia, Botswana, Finland, Indonesia and Zimbabwe, while the laterites are the source in New Caledonia and Cuba, as well as Brazil, Dominica, Greece, Guatemala, Indonesia, Phillipines, USA, USSR and Venezuela.

The sulphides are concentrated and segregated by flotation and magnetic separation and are processed by combinations of pyro-, hydro-, electro- and vapo-metallurgy on the route to high-purity metal. The oxide ores are low-grade, with about 2% Ni, and are not susceptible to mineral processing. They can either be treated as sulphides, after a sulphidising smelt, or they can be smelted or leached as such.

PYROMETALLURGY OF OXIDE/SILICATE ORES

One method utilised in New Caledonia by the Société Le Nickel is to sinter the ore and smelt it with gypsum, coke and limestone in a blast furnace to produce an Fe/Ni/S matte and a waste slag. The matte is blown in a converter to raise the Ni content to 77%, with 1% Fe + Co and 22% S. This is crushed and roasted to oxide in a fluidised bed reactor, before forming into cylinders and reducing to metal for sale, with wood charcoal in shaft furnaces.

An alternative consists in smelting to ferro-nickel in an electric arc furnace, after heating to dehydrate and partially reduce in a rotary kiln. The ferro-nickel (23% Ni) is desulphurised by a basic slag in a Top Blown Rotary Converter and the residual C and Si are removed. Higher grade ferro-nickel (50% Ni) can be made, using the Ugine-Perrin process, in which melted ore is caused to react with a controlled quantity of ferro-silicon. Final purification is effected by contact with a slag rich in iron oxide.

HYDROMETALLURGY OF ROASTED SULPHIDE OR OXIDE/SILICATE ORE

When the oxides arise from the roasting of crushed and ground matte at about 800°C, they generally contain copper oxide. At Falconbridge, leaching is carried out in dilute H_2SO_4 and copper oxide is more readily dissolved than NiO. The leach residue is dried and melted in an arc furnace to produce high-Ni anodes. Cu is recovered from the leach liquor by electrowinning, after purification and concentration.

The solvent used for the recovery of nickel from ore will depend on the nature of the gangue. If it is high in MgO, the solvent will be an ammoniacal ammonium carbonate liquor, but if it is mainly iron oxide, H_2SO_4 can be used in a variety of ways. $CuSO_4$ solution and spent $CuSO_4$ electrolyte can also be used, under appropriate conditions. Cuban laterite with basic gangue is ground to -200 mesh and partially reduced in a rotary kiln or multi-hearth furnace with coke breeze, so as to produce Ni from NiO and Fe_3O_4 from the iron oxides. This mixture is leached with an ammoniacal liquor, which is subsequently heated and aerated with air and CO_2 to remove NH_3 and precipitate $NiCO_3$. This solid is roasted to NiO and reduced to metal.

Ore suitable for acid leaching can be mixed with the acid, either by making a paste of ore and 98% H_2SO_4 and heating it to about 700°C, or by conventional mixing and reaction at high temperature and pressure. The acid may be generated *in situ* by reaction between water, pyrite and oxygen at high temperature and pressure, or the ore may be pre-reduced and leached with dilute H_2SO_4 at about 80°C. The pressure processes give $NiSO_4$ and $CoSO_4$ in solution and the iron as a hydrated oxide residue. NiS and CoS are precipitated by treatment with H_2S under pressure and filtered off. They are then redissolved by heating in water under oxygen pressure and the solution is purified before Ni powder is precipitated by H_2.

Copper sulphate solution can be used on basic laterites after partial reduction.

The gangue is not attacked and the Ni can be reduced from solution by H_2. Ni and Co can also be solubilised by heating the ore in $SO_2 + O_2$ and then leaching the sulphates out with water. Chlorination can also be used as a preliminary to leaching.

PYROMETALLURGY OF SULPHIDES

Nickel sulphide concentrate obtained either by flotation or magnetic separation, followed by differential flotation to remove copper sulphide, still contains Cu, Co and Fe. Pyrometallurgy makes use of the differences in affinity for oxygen and sulphur between them to obtain the separate metals efficiently and in the required state of purity. The nickel sulphide concentrate is partially oxidised and smelted in a reverberatory, flash, or electric arc furnace and then further oxidised in a converter to leave a Ni-Cu-S matte, while the iron is oxidised and combined with added silica to form a slag. The metallic output can range from matte to almost pure Ni_3S_2 and, with suitable feed, low-S nickel can be produced from a Top Blown Rotary Converter. Sulphide matte can be cast into anodes for nickel electrodeposition (see Electrowinning), ground and roasted to oxides and treated by leaching (see Hydrometallurgy) and if the product is low-S Ni it can be electrorefined (see Electrorefining) or converted to high purity metal by the Langer-Mond process operated by INCO.

This vapo-metallurgy process is based on the formation of nickel carbonyl from metal and CO:

$$Ni + 4CO = Ni(CO)_4 \qquad (5.90)$$

Fe in the input will form $Fe(CO)_5$, but Cu and Co do not react. The standard free energies of formation of $Ni(CO)_4$ and $Fe(CO)_5$ differ considerably. From Fig. 74 it can be seen that $Ni(CO)_4$ is more stable than $Fe(CO)_5$ and Table 11, calculated from the plots of Fig. 74, gives the percentage of carbonyl in the gaseous phase in equilibrium with the pure metals at different temperatures and pressures. The carbonyls form most readily at low temperatures (50-75°C) and high pressures (20 + bar) and decompose to metal and CO on lowering the pressure to atmospheric and raising the temperature to 180-220°C. Because $Ni(CO)_4$ boils at 43°C and $Fe(CO)_5$ at 103°C, they can be separated efficiently by fractional distillation. In industrial practice, nickel is obtained from roasted sulphide concentrate by reducing the oxide with H_2 at about 400°C and then volatilising $Ni(CO)_4$ as a dilute vapour in a stream of CO, treated to inhibit Fe volatilisation, at about 50°C and atmospheric pressure. An alternative is to granulate low-S metal prepared in the TBRC by pouring at 1650°C into high pressure water jets and drying the granules before subjecting them to reaction with CO at about 70 bar and 175°C to form liquid $Ni(CO)_4$ and $Fe(CO)_5$. The $Ni(CO)_4$ is separated by fractional distillation and decomposed in contact with circulating heated "seed" pellets which grow from about 1-2 mm to 6 mm diameter. The separated $Fe(CO)_5$ is decomposed to make Fe powder. If the CO pressure is raised to 200 bar, Ni can be volatilised from Ni-Cu-S matte, provided that the Cu is sufficient to form Cu_2S, otherwise COS will form and decompose to give contaminated Ni.

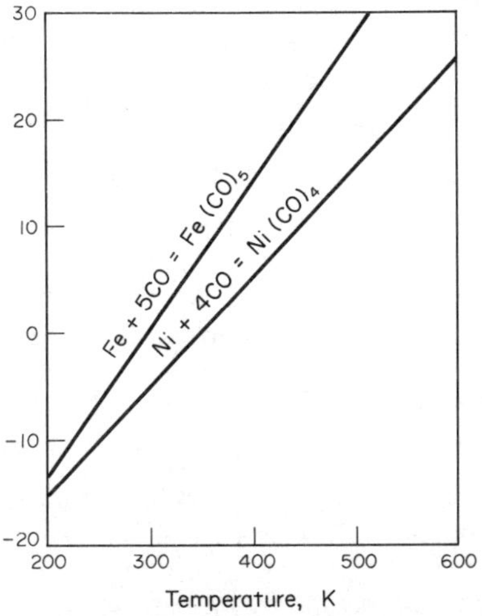

Fig. 74. Free energy of formation of nickel and iron carbonyl

TABLE 11 Fraction of Carbonyl in the Gas Phase (%) in Equilibrium with pure Ni and Fe at Different Temperatures and Pressures

Pressure bar	Temperature (K)					
	300		400		500	
	$Ni(CO)_4$	$Fe(CO)_5$	$Ni(CO)_4$	$Fe(CO)_5$	$Ni(CO)_4$	$Fe(CO)_5$
1	81.6	25.2	0.05	4×10^{-4}	9×10^{-6}	1×10^{-10}
10	96.6	84.9	19.7	0.04	0.009	1.3×10^{-4}
20	98.0	91.2	42.3	0.59	0.07	2.1×10^{-4}
100	99.4	97.5	79.8	37.2	6.9	0.013
200	99.6	98.6	87.7	60.3	24.2	0.204

HYDROMETALLURGY OF SULPHIDE ORES

The Sherritt-Gordon ammonia pressure leach process is the most widely applied in this field. Finely ground concentrate is reacted as a slurry with water, O_2 and NH_3 at about 100°C and 10 bar pressure. The Ni, Co and Cu sulphides are oxidised and dissolved as ammines and the FeS is oxidised and hydrolysed to insoluble hydrated iron oxide. The balance of the NH_3 is converted to ammonium sulphate. Trithionates, thiosulphates and sulphamates are also formed in solution. When reaction is complete, the solution is filtered off and heated to distil off part of the NH_3 and cause the unsaturated sulphur compounds to react with Cu^{2+} and precipitate as Cu_2S. After separation of this Cu_2S, the solution is heated to 200°C under H_2 gas at 30 bar in order to reduce the Ni ammines to metal powder. The Ni is filtered off and the residual liquor is treated with H_2S to precipitate the last of the Ni and the Co as sulphides. This precipitate is separated and redissolved to recover the metals separately.

Acid pressure leaching can be operated with H_2SO_4 added to the charge, or it can be generated by oxidation and hydrolysis of iron sulphides. The conditions of temperature, pressure and atmosphere are similar to those for ammonia leaching, as is the treatment of solutions subsequent to any necessary adjustment of pH . Ni is recovered as metal and Co can be precipitated as metal or sulphide. H_2 reduction of Co^{3+} is facilitated by preliminary reduction to Co^{2+} by the addition of Co powder.

ELECTROWINNING OF NICKEL

This may be carried out in the conventional manner by electrolysing purified solutions of mixed nickel chloride/sulphate at about pH 4, using lead anodes and thin nickel cathode starter sheets, but the more general method is to use sulphide anodes, which may have significant contents of Cu_2S. When polarised anodically, the anode sulphides will decompose according to

$$MS = M^{2+} + S^{o} + 2e^{-} \qquad (5.91)$$

The sulphur will remain as a porous skin, which may disintegrate and fall in the cell, or remain adherent to the metal-containing core. In the latter case, it will raise the cell voltage because of its insulating properties. The M^{2+} ions will discharge normally at the cathode and because the deposition potential of Ni is electronegative to those of Cu and H_2, the electrolyte from which the metal is deposited must be free from Cu^{2+} and the pH must be raised, so as to lower the activity of H^+. The electrolyte around the anode will be impure by virtue of solution from it and the aim of producing 99.95% pure Ni is achieved by having separate compartments for anolyte and catholyte with a permeable membrane in between. The anolyte is taken from its compartment and purified before it is returned to the catholyte compartment within which the cathode is hung. The anolyte first flows to a tank where Cu and As are removed by H_2S and from there to a leach tank where the Ni content of the electrolyte is raised by aeration in the presence of disintegrated anode scrap. Fe and Co are then removed in sequence by adding $NiCO_3$ and bubbling with Cl_2. Dissolved Cl_2 is removed by aeration and a final treatment with H_2S and pH adjustment with H_2SO_4 complete the preparation of catholyte. This liquid is pumped into the cathode compartment and flows out through the membrane. The velocity of flow is controlled so that no metal ions can pass directly from the anode compartment into that containing the cathode, but have to pass out into the purification system.

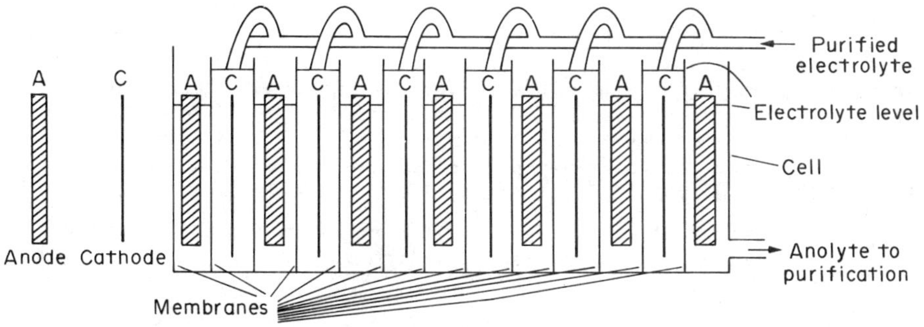

Fig. 75. Nickel electrolyte circulation

ELECTROREFINING OF NICKEL

This is carried out using metallic nickel anodes which may contain over 15% of Cu and the process is therefore subject to the same requirements for electrolyte purity as in electrowinning. Anodes are prepared from sulphide by roasting, reduction, controlled deoxidation and casting to shape with a satisfactory flatness and surface smoothness. Cathode starter sheets are fabricated by electrodeposition on stainless steel blanks and the electrolyte is a sulphate/chloride solution free from Cu^{2+} and of pH 4. The purification circuit is simpler than for electrowinning because it is not necessary to compensate for the relatively low anode efficiency of sulphides. Copper is cemented out by Ni and the efficiency of removal is increased by adding S with the Ni to precipitate Cu_2S . Fe and Co are precipitated by oxidation and hydrolysis and the catholyte is pure enough to give deposits of 99.95% Ni.

5.5.3. METALLURGY OF LEAD (at.wt. 207, density 11.4 g cm^{-3}, T_f 327°C)

This metal is resistant to many forms of corrosion and it can be easily cast, rolled and extruded into useful shapes. It is opaque to radiation and is widely used for screening purposes as well as in the manufacture of chemical plant, lead-acid accumulators and cable sheathing. It suffers from a tendency to creep and deform, even at higher atmospheric temperatures and multiple supports are necessary for extended lengths or areas.

The principal mineral is Galena, PbS, often found in massive deposits associated with ZnS and FeS and containing As, Sb, Bi, Sn, Ag and Au. Cerussite, $PbCO_3$, and Anglesite, $PbSO_4$, are the oxidised forms found in surface deposits. Lead ores contain up to 10-12% of the metal and the minerals are readily concentrated by gravity methods. The finely disseminated or crushed and ground minerals are readily floated.

EXTRACTION OF LEAD

PbS is reducible by Fe and PbO is readily reducible by C, CO and H_2. In common with the sulphides of Cu and Bi, PbS will react with PbO and $PbSO_4$ to produce metal. Early methods of production made use of this roast and reaction process, but modern plants are based on reduction of the oxide by carbon in a blast furnace. Sulphide flotation concentrates can be produced which are almost

100% mineral and these are oxidised and agglomerated into strong, permeable sinter, containing about 1% S by roasting on a Dwight-Lloyd pattern sintering machine. The charge is made up of concentrates, fluxes, recycled sinter fines and furnace dust to contain about 6% S. This will oxidise to provide the heat necessary to agglomerate the remainder. The raw sinter is crushed and screened to provide the correct proportions of 50-75 mm furnace feed and return fines at about 6 mm, which will become coated with concentrate to give a readily combustible sinter feed mixture. The coarse lumps, containing about 50% Pb, 1-1.5% S, 10% SiO_2, 8% CaO and 15% Fe_2O_3, constitute the blast furnace feed, mixed with about 20% of its weight of coke. The lead content is maintained by up-draught sintering so minimising deleading resulting from the reaction:

$$PbS + 2PbO = 3Pb + SO_2 \qquad (5.92)$$

The blast furnace is rectangular, with upper walls of brickwork, lower sides and ends of water jackets and a very strongly reinforced hearth lined with magnesite. An air blast through the tuyeres generates a temperature of about 1200°C and a large volume of CO-rich gas which passes upward through the charge. PbO is reduced in the shaft by this CO and the residual quantity which has become incorporated in the slag by reaction with the gangue is reduced by solid C in the hearth. Cu, Fe, S and As form matte and speiss. Lead may be separated from slag, matte and speiss, either by use of a siphon taphole, or by passage through a forehearth. It is then cooled in stages to cause separation of entrained oxides, speiss and matte.

REFINING

The lead from this separation is treated with sulphur at below 350°C in order to reduce the Cu to below 0.005%. Other impurities, As, Sb, Sn, can be removed, either by further oxidation, or electrolysis. In the Harris process these elements are oxidised and slagged by a mixture of NaOH and $NaNO_3$ at 400°C to form sodium arsenate, antimonate and stannate. Oxygenated air can be used to give the same selective oxidation, without alkalis, at 650°C and it can be done by air blowing at 800°C. If the lead ore is low in Bi, Au and Ag, the metal can be sold at this stage. Bi can be removed by the Kroll-Betterton process which involves the addition of Ca and Mg to form insoluble compounds with it at about 380°C and scraping off the bismuthide crusts. Excess of Ca and Mg is removed by chlorination and oxidation.

Most lead ores contain silver and recovery of this metal is generally profitable. It is done using the Parkes process, which depends on the formation of a series of compounds Ag_nZn_m when zinc is added to molten lead containing silver. The lead must be "soft", or free from other impurities, so the feed is usually lead which has been refined by treatment with sulphur and oxygen. In the batch process, solid zinc is added to lead at about 500°C at the rate of about 10 kg per tonne and the bath is allowed to cool to allow the Ag_nZn_m crystals to separate. These are recovered by squeezing the semi-solid mass to expel the liquid lead.

A further zinc addition is made before cooling to about 330°C and repeating. The crystals are enriched by liquation and then heated to distil the zinc and leave a predominantly Pb-Ag alloy. This is cupelled, i.e. oxidised by a blast of air at such a temperature that the PbO is liquid and flows away, ultimately leaving a pool of silver. The PbO is recycled to the blast furnace and the lead squeezed from the crystals is heated under vacuum and blown with air under molten alkali to remove zinc and make it saleable.

The continuous lead refining plant at Port Pirie carries out the silver extraction process on a flowing stream of metal, incorporating a desilverising vessel in which the Ag/Pb liquid is contacted with zinc at about 600°C, cooled to 330°C to effect maximum Ag_nZn_m separation and then reheated by heat exchange before passing to the vacuum dezincing vessel.

When lead has to be electrorefined, anodes with 0.005% Cu are cast on a wheel to give slabs about 1 m square and 3 cm thick. Cathode lead is melted and cast on a water-cooled drum to make starter sheets about the same size and 6 mm thick. The electrolyte is an acid solution of lead fluosilicate at 40-50°C containing about 10% H_2SiF_6 and 8% Pb, with glue added to give a smooth deposit. Electric current at 0.4-0.5 V per cell and 250 A m^{-2} causes the lead to dissolve and deposit on the cathode, leaving an anode mud containing the Cu, As, Sb and Ag, while Sn, Zn and Fe pass into a solution, any Sn in the anode will pass to the cathode. The refined lead contains about 20 ppm impurities. The anode mud is roasted and smelted with an alkaline slag to leave the precious metals, mainly silver, which are melted and cast into anodes for electrolytic separation.

5.5.4. METALLURGY OF ZINC (at.wt. 65.4, density 7.13 g cm^{-3}, T_f 419.6°C)

Zinc is resistant to atmospheric attack, but not to solutions of acids or alkalis. The principal uses are for dry battery cells, photolithography plate and thin sheet for roofing. Steel sheet and fabrications are given a protective coating of zinc, either by hot dip galvanising or electrodeposition. Zinc-aluminium alloys form the basis of the die casting and "superplastic" alloys and the alloys with copper and nickel are familiar. Zinc dust is used to cement out gold from solution and as a chemical reagent in dyestuff production. The oxide and sulphide can be used as pigments and zinc chloride is widely used as a flux in soldering and tinning.

The principal mineral is the sulphide, zinc blende, associated with FeS_2, PbS, Cu_2S and CdS. There are also oxidised minerals, silicate, carbonate, oxide and a zincate of iron. The sulphides are concentrated by flotation and the oxidised compounds are either gravity concentrated or leached in sulphuric acid. ZnO is reducible by C, but ZnS is not and must be roasted to ZnO + SO_2, the gas being made into H_2SO_4. While reduction of the oxide was the method of producing the metal for some hundreds of years, about 80% of current production is by leaching and electrowinning. Roasting for smelting is carried out on a Dwight-Lloyd pattern sintering machine at 1200-1300°C, using a feed mixture of five parts by weight of return sinter fines to one of flotation concentrates, in order to produce a strong, permeable and completely oxidised product. Oxide for leaching is produced by either flash (suspension) roasting at 900-1000°C, to minimise insoluble ferrite formation by reaction between ZnO and Fe_2O_3 from pyrite in the concentrate, or by fluidised-bed roasting at 750-800°C to reduce the sulphide sulphur to about 0.5% and retain about 2.5% S as $ZnSO_4$.

PYROMETALLURGY

Thermodynamic studies of the Zn-O-C system show that ZnO can be reduced by C as well as CO according to the reactions:

(a) $$ZnO + C = Zn + CO \quad (5.93)$$

(b) $$ZnO + CO = Zn + CO_2 \quad (5.94)$$

Reaction (a) between ZnO and solid C can best be represented by the sum of (b) and the Boudouard reaction, the kinetics of which are rapid only above about 950°C.

(c) $$CO_2 + C = 2CO \quad (5.95)$$

Reactions (b) and (c) are endothermic and a high temperature, 1000°C, is necessary for them to proceed. At this temperature, zinc is a vapour at atmospheric pressure (b.p. 906°C) and has to be recovered by condensation from the reaction products. However, reaction (b) is reversible and when the temperature falls, Zn can be oxidised by CO_2; for example, starting with a quantity of reaction products at Point X on Fig. 66a and moving horizontally to the left to represent cooling, it can be seen that the system becomes oxidising toward Zn (P_{Zn} decreases). For this reason, the process must be carried out in a closed system in which there is precise control of temperature and atmosphere composition. In the retort and electrothermic processes the temperature is virtually constant throughout and the atmosphere is kept low in CO_2 by the presence of a considerable excess of C. The time to pass to the condenser is so short that reoxidation of Zn is insignificant and if $P = 1$ atm, $P_{Zn} \simeq P_{CO} \simeq 0.5$ atm. In the blast furnace it is necessary to accommodate the need for an atmosphere low in CO_2 while achieving the necessary temperature by burning C and CO to CO_2. The distance and time between the hearth and the condenser are such that considerable reoxidation could occur without special provisions during transit to keep the temperature up and then to cool the gas very rapidly. Iron oxides can be reduced in both types of process if the $CO/CO + CO_2$ is above the Fe-FeO curve of Fig. 66a. In the retort and electrothermic processes this is the case, but the temperature is such that the residue is solid and there is no problem if metallic iron is produced. In the blast furnace, the slag is liquid and the reduction of FeO must be avoided, (because the solid Fe would accumulate and fill the hearth). A careful examination of Fig. 66a will reveal that there is an area to the right of the Zn-ZnO curve at 1 atm, and below the Fe-FeO curve, including point X, in which the CO-CO_2 mixture is, at the same time, in equilibrium with Zn vapour and with FeO, but not with C. The displacement from equilibrium with C is obtained by blowing in an appropriate excess of air.

Early furnaces operated on a 24- or 48-hour cycle of charging a mixture of oxide and carbon into a large number of small retorts, set horizontally and heated externally by the combustion of fossil fuel. Each retort had a refractory condenser sealed into the open end and a sheet iron prolong on the open end of that. These retorts were heated to about 1300°C on the outside and reached about 1100°C inside. With three times the theoretical requirement of C present, in the form of coke or anthracite coal, relative to the ZnO, the atmosphere at the operating temperature was essentially 50% CO and 50% Zn vapour. As this passed through the condenser, it was cooled to about 550°C and the Zn condensed to liquid, about 5% of the input Zn was collected as dust "blue powder" in the prolongs and about 2% was lost. The liquid zinc was scraped from the condensers at intervals and cast into slabs. It was relatively impure, containing 1.0-2.0% Pb, 0.05-0.1% Fe and 0.3-0.8% Cd.

Between 1920 and 1930, two patterns of continuous vertical retort were brought into use, The New Jersey Zinc Company version was externally heated and the briquetted and preheated charge of oxide and carbon moved continuously from top to bottom of the rectangular silicon carbide slab shaft. A splash condenser and gas exhaust system were fitted to the top and cooling and screw discharge to the bottom. Metal collected in the condenser was tapped out and cast and any dust in the gas was scrubbed out and returned to the charge. Each retort produced 8-9 tonnes of metal a day. The St. Joseph electrothermic retort is circular, with an external annular vapour collection ring leading to a condenser in which effluent gases have to bubble through a shallow layer of condensed liquid. Heat is generated by passing a heavy current through the coke of the charge, using graphite inserts in the wall at the top and bottom as connectors. A rotary discharge removes the dry residues at the bottom and metal flows from the condenser to a collecting sump. One furnace can produce 50 tonnes of metal per day.

It was apparent that the output of the externally heated retorts was limited by heat transfer through the walls and that the labour and fuel demands of the horizontal retort furnaces, with each retort producing about 35 kg per day, were very high. Few plants could expect power costs low enough to employ electrothermic heating and after 1935 much work was carried out on producing zinc from a unit based on the lead blast furnace. The hearth temperature was raised to 1200-1300°C by burning coke in air raised to about 800°C in a stove and supplied in such quantities as to give a $CO/CO + CO_2$ ratio of about 0.65. As this gas, with increasing Zn vapour content rises, it reacts with C and cools to about 1050°C and remains at this temperature until it is near the condenser offtake at the top of the shaft. The temperature near to and at the top is maintained by either admitting air and burning some CO to CO_2, or by exothermic reduction of PbO in the charge in which the coke is preheated to about 400°C, by CO. The problem of achieving very rapid cooling below the inversion temperature has been solved by passing the gases, containing about 5% Zn, 11% CO_2 and 23% CO into a lead splash condenser, where they are cooled very quickly to about 550°C and the zinc is dissolved after being condensed by flying drops of liquid lead at this temperature. About 400 tonnes of lead has to be circulated for each tonne of zinc produced. The zinc is separated from the lead by cooling to about 440°C and it contains about 1.3% Pb, 0.05% As, 0.02% Fe and 0.08% Cd. The As is removed by treatment with Na and the zinc upgraded to 99.99% purity by two-stage distillation, first separating Zn + Cd from Fe + Pb and then Zn from Cd.

HYDROMETALLURGY

Zinc is the most electronegative metal capable of being electrodeposited from an aqueous solution. Any other depositable metal will therefore be deposited preferentially if it is in the electrolyte and will tend to cause solution of cathode zinc by galvanic action. At first sight it appears impossible to electrodeposit zinc from an acid solution because H^+ is strongly electropositive to Zn^{2+}, but the overvoltage for H_2 formation on pure Zn increases so rapidly with the current density that there is a reversal of deposition potentials at high values (Fig. 76).

Zinc production by electrolysis was established in about 1920, with the immediate advantage of producing 99.99% pure metal. The initial problem of poor yield from concentrates containing Fe and electrochemical losses from trace impurities in solution were overcome and recent improvements in fluidised-bed roasting, more efficient leaching and residue treatment, together with improved electrical plant, a wider use of mechanisation in cathode handling and electrical melting of cathodes have so improved the actual operation, as well as the economics, that the proportion of the total produced by this method is increasing continuously.

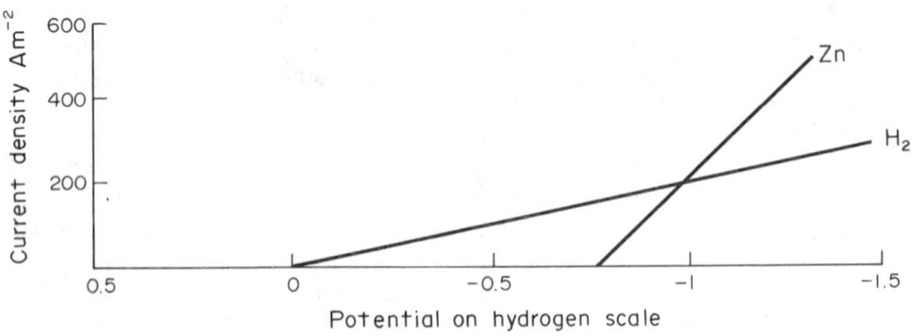

Fig. 76. Polarization curves for H_2 and Zn on pure Zn

Oxidised ores and roasted concentrates are leached in sulphuric acid, often in two stages so as to minimise the loss as insoluble ferrite formed during roasting and the solution is then exhaustively purified. The first stage consists of iron removal and this is now effected with much lower loss of occluded zinc and fewer problems in handling the ferric hydrate, by forming the less absorbent and easily filterable compounds Jarosite or Goethite. Jarosite is a crystalline double salt of iron and an alkali, based on $Fe_2(SO_4)_3$ and extensively hydrolysed, $Na_2Fe_6(SO_4)_4(OH)_{12}$ and $(NH_4)_2Fe_6(SO_4)_4(OH)_{12}$ are sodium and ammonium jarosites. They are precipitated from relatively acid solution, pH 1.5. Goethite is a hydrated ferrous oxide, FeO . OH, but $Fe(OH)_3$ and Fe_2O_3 may also be present. Precipitation is effected at 80-90°C and pH 2-3.5 while the ferrous solution is being oxidised. After the iron removal, there are three or four stages of cementation with zinc dust to remove Cu, Ni, Cd, Sb, Co and Ge. The first will remove Cu and Ni and some Cd and the second the balance of the Cd. Excess of Zn, activated by $CuSO_4$ will also remove Ge and Sb. Co can be removed by Zn with As_2O_3, by Cl_2, or by α-nitroso-β-naphthol. Mn can be tolerated in the final liquor, since it forms MnO_2 on the anodes. The Cu, Co, Cd and other metals can be recovered as by-products. After filtration, the final liquor is a solution of $ZnSO_4$ of pH 5 and this is fed to the electrolysis tanks at the rate necessary to maintain the specified acidity, spent electrolyte containing about 60 g l^{-1} Zn and 150-200 g l^{-1} H_2SO_4 being withdrawn continuously. The anodes are of Pb containing 1% Ag and the cathode blanks are of C.P. Al, each about 1 m^2 in area. Deposits are stripped every 24 or 48 hours and are melted and cast into slabs. Additives, such as glue and β-naphthol promote the growth of smooth deposits. The current density increases with the acidity of the electrolyte and can be up to 1000 Am^{-2}. Under these conditions, the electrolyte must be cooled by immersed lead pipes carrying cold water to keep it at 30-35°C. Spent electrolyte is returned to the second stage of leaching and the resultant solution is passed through the purification system before it is returned to the cells.

5.5.5. THE METALLURGY OF TIN (at.wt. 119, density 7.5 g cm^{-3}, T_f 231.9°C)

The properties of tin which are most frequently quoted are those of the β (white) allotrope, silvery white, soft, malleable but not ductile and stable above 13.2°C. The α allotrope, stable below this temperature, is usually found as a grey powder,

$d = 5.77\ g\ cm^{-3}$ and transformation to this form at low temperature is accelerated by mechanical working and by contact of β with α. Pure tin has a high resistance to attack by a wide range of chemicals and is almost unique in the length of the temperature range between the melting and boiling points. This facilitates refining, just as the low boiling points of SnO, SnS and Sn chlorides facilitate extraction. The principal use of tin is as the thin coating on steel in tinplate, but there are many other examples of tin coatings on copper, steel and cast iron. Pure tin has been widely used for the manufacture of collapsible tubes and substantial quantities are essential for the manufacture of float glass. Important alloy systems are formed with Pb (soft solders), Pb and Sb (type metal and bearing metal), Cu (bronze) and Ni and Zn. Organic and inorganic compounds of tin have wide uses as catalysts, pigments and enamels and there are numerous bactericidal applications.

The principal mineral of tin is cassiterite, SnO_2, found in association with Fe_3O_4, $FeWO_4$ and FeS_2, as well as sulphides of Cu, Ni, Bi, As, Sb and Pb. The alluvial deposits of S.E. Asia are of much higher purity than the vein deposits of S. America, Australia, Africa and the U.K. Because of the volatility of some compounds of tin and the amphoteric character of SnO_2, concentrates for smelting must be of high purity, in order to limit losses by volatilisation and slagging. However, this volatility can be of use in aiding the recovery of tin from slags and low grade materials. Primary concentration is generally by gravity, the plant ranging from simple sluices to complex assemblies of tables, jigs, cones and other specialised washing equipment. Magnetic separation, roasting, leaching and flotation are used to produce concentrates which are about 90% SnO_2 (70-74% Sn).

EXTRACTION

Because SnO_2 and FeO are very similar in reducibility by C and CO and tin ores always contain small amounts of iron oxide, there is an equilibrium between SnO_2 and FeO in the slags and the Sn and Fe in the metal. This can be expressed in the form:

$$\frac{a^2_{(Fe)_m}}{a_{(FeO)_s}} = \frac{1}{K} = \frac{a_{(Sn)_m}}{a_{(SnO)_s}} \tag{5.96}$$

for the reaction:

$$(Fe)_m + (SnO)_s = (Sn)_m + (FeO)_s$$

K = 100 at 1400 K and the two desirable objectives, tin low in iron and slag low in tin, cannot be achieved simultaneously. Tin smelting is therefore carried out in two stages. In the first or "ore smelt", concentrate is heated to about 1150°C in contact with carbon and an iron-tin alloy recovered from the second "slag smelting" stage. Smelting can be carried out in fuel-fired shaft, reverberatory or rotary furnaces, or in submerged-arc electric furnaces. About 80% of the SnO_2 is reduced by C and Fe to make tin metal, about 99% pure and a slag containing 20-30% Sn combined with SiO_2 as stannous silicate accompanied by 20-40% FeO as Fayalite. In the "slag smelt", CaO is added to combine with the SiO_2 as

$2CaO \cdot SiO_2$, releasing SnO_2 and FeO to be reduced by C at 1350-1400°C to leave a slag with about 1% Sn and metal containing about 25% Fe. This alloy is granulated for return to the ore smelt. If electric smelting is used, ferrosilicon can be added to produce a discardable slag in one step. The slags produced in the "slag smelt" can be smelted with pyrite or gypsum and coke to reduce the tin content to less than 0.1% by volatilisation of SnS. These slags can also be smelted to recover W as an alloy with Fe, or they can be processed to recover Ta_2O_5 or Nb_2O_5. All furnaces must be fitted with efficient fume collection equipment because of the volatilisation of SnO which occurs. Tungsten also accumulates as a refractory alloy with iron on the furnace hearth and this is an important factor in limiting the length of a campaign.

REFINING

This is necessary to remove Fe, As, Sb and Bi. Fe can be reduced to 0.01% by liquation at 240°C to leave crystals of $FeSn_2$, which are returned to the ore smelt: tin ingots are stacked on inclined cast iron trays and heated by "soft" flames to cause tin low in iron to flow into the collecting "kettle". This metal is then heated to about 440°C and air is blown through it to form an Fe, Sn oxide dross which is also returned to the ore smelt. As and Sb are removed by adding Al to form compounds which float on the surface and oxidise before they are skimmed off. Excess Al also oxidises and is skimmed away. Cu is drossed as Cu_2S and Bi is removed as a Ca-Mg-Bi alloy/oxide dross. All of the solid phases skimmed off are wet with tin and provision must be made for it to be recovered.

Liquated tin can be cast into anodes and refined by electrolysis in a fluosilicate, fluoborate or sulphamate electrolyte. Because Pb is also soluble in these liquors and its electrode potential (-0.122 V on the hydrogen scale) is very close to that of Sn (-0.136 V), H_2SO_4 is added to precipitate $PbSO_4$. As, Sb, Bi and Ag will collect in the anode mud and Fe and any Ni partly in the mud and partly in the electrolyte. Glue and cresol are added to the electrolyte to give a smooth deposit. Anodes with up to 25% Pb can be refined in a thiostannate bath, the Pb reporting as PbS in the anode mud. The cathodes are 99.93-99.98% Sn.

Significant quantities of tin are recovered by treatment of tinplate and tin-can manufacturing scrap and increasing amounts by treatment of tin-cans recovered from domestic waste. The tin is separated from the steel base by electrolytic and chemical processes in alkali and gaseous media and is obtained as metal, or chemical compounds which can be readily reduced.

5.5.6. MINERAL PROCESSING

GRAVITY CONCENTRATION

Separation of mineral and gangue is effected according to the response of these materials of different density when a quantity of closely sized ore is either immersed in a fluid of density intermediate between those of these constituents, or allowed to settle in a stream of water flowing either vertically or substantially horizontally.

In the first case, the heavy mineral will sink and the light gangue will float and the method is more generally described as either "Sink and Float" or "Dense Medium" separation. The ore particles can be as large as the plant can handle, subject to satisfactory release of mineral from gangue, down to about 0.5 mm, but slime must have been removed. The dense medium consists of a suspension of up to about 30% by volume of solid, kept as such by agitation or circulation. The solids most widely used are ferrosilicon and magnetite with particle sizes between about 150 and 50 μm. They have the advantage that they can be recovered magnetically after washing from the separated products and reused after demagnetisation. Fine silica sand can be used for coal washing. The operation can be carried out in cones, drums and cyclones and, by variation of medium density, the concentrate can either be of low yield and high grade, or high yield and low grade.

The sedimentation rate (limiting velocity of descent) of a particle is a function of the apparent mass m_a ($m_a = (\rho - \rho_f) \times$ volume) and of its cross-section S:

$$V(\text{velocity}) = Km_a/S$$

The separation of two particles having limiting velocities V_1 and V_2 can be effected by water moving upward at a velocity between these values.

When particles of different terminal velocities are fed into a horizontally flowing stream, they will be carried in that direction for such a distance as a floating body would move before they reach the bottom and come to rest in the very slowly moving liquid. In feed of mixed sizes, the large particles will fall quickly and the small ones more slowly, so that there will be a "middling" zone where coarse gangue is mixed with fine mineral, downstream from the coarse mineral, but upstream from the fine gangue. This simplest separation is exemplified in the recovery of cassiterite in long sluices or palongs. Coarse material of 5-15 mm particle size can be separated in the alternating vertical currents of a jig, while 5-50 μm slimes can be processed in tilting multi-deck tables. In between there are drums, cones, spirals, hydrocyclones and shaking tables in many modifications, all capable of effective operation on suitably and closely sized feed.

MAGNETIC SEPARATION

In the same way that all materials have a characteristic density, they have a property defined as magnetic susceptibility and all materials can be put into one of three classes. The diamagnetic materials have susceptibility values $\kappa \lessgtr 0$ and tend, in a magnetic field, to move to the point of minimum field strength. Paramagnetic materials are still only weakly magnetisable but are attracted to points of highest field strength, $\kappa \gtrless 0$. Ferromagnetic materials are strongly attracted to points of high field strength and display remanence or residual magnetism, $\kappa >>> 0$. The force acting on a particle is a function of the field strength and the field gradient:

$$F = kHdH/dl$$

and k depends on the volume and susceptibility.

Separation can be effected by subjecting the mixture of crushed gangue and mineral to a magnetic field, either dry, or as a pulp in water and collecting the different fractions. Metallic iron and magnetite are the strongest ferromagnetics and require only low intensity fields for concentration. Tramp iron is lifted from ore on conveyor belts by electromagnets suspended above and these are periodically moved, switched off and cleaned, before return. Fine mineral can be concentrated by devices ranging from magnetic pulleys to drums, belts, discs and the high intensity separators, of which the Jones is an important example. Stronger fields are needed for haematite, wolframite, chromite and the sulphides of iron, nickel and cobalt and these are concentrated by diversion or withdrawal from moving streams of crushed material.

FLOTATION

Mineral separation by flotation depends on the fact that some materials are wetted by water and some are not. In a suspension of water, air bubbles and small particles, those that are not wetted will attach themselves to the bubbles and rise to the surface. If a substance has been added to the water which stabilises the bubbles into a froth, this will float and can be swept off. Modern flotation technology depends on the availability of reagents which can be used to render minerals non-wetting on a highly selective basis, which can form froths of widely differing characteristics and which can depress and activate selected species at will. Refinements of selectivity can be achieved by adjustment of pH and other ion contents.

Attachment of solid to bubble can be expressed in terms of the surface tension values between liquid, solid and gaseous phases (γ_{LS}, γ_{SG} etc.):

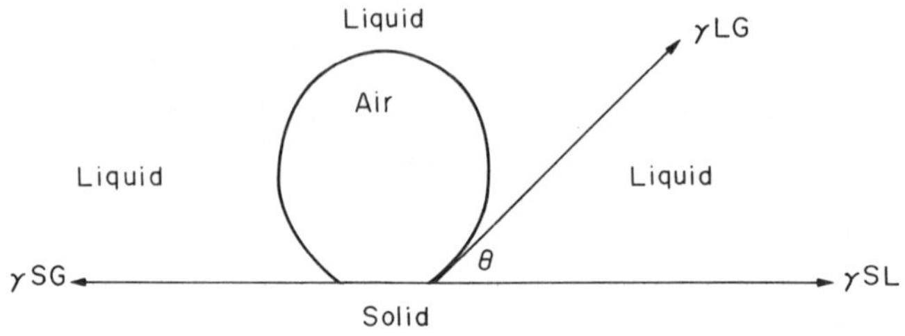

Fig. 77. Bubble attachment to solid mineral

$$\gamma_{SG} = \gamma_{LS} + \gamma_{LG} \cos\theta$$

where θ is the angle of contact between the bubble and the solid. The work of attachment of the bubble is given by:

$$W_{SG} = \gamma_{LG} + \gamma_{SL} - \gamma_{SG}$$

Combining these equations:

$$W_{SG} = \gamma_{LG} (1 - \cos\theta)$$

and the value increases with the value of θ.

The reagents which confer the non-wetting properties are described as promoters or collectors and can range from virtually insoluble hydrocarbons to ionising solutes consisting of a combination of hydrocarbon and polar groups which may be anionic or cationic and adsorb as virtually monomolecular layers on specified minerals.

Ore pulps are conditioned by mixing with reagents which control pH and with collectors which may take up to an hour to distribute on the mineral. Frother is then mixed in and the pulp is passed to the flotation machine, where high speed agitation and aeration combine to form bubbles and cause them to impact with mineral particles before rising to the surface. The froth flows, or is swept to the concentrate launders, where the bubble structure is broken down before the pulp is filtered to recover the concentrate. There is an initiation period between entry into the machine and the appearance of mineral in the froth and differences in this property can be utilised by passing a pulp through several machines in series, so as to collect a succession of different minerals.

GENERAL READING

Biswas, A.K. and W.G. Davenport, *Extractive Metallurgy of Copper*, Pergamon Press, Oxford (1980).

Boldt, J.R. and P. Queneau, *The Winning of Nickel*, Longmans, Toronto (1967).

Dennis, W.H., *Metallurgy of the Non-Ferrous Metals*, Pitman Bros., London (1954).

Gilchrist, J.D., *Extraction Metallurgy*, Pergamon Press, London (1967).

Hopkins, D.W., *Physical Chemistry and Metal Extraction*, J. Garnet Miller Ltd., London (1954).

Institution of Mining and Metallurgy, *Advances in Extractive Metallurgy, Proc. Symposium*, London, 1967. The Institute of Mining and Metallurgy, London (1968).

Kubaschewski, O., E.Ll. Evans and C.B. Alcock, *Metallurgical Thermochemistry*, Pergamon Press, London (1967).

Parker, R.H., *An Introduction to Chemical Metallurgy*, Pergamon Press, London (1967).

Peacey, J.G. and W.G. Davenport, *The Iron Blast Furnace*, 2nd Ed., Pergamon Press, Oxford (1980).

Pehlke, R., *Unit Processes of Extractive Metallurgy*, Elsevier, N.Y. (1973).

Rosenqvist, T., *Principles of Extractive Metallurgy*, McGraw-Hill, N.Y. (1974).

St. Pierre, G.R., Ed., The Met. Soc. Conferences, *Physical Chemistry of Process Metallurgy*. Am. Inst. of Mining. *Metallurgical and Petroleum Eng.*, Vols. 7 and 8, Interscience Publishers, N.Y. (1959).

TABLE 12 Summaries of Processes for Extraction of Metals from Ores (Avnex)

Metal	Principal sources	Preliminary treatments	Extractive processes
Al	Oxide	Solution in NaOH, precipitation, calcination	Electrolysis of Al_2O_3 in fused cryolite, electrolysis in fused chloride
Sb	Sulphide	Liquation of rich ore. Gravity or flotation, volatilising roast	Reduction of Sb_2S_3 by Fe. Reduction of Sb_2O_3 by C
As	By-product of Pb, Cu, Co, Zn	Volatilising roast	Reduction of As_2O_3 by C
Be	Complex silicate	Alkali sintering, chemical purification, BeF_2. Sintering, purification $BeCl_2$, distillation	Reduction of BeF_2 by Mg. Electrolysis of $BeCl_2$/NaCl fused salt
Bi	Native, sulphide, oxide	Liquate rich native metal ore. Flotation, roast	Reduce Bi_2S_3 with Fe. Reduce Bi_2O_3 with C
Cd	By-product of Zn, Pb, Cu	Recycle dust to concentrate Cd, leach with H_2SO_4, purify solution	Cement with Zn or electrolyse $CdSO_4$ solution
Ca	Limestone, gypsum brines	Calcination. Evaporation, crystallisation, fusion $CaCl_2$	Reduction of CaO with Al *in vacuo*. Electrolysis of fused $CaCl_2$. Distillation of Ca from cathode of Cu
Cr	Iron chromite	Gravity concentration for lean ore. Sinter with Na_2CO_3, leach, purify, convert to Cr_2O_3. Sinter, leach, convert to $CrCl_3$	Reduce to high-C ferrochrome with C Reduce Cr_2O_3 with Al in electric furnace. Electrolyse $CrCl_3$ solution Reduce molten chromite with molten FeSi
Co	Co-S-As minerals, by-product of Ni and Cu oxides	Matte or speiss, roast, leach, precipitate, leach with H_2SO_4, precipitate as $Co(OH)_3$, calcine to Co_2O_3. Roasting and pressure leaching, purification	Electrolysis of neutralised sulphate solution. Reduction of oxides by C. Reduction of ammines by H_2 under pressure

Metal	Principal sources	Preliminary treatments	Extractive processes
Cu	Oxides, sulphides	Leaching, solvent extraction, purification. Flotation, roasting, matte smelting, conversion, anode production or tough pitch	Electrowinning (remelting and electrorefining). Electrorefining
Ga	By-product of Al	Hg-cathode deposition, solution in NaOH	Electrodeposition, purification
Ge	Sulphide, by-product of Zn, coal flue dust	Production of $GeCl_4$ solution, precipitation of As, distillation of $GeCl_4$	Reduction by H_2, zone refining
Au	Native, by-product of Cu, Pb, Ag, telluride	Cyanidation. Amalgamation. Anode slimes, bullions, precipitates roasted, leached. Dusts leached, cupelled. Roasting, cyanidation or amalgamation	Cementation by Zn. Distillation of Hg. Cupellation, chlorination. Cementation or distillation
Fe	Oxides	Sizing by crushing and agglomeration	Reduction by C, refining by oxidation. Reduction by $CO + H_2$
Pb	Sulphides, oxides	Gravity, flotation, roasting, sintering. Gravity, roasting or direct smelting	Reduction of PbO by C and CO, desilverising, debismuthising. Electrorefining
Mg	Carbonate, hydrate, brines, sea water	Calcination, chlorination $MgCl_2$ production, crystallisation	Electrolysis of $MgCl_2$ $CaCl_2$ NaCl electrolyte. Reduction of MgO by Si (as FeSi) *in vacuo*
Mn	Oxides	Leach in H_2SO_4, purify	Electrolyse in diaphragm cell. (Ferroalloy by smelting with iron oxide in blastfurnace or electric reduction furnace
Hg	Sulphide	Calcination/distillation	Condensation, filtration, redistillation

Metal	Principal sources	Preliminary treatments	Extractive processes
Mo	Sulphide	Flotation, roast to MoO_3, sublime	Reduce with H_2. (Ferroalloy by reduction with C and Fe)
Ni	Sulphide	Flotation, calcination, matte smelting, FeS oxidation, slow cooling, flotation and magnetic separation, roasting. Pressure leaching, purification.	Reduction of oxide by C, reduction of oxide by H_2, carbonyl formation and decomposition. Sulphide electrolysis. Metal anode electrolysis. H_2 pressure reduction.
	Oxide	Sulphidisation, then as above. Partial reduction leaching, purification	Melting, H_2 reduction, matte smelting, etc.
Nb and Ta	Complex oxides	Gravity concentration, alkali fusion, leaching, solvent extraction, water leach	Reduction with C, carbide/oxide reaction to metal
Pt Group	Native alloy, by-product of Ni. Arsenide in CuNi ore	Collect in lead bullion, cupel. Chemical and smelting treatment. Recover NiPt alloy magnetically, treat aqua regia, then as above. Matte smelt, roast, leach, as above	Electrodeposit Au. Decompose $(NH_4)PtCl_6$ by heat. Decompose $(NH_3)_2PdCl_2$ by heat. Reduce $K_3(RhCl_6)$ in H_2. Reduce $RuCl_3$ in H_2. Reduce $(NH_4)_2(IrCl_6)$ in H_2. Reduce OsO_4 with H_2
K	Salt deposits, brines	Leaching, flotation, crystallisation. Evaporation, flotation	Electrolysis of fused KCl
Si	Oxide	Washing, drying. Chlorination	Reduction by C in arc furnace. Reduction of $SiCl_4$ by H_2
Ag	Sulphide, by-product of Cu, Pb, Zn and Au	Chloridizing roast, cyaniding. Anode slimes roasted, leached, smelted.	Cementation.
		Parkes process for concentration in Zn, dezincing, cupellation	Electrodeposition
Na	Sea water and brines	Evaporation, crystallisation	Electrolysis of fused NaCl

Metal	Principal sources	Preliminary treatments	Extractive processes
Th	Beach sands	Gravity, magnetic concentration, H_2SO_4 or NaOH digestion, conversion to oxide or halide. Conversion to nitrate, solvent extraction	Reduction of ThF_4 by Ca. Electrolysis of fused fluorides or chlorides
Sn	Oxide	Gravity, magnetic separation, flotation	Ore smelt and slag smelt, liquation, oxidation, electrolysis
Ti	Oxide, Iron titanate	Gravity concentration, chlorination. Smelting, TiO_2-rich slag, HCl leach	Reduction by Mg or Na. Ferroalloy by reduction with Si or Al
W	Complex oxides	Gravity, magnetic concentration, alkali fusion, leaching, purification, WO_3	Reduction of WO_3 by H_2, vacuum arc melting. Ferroalloy by reduction of concentrate with C
U	Oxide	Gravity concentration, leaching, solvent extraction, precipitation, calcination, solution, precipitation, UO_2, UF_4	Reduction of UF_4 by Mg
V	Complex oxides	Gravity concentration, Na_2CO_3 sintering, leaching, precipitation of V_2O_5	Reduction by Ca in bomb. Ferroalloy by reduction with C or Si (+ Fe)
Zn	Sulphides	Flotation, roasting or sintering. Roasting, leaching, purification	Reduction of ZnO by C in retort or ISC furnace. Distillation. Electrolysis of acid sulphate solution
Zr	Silicate, oxide	Gravity, magnetic concentration, smelt to carbide, chlorinate, condense $ZrCl_4$	Reduce $ZrCl_4$ by Mg in Kroll process

Chapter 6

SLAGS

INTRODUCTION

Slag is a generic term used to designate a great variety of simple or complex compounds, which may be solutions of oxides from different sources, sulphides from the charge or fuel and, in some cases, halides such as CaF_2 which are added as flux. Slag forms a separate phase from the metallic bath, due to its immiscibility and low density (3 to 4 for slags compared with 5.5 for sulphide matte and 7.8 for iron or steel).

Slags play a most important role in pyrometallurgy. They perform a wide variety of chemical and physical functions ranging from being the receptacle for gangue and unreduced oxides in primary extraction, to the reservoir of chemical reactant and absorber of extracted impurities in pyrometallurgical refining processes. The slag cover also protects the metal or matte from oxidation and diminishes heat losses. In the electric smelting furnace, the slag is frequently made use of as a heating resistor. To achieve these functions, slags must have certain physical properties (melting point, viscosity) and chemical properties (basicity, oxidation potential, thermodynamic properties), the actual values of which are controlled by variation of the composition and structure.

6.1. STRUCTURE OF SLAGS

6.1.1. STRUCTURE OF PURE OXIDES

The structures of pure oxides vary according to the respective dimensions of cations and anions and the type of bonds between them.

(A) INFLUENCE OF THE ION DIMENSION ON THE STRUCTURE

The solid oxide structure is relatively simple: metallic cations are surrounded by oxygen ions in a three-dimensional crystalline network. Pauling's first law establishes that each cation shall be surrounded by a maximum number of oxygen anions in a close-packed structure. This number is called the "coordination number" and depends only on the respective sizes and charges on the anion and cation. In Fig. 77a are shown examples of stable and unstable structures.

For a given column of Mendeleev's periodic table, the atomic radii of the elements increase from the top to the bottom, since the number of electron shells increases. For a horizontal row, the radii decrease from left to right, due to the fact that each electron of a given shell is attracted by a greater number of protons in the nucleus. In the case of ions, the radius depends not only on the radius of the corresponding atom, but also on its charge. Thus, for example, the radii of the anions are much greater than those of the cations. The ion Fe^{3+}, which is an iron atom which has lost three electrons, is smaller than the ion Fe^{2+}, which has lost two electrons. In Table 13, the radii of the most common ions are listed.

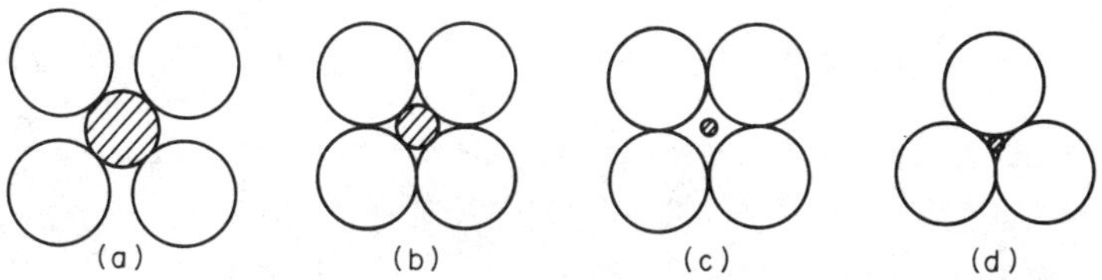

Fig. 77a. Example of structures. (a), (b), (d): Stable; (c): Unstable

TABLE 13 Radii of Cations R_c and Anions R_a

Cations	Ca^{2+}	Mn^{2+}	Fe^{2+}	Fe^{3+}	Mg^{2+}	Al^{3+}	Si^{4+}
R_c (Å)	0.93	0.80	0.75	0.60	0.65	0.50	0.41
Anions	O^{2-}	S^{2-}	F^-				
R_a (Å)	1.40	1.84	1.36				

The limiting ratio R_c/R_a for any stable structure can be determined from geometrical considerations. For an octahedral limiting structure, i.e. one in which all the ions are in contact, Fig. 77b. indicates that in the crystalline network, we have:

$$b = 2(R_c + R_a) \qquad \text{and} \qquad b\sqrt{2} = 4\ R_a$$

The limiting value of R_c/R_a will then be equal to 0.414. With a smaller ratio, i.e. with a cation of smaller size, the octahedral structure will not be stable. Table 14 indicates the possible structures with the corresponding co-ordination numbers and R_c/R_a ratios. Cations of greater radius (Ca^{2+} up to Mg^{2+}) in oxides have an octahedral structure with a co-ordination number of 6, whereas oxides of smaller cations (Si^{4+}, P^{5+}, Al^{3+}, etc.) have a tetrahedral structure.

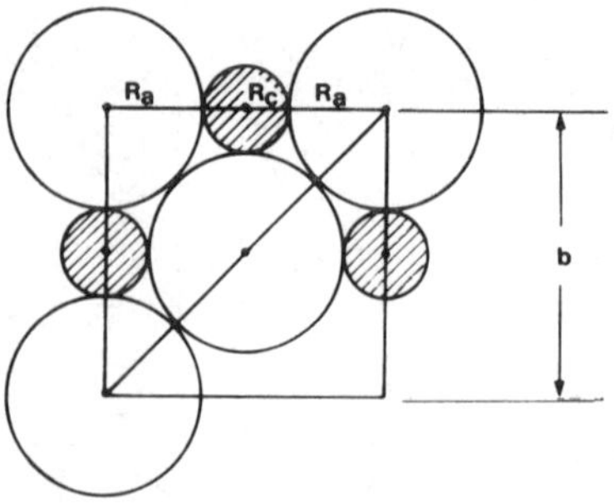

Fig. 77b. Octahedral limiting structure in which each cation is surrounded by six anions

TABLE 14 Structure of Solid Oxides

Structure	Co-ordination number	R_c/R_a	Examples
Cubic	8	1-0.732	
Octahedral	6	0.732-0.414	CaO, MgO, MnO, FeO
Tetrahedral	4	0.414-0.225	SiO_2, P_2O_5
Triangular	3	0.225-0.155	

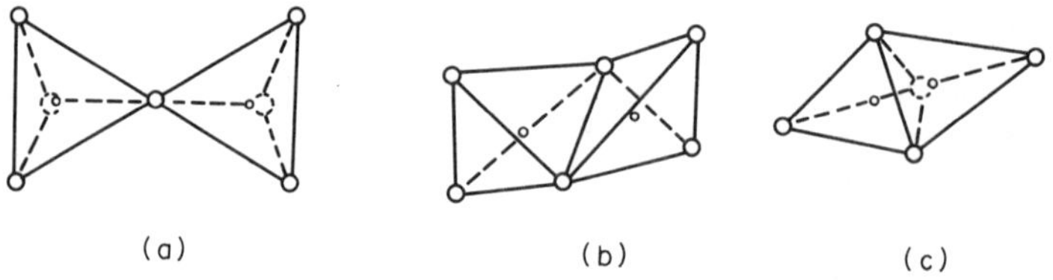

Fig. 78a. Possible methods of joining between silica tetrahedra: (a) at a vertex; (b) along a side; (c) over a face

There are three possible ways in which the elemental tetrahedra of silica could combine; at a face, a side, or a vertex (Fig. 78a). Pauling's second law establishes that the interval between two neighbouring cations must be the maximum, because they are mutually repellent. Therefore, the tetrahedra must be joined at the vertices to give the hexagonal configuration in three dimensions of Fig. 78b. In this way, each atom of silicon has bonds with four atoms of oxygen, and each atom of oxygen has bonds with two atoms of silicon, which results in a structure formula $(SiO_2)_n$, or simply SiO_2.

During melting, the crystalline network of oxides is destroyed, and in the liquid phase the bonds between ions are broken by thermal agitation. Under these circumstances, the cations are still surrounded by anions, but in a less rigid form, and the coordination number may vary. When the bonds between cations and anions are very strong, as is the case in silica, the tetrahedral crystalline network remains unaltered during melting. As the temperature is increased beyond this point, the tetrahedral joins are broken to an increasing extent. Initially, there are large ions, such as $(Si_9O_{21})^{6-}$, and a small amount of Si^{4+}. Later, the size of the complex ions decreases as the joins in the crystalline network are destroyed. For example, the $(Si_3O_9)^{8-}$ ion may be found, further increasing the number of cations (Si^{4+}). At a very high temperature, each tetrahedron would be separated from the others and there would be an equal number of anions (SiO_4^{4-}) and Si^{4+} cations. Passage from completely solid silica to a viscous liquid takes place in several stages with no well-defined fusion point; merely a gradual change from infinite viscosity to measurable values.

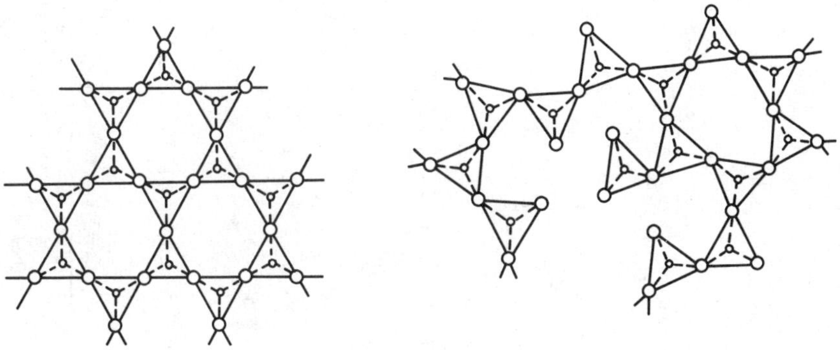

(a) (b)

Fig. 78b. Structure of silica: (a) solid; (b) molten

(B) METAL-OXYGEN BONDING FORCES

Two principal types of bonds exist in crystals: electrovalent and covalent. In electrovalent bonds, the metal atom transfers one or more electrons to the oxygen atom, transforming itself into a cation and the oxygen into an O^{2-} anion. The attraction force between anions and cations, or Coulomb's force, increases as the charges ze and $z'e$ on cations and anions increase, and as the distance between ions of opposite charge is decreased. As the charge on oxygen ions is $2e$ and the distance between ions is $R_a + R_c$, the Coulomb law can be expressed in the form:

$$F = 2ze^2/(R_c + R_a)^2$$

The elements in the first column of Mendeleev's Table, i.e. the alkali metals, have a smaller number of unit charges than the elements in the later columns, and their ionic radii are larger. The electrovalent bonding forces between these cations and oxygen anions will then be relatively small. The fact that these cations have large coordination numbers, because of their size, decreases the bond strength still further because it must be shared between a larger number of oxygen ions. In the second column of Table 15 the ratio $z/(R_c + R_a)^2$, which is proportional to the attraction forces between ions, is given for a number of oxides.

TABLE 15 Types of Bonding and Attraction Forces between Cations and O^{2-} Anion

Oxide	$z/(R_c + R_a)^2$ ($\AA^{-2}$)	Ionic fraction of bond	Coordination number Solid	Coordination number liquid	Character of the oxide
Na_2O	0.18	0.65	6	6 to 8	Network Breakers or Basic Oxides
BaO	0.27	0.65	8	8 to 12	
SrO	0.32	0.61	8		
CaO	0.35	0.61	6		
MnO	0.42	0.47	6	6 to 8	
FeO	0.44	0.38	6	6	
ZnO	0.44	0.44	6		
MgO	0.48	0.54	6		
BeO	0.69	0.44	4		
Cr_2O_3	0.72	0.41	4		Amphoteric Oxides
Fe_2O_3	0.75	0.36	4		
Al_2O_3	0.83	0.44	6	4 to 6	
TiO_2	0.93	0.41	4		
SiO_2	1.22	0.36	4	4	Network Formers or Acid Oxides
P_2O_5	1.66	0.28	4	4	

The central elements of Mendeleev's table, of intermediate size between the alkali metals and the halogens, neither lose nor gain peripheral electrons as easily as the elements at the extremities. Their bonds with oxygen are mainly covalent, both elements sharing their peripheral electrons. The binding force between atoms that form a covalent molecule is very large, and high temperatures are needed to destroy such bonds. Nevertheless, the binding forces between molecules are smaller, which explains why the purely covalent oxides (CO_2, SO_2) are gaseous.

Bonds of purely ionic or purely covalent type do not exist in the oxides in slags. There is a certain proportion of binding of both types; the ionic fraction of the bond decreases from sodium oxide to phosphorus oxide. This fraction is a measure of the tendency to dissociate into simple ions in the liquid state. In the third column of Table 15, the ionic fractions of the bonds between various cations and oxygen anions are given for the most common oxides in slags. It should be noted that this fraction decreases at almost the same rate as the Coulomb attraction between the ions increases. In the last group of oxides bonding is mainly covalent, and that part which is electrovalent is strong, because the cation is small and carries a large charge, with a coordination number of 4. For these reasons, the bonding between cations and O^{2-} anions is strong, and these simple ions group to form complex ions such as SiO_4^{4-} and PO_4^{3-}. In slags, these tend to form a stable hexagonal network (Fig. 78b). They are, therefore, called "network formers", or acids, as will be seen later in the study of basicity of slags. When members of the first group of oxides in Table 15 are heated to the melting point, or are incorporated in a liquid slag, they form simple ions such as Ca^{2+} and O^{2-}. They are called "network breakers", since they destroy the hexagonal network of silica by reacting with it. All of them are basic oxides. The following comments can be made on the values given in Table 15:

When a metal has two valencies and can therefore form two oxides, the cation of lower valency is the larger. The metal-oxygen bonding in the corresponding oxide is then more electrovalent, and therefore this oxide is more basic than that of higher valency. Thus, iron has two oxides, Fe_2O_3 (acid), and FeO (basic). Magnetite, Fe_3O_4, results from reaction between the basic oxide FeO and the acid Fe_2O_3. The increase of acidity of the oxides with the valency of the cation is equivalent to what happens in an aqueous medium: chrome III is amphoteric, whereas chrome VI is acid.

Table 15 gives a scale of attraction between the elements and oxygen. This attraction is a function of the size of the ions and their charges, and should not be mistaken for the affinity of an element for oxygen, which is a thermodynamic property of the molecule, and not of the crystal.

The ionic fraction of the metal-oxygen bonding does not indicate the dissociation percentage of oxides in those liquid solutions which are slags. In fact, silica is an ionising solvent in that it increases the ionised fraction of the oxide in solution. A similar situation occurs in aqueous solutions where a covalent compound such as gaseous hydrogen chloride is completely dissociated into ions when it dissolves. In slags, the dissociation of basic oxides into simple ions is not complete. Nevertheless, dissociation of the oxides of the first group of Table 15 is more complete than that of the later groups, and depends greatly on the ionising power of the acid solvent.

6.1.2. STRUCTURE OF SLAGS

As has been seen, most slags are silicates. If a basic oxide is added to the hexagonal network of silica, it will form two simple ions and, in the case of CaO:

$$(CaO) \rightleftarrows (Ca^{2+}) + (O^{2-})$$

it will introduce O^{2-} into the three-dimensional Si-O network, breaking it according to the following diagram:

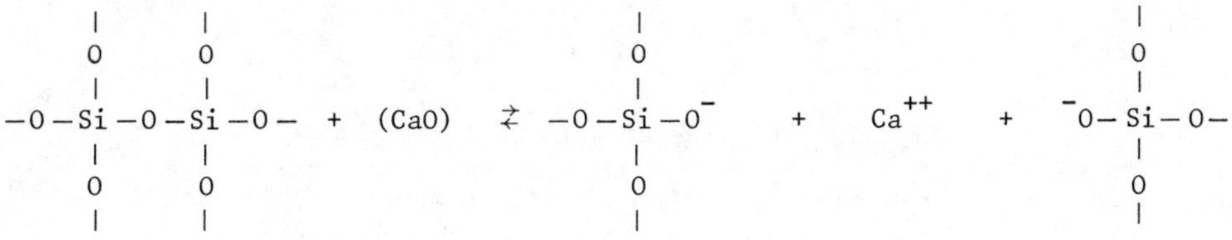

with an alkaline base:

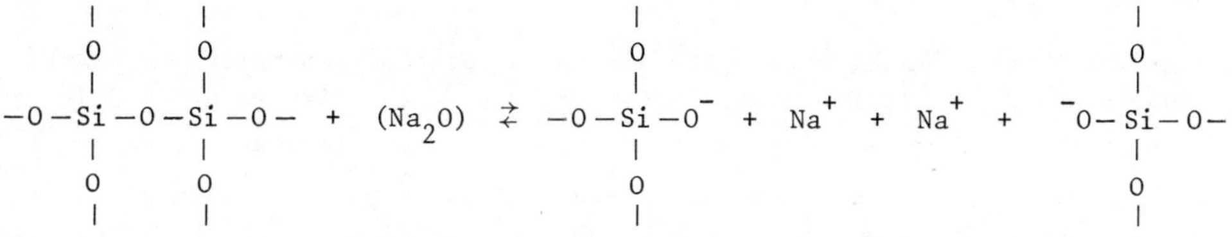

The number of joints destroyed depends on the fraction of basic oxide, i.e. the ratio O/Si, as indicated in Table 16.

TABLE 16 Influence of the Addition of a Basic Oxide on the Structure of Silicates

O/Si	Formula	Structure
2/1	SiO_2	The tetrahedra form a perfect hexagonal network.
5/2	$MO \cdot 2SiO_2$	One vertex joint breaks.
3/1	$MO \cdot SiO_2$	Two vertex joints are broken.
7/2	$3MO \cdot 2SiO_2$	Three vertex joints are broken.
4/1	$2MO \cdot SiO_2$	Four vertex joints are broken.

Different structures will be found in the solid state, depending on the ratio O/Si of the natural silicates. In kaolinite, a hydrated silicate of aluminium, $Al_2Si_2O_5(OH)_4$ (O/Si = 5/2), the elementary tetrahedra of silica are united in a plane (two-dimensional or lamellar structures). The vertices of the tetrahedra, which are not joined together, have combined with aluminium octahedra. The structures of the micas, also lamellar, explain why it is easy to separate them into layers.

With two vertices destroyed in each tetrahedron, the fibrous structure of the pyroxenes, $MgO \cdot SiO_2$, will be obtained, as shown in Fig. 79a. Asbestos is of this type.

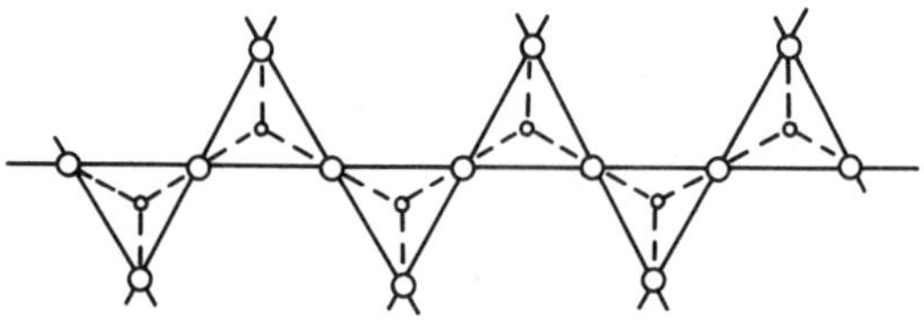

Fig. 79a. Fibrous structure of a pyroxene

Fayalite, Fe_2SiO_4 (O/Si = 4), is an iron silicate where all the vertices of the tetrahedral of the hexagonal network are affected. This compound crystallises in the form indicated in Fig. 79b. It can be observed that all the tetrahedra are separated by Fe^{2+} ions.

In contrast to silica, the melting of silicates results, generally, in a well defined fusion point, with no high viscosity range. The electrovalent bonds between simple cations and complex anions, SiO_4^{4-}, which are weaker, are the first to be broken. As the temperature increases, the solid silicate structure completely disappears (Fig. 79b).

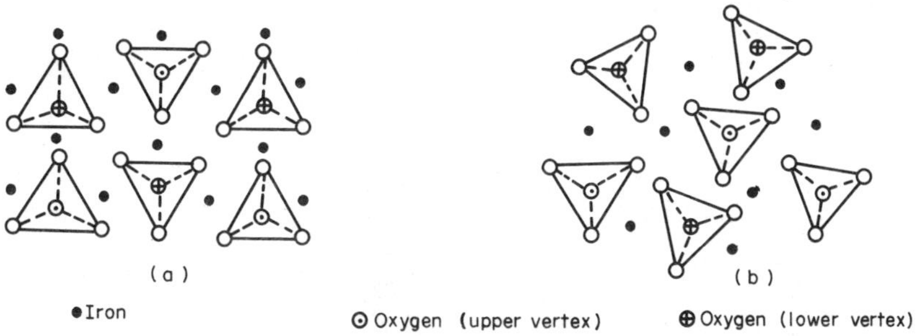

Fig. 79b. Fayalite[2] structure: (a) solid; (b) molten

6.2. PROPERTIES OF SLAGS

6.2.1. ELECTRICAL AND THERMAL CONDUCTIVITIES

Molten silica is a weak electrical conductor, but by adding a flux such as CaO or, more especially, FeO or MnO, the electrical conductivity of the slag is increased and is generally considerably higher than that of the pure oxides. This higher conductivity can be explained by the formation of ions, silica being an ionising solvent. Conductivity values can be used to measure the extent of ionisation of a slag. For iron and manganese silicates, the conductivity is too large to be explained by ionic conduction only, and it must be assumed that the conduction is partly electronic. The electrical conductivity of slags depends not only on the number of ions present, but also on their mobility, which is a function of their size and of the viscosity of the slag in which they move. The conductivity will then be greater in the liquid state, and increases with the temperature.

The thermal conductivity of slags is generally very low, but because the bulk of heat transfer through liquid slag is due to convection, the heat losses are far greater than those which could be calculated starting only with thermal conductivity data.

6.2.2. VISCOSITY OF SLAGS

The viscosity of a slag depends mainly on two factors: the composition, and temperature. When the temperature rises, the viscosity, η, of a slag of given composition decreases in an exponential form (Figs. 81 and 82), according to:

$$\eta = A \exp(E_\eta/RT)$$

A is a constant, and E_η is the activation energy of viscous flow of the slag, which depends on its composition. The decrease of viscosity of molten silica as the temperature is raised is small, an indication that the activation energy for viscous flow is large. The value of this activation energy rapidly decreases with the addition of a flux such as a basic oxide, which, as has been seen, breaks the bonds between the tetrahedra of silica. Since it is enough to cut a small number of bonds to cause a drastic decrease in the size of unit in the silica network, and consequently its viscosity, the first addition of flux to silica (under 15%) will have a far greater proportional effect in decreasing the activation energy, as is shown in Fig. 80. The effect is much larger the smaller the binding forces between cation and anion in the flux, that is, the more ionic are the bonds, the smaller the size of the ions. This explains why sodium oxide and calcium fluoride, for example, are energetic fluxes for silica and acid slag.

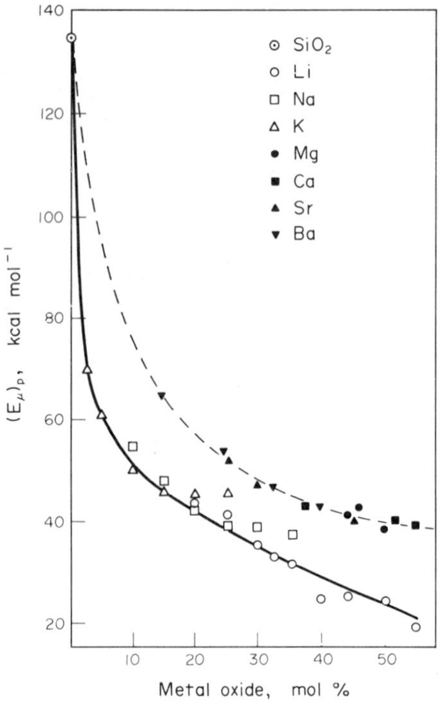

Fig. 80. Influence of the addition of flux on the flow activation energy for liquid slags[34]

(Acknowledgement to F.D. Richardson, *Physical Chemistry of Melts in Metallurgy*, Academic Press, London (1974)

In the same way that basic oxides are used as fluxes for acid slags, silica serves as flux for basic slags. In Figs. 81 and 82 it can be seen how the viscosities of acid and basic blast furnace slags of different compositions vary with the temperature. These two diagrams show the exponential decrease of viscosity with rising T. It can also be noted that the basic slags, in spite of having a much higher melting point, have viscosities which decrease rapidly at high temperatures. The activation energies for viscous flow of these slags are much smaller than for the acid slags.

The lowest viscosity is obtained with a "basicity index", i (i = (% CaO)/(% SiO_2)), of 1.35. Slags of this composition have a lower melting point than slags of any other composition, as is shown in the binary $CaO-SiO_2$ phase diagram of Fig. 86.

Calcium fluoride (CaF_2) or fluorspar serves as flux for acid and basic slags, but its effects are much more pronounced for basic slags than for acid (Figs. 83 and 84). The influence of CaF_2 is assumed to be based on the effect of F^- ions in breaking networks and on the low melting point of undissociated CaF_2.

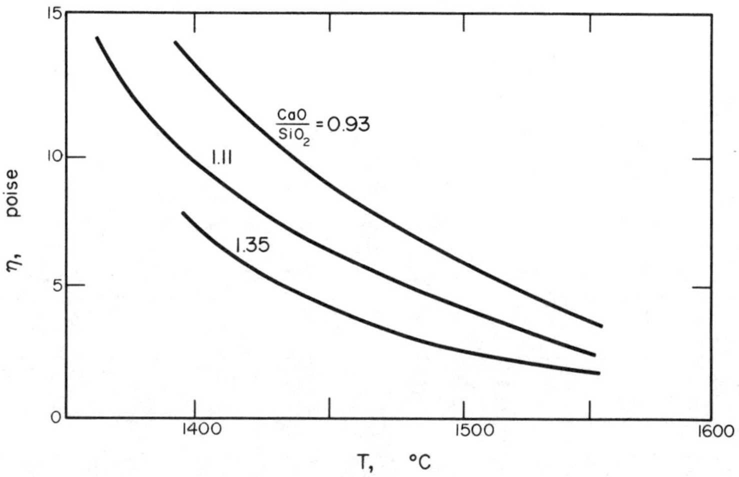

Fig. 81. Viscosity of acid blast furnace slags

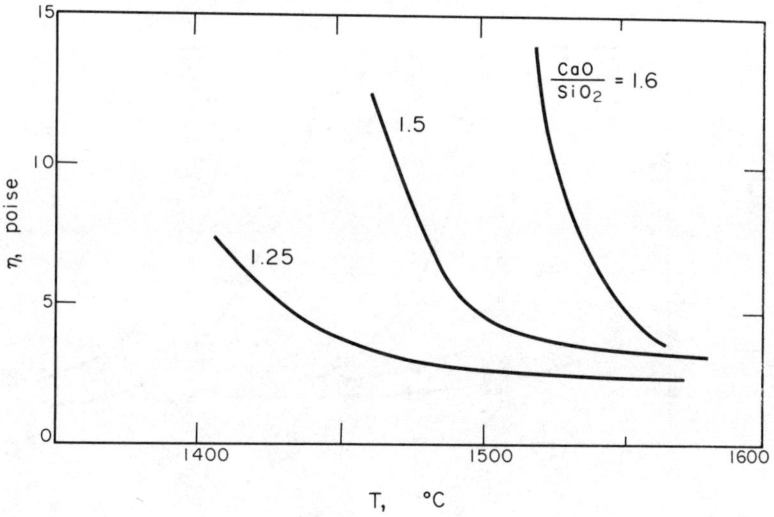

Fig. 82. Viscosity of basic blast furnace slags

Alumina fulfils the role of network breaker in acid slags, and as a network former for basic slags. In the latter case, aluminium has a coordination number of 4, and alumina tetrahedra replace those of silica. The addition of alumina, therefore, increases the viscosity of a basic slag and decreases that of an acid combination. It has less effect on weakly acid slags, as can be deduced from the viscosity curves of Fig. 85.

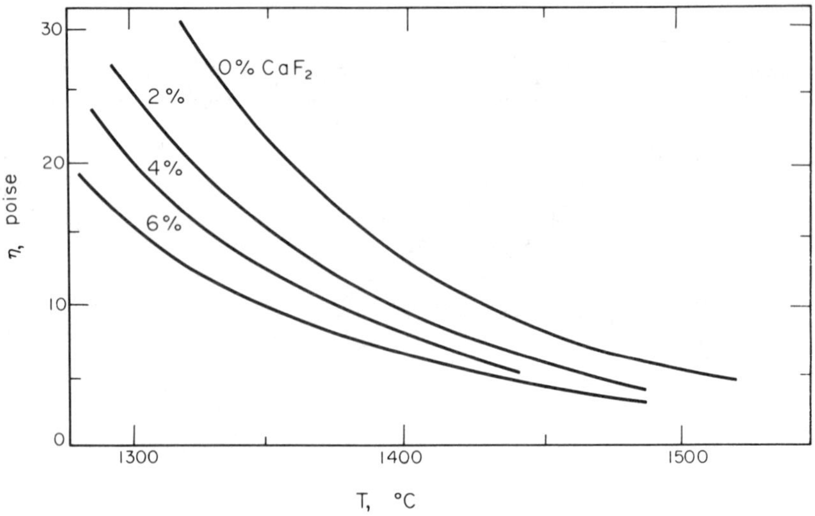

Fig. 83. Influence of the addition of CaF_2 on the viscosity of acid slags. Slag with 44% SiO_2, 12% Al_2O_3, 41% CaO and 3% MgO.(9, 10)

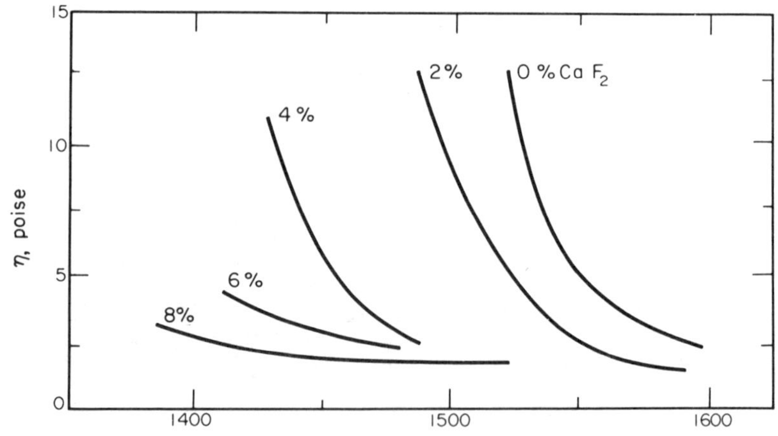

Fig. 84. Influence of the addition of CaF_2 on the viscosity of basic slags. Slag with 32% SiO_2, 13% Al_2O_3, 52% CaO and 3% MgO.(9, 10)

In industrial practice, when the viscosity of a slag decreases slowly with a rise in temperature, it is called "long" slag. The "short" slags, on the contrary, are those in which the viscosity decreases rapidly with the temperature. Acid slags are "long", and basic are "short".

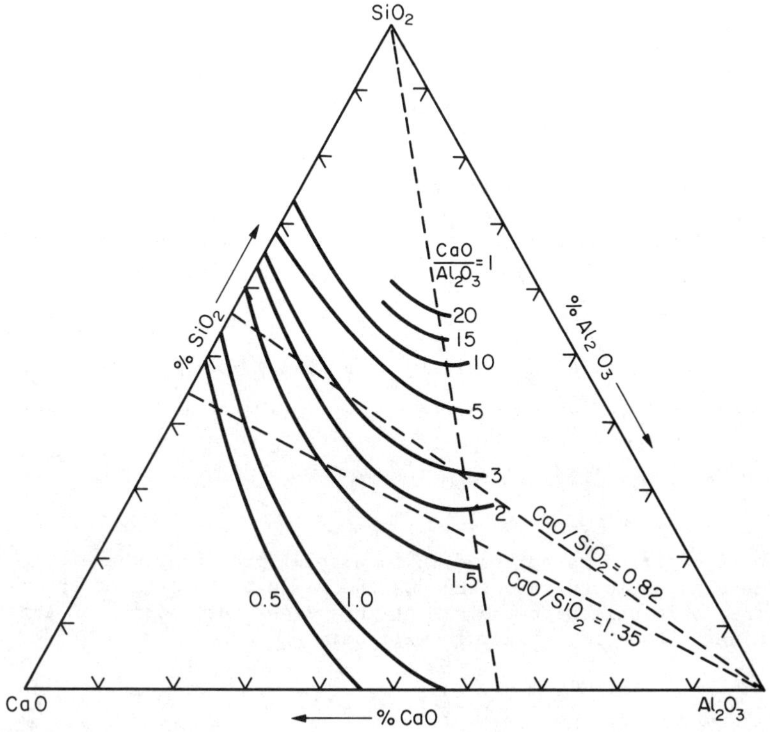

Fig. 85. Viscosity of $CaO-SiO_2-Al_2O_3$ slags at 1600°C. (Viscosity in poises; composition in weight percent)[(6-10)]

6.2.3. BASICITY OF SLAGS

compare to Lewis acids + bases

According to Bronsted's theory, an acid is a compound able to provide one or more protons, i.e. H^+, in aqueous solutions, while a base can accept this proton. In slags in which most of the constituents are oxides, an acid can accept one or several O^{2-} anions to form a complex ion, whereas a base is a source of O^{2-} ions:

$$\text{Base} \rightleftarrows \text{Acid} + O^{2-} \tag{6.1}$$

Examples of acids are the oxides P_2O_5, SiO_2, CO_2, SO_3, etc.:

$$(SiO_2) + 2(O^{2-}) \rightleftarrows (SiO_4^{4-}) \tag{6.2}$$

Bases that supply O^{2-} ions are in the first group of oxides in Table 15: CaO, Na_2O, MnO, etc.

$$(CaO) \rightleftarrows (Ca^{2+}) + (O^{2-}) \tag{6.3}$$

Finally, the amphoteric oxides behave as bases in the presence of an acid or as acids in presence of a base:

$$(Al_2O_3) + (O^{2-}) \rightleftarrows 2(AlO_2^-) \quad (\text{or} \quad Al_2O_4^{2-}) \tag{6.4}$$

$$(Al_2O_3) \rightleftarrows 2(Al^{3+}) + 3(O^{2-}) \tag{6.5}$$

A neutral slag is one which contains enough O^{2-} ions to ensure that each tetrahedron of the acid oxide is independent of the rest. In the binary system, CaO-SiO_2, neutrality will be reached at the composition 2CaO . SiO_2, i.e. with a slag that contains 33.3 mol. % of SiO_2. Accordingly, as the percentage of SiO_2 is less, equal or more than 33.3 mol. %, the slag will be basic, neutral, or acid.

Various scales exist to measure the basicity of a slag. According to the ionic theory, basicity is expressed as the excess of O^{2-} ions in 100 g of slag:

$$n_{O^{2-}} = n_{CaO} + n_{MgO} + n_{FeO} + n_{MnO} + \ldots - 2n_{SiO_2} - n_{Al_2O_3} - 3n_{P_2O_5} \tag{6.6}$$

This ionic scale of basicity is not used in industrial practice, where the "Index of Basicity" generally employed is defined as the ratio of the sum of the weight percentages of the basic oxides to the sum of the weight percentages of the acid oxides. For a binary CaO-SiO_2 slag, it will be:

$$\text{wt. \% CaO/wt. \% } SiO_2 \tag{6.7}$$

In a complex slag, the basicity index takes into account the difference in strength between the bases. For example: in a slag composed of CaO, MgO, SiO_2, and P_2O_5, which is used in the dephosphorisation of steel, magnesite is a weaker base than lime and, in this case, the ratio is taken as:

$$i = (\text{wt. \% CaO} + \frac{2}{3}\text{ wt. \% MgO})/(\text{wt. \% } SiO_2 + \text{wt. \% } P_2O_5) \tag{6.8}$$

In slags with compositions near to neutral, as generally occurs with metallurgical slags, the amphoteric oxides are not included when calculating basicity.

6.2.4. OXIDIZING AND REDUCING CAPACITY OF SLAGS

The oxidizing or reducing power of a slag refers to the capacity to participate in the transfer of oxygen to and from the metallic bath. In iron and steel making, the metallic bath contains iron as the principal component and other elements more or less noble than iron. The first group will oxidise less readily than iron during the refining of pig iron and, consequently, they will only be found in the metallic phase. The elements which are easier to oxidise than iron will be found to a large extent in the slag as oxides. However, these oxides may be so stable that they are not able to supply oxygen to the metallic bath, and the oxidising capacity of the slag then depends on the activity of the iron oxide which is present in the slag. The equilibrium between the oxygen dissolved in the metal and iron oxide in the slag can be written:

$$(FeO)_s = (\underline{Fe})_m + (\underline{O})_m \tag{6.9}$$

In steel-making the activity of the iron in a low alloy melt is assumed equal to one, and the equilibrium constant between oxygen in the metal and iron oxide in the slag will then be equal to:

$$K = \frac{a_{(O)_m}}{a_{(FeO)_s}} \tag{6.10}$$

The activity of oxygen in the metallic phase is then proportional to the activity of FeO in the slag, and the equilibrium constant is equal to the activity of oxygen in a steel in equilibrium with pure FeO: $K = a_{(O)_{sat}}$

6.3. ACTIVITIES OF SLAG COMPONENTS

The thermodynamic behaviour of slag components has been the subject of many studies, and several theories exist that try to explain such behaviour. Among them, the most important are the ionic theories and the molecular theory. Nevertheless, no theory represents the reality completely, and they are often difficult to apply. For these reasons, the activity of a component of a slag should be determined experimentally by one of the methods described in Chapter 3, and particularly by methods based on the equilibrium between metal, slag, and atmosphere.

6.3.1. IONIC THEORIES

(A) TEMKIN'S THEORY

Temkin's Theory[11] postulates that slags are solutions which are completely dissociated into ions, with no interaction between ions of the same charge, the state being that of complete randomness. Solutions of salts or oxides may then be assumed to be made up of two ideal solutions: one of cations, i^+, and the other of anions, j^-. Accepting this hypothesis, we can write:

$$a_{i^+} = N_{i^+} = \frac{n_{i^+}}{\Sigma n_{i^+}} \tag{6.11}$$

and

$$a_{j^-} = N_{j^-} = \frac{n_{j^-}}{\Sigma n_{j^-}} \tag{6.12}$$

To define the standard state of a component ij of this solution, the following equilibrium is considered:

$$ij = i^+ + j^- \tag{6.13}$$

the free energy of which is nil, and therefore: $\Delta G^o = -RT \ln K$. If the pure component in equilibrium with its ions is taken as the standard state, the standard free energy will be nil, and $K = 1$.

$$K = \frac{a_{i^+} \cdot a_{j^-}}{a_{ij}} = 1 \qquad (6.14)$$

or, according to (6.11) and (6.12):

$$a_{ij} = N_{i^+} \cdot N_{j^-} \qquad (6.15)$$

It is assumed that the latter relation holds, whatever the molar fraction of the component ij in solution.

The ions that are usually in solution in slags are Ca^{2+}, Mg^{2+}, Fe^{2+}, SiO_4^{4-}, PO_4^{3-}, $Al_2O_4^{2-}$, $Fe_2O_5^{4-}$, O^{2-}, etc. The fraction of the anions represented by O^{2-} in a slag can be calculated by means of (6.6), and the activity of an oxide by (6.15).

Example

Calculate the activity of FeO in a slag of the following molar composition:

CaO : 0.339	SiO_2 : 0.016
MgO : 0.114	Fe_2O_3 : 0.064
FeO : 0.467	

The solution, assumed perfectly ionised, will contain the cations Ca^{2+}, Mg^{2+}, and Fe^{2+}, and the anions SiO_4^{4-}, $Fe_2O_5^{4-}$ and O^{2-}. The activity of FeO, according to (6.15) is expressed by:

$$a_{FeO} = N_{Fe^{2+}} \cdot N_{O^{2-}}$$

The cationic fraction of Fe^{2+} is equal to:

$$N_{Fe^{2+}} = \frac{n_{Fe^{2+}}}{n_{Ca^{2+}} + n_{Mg^{2+}} + n_{Fe^{2+}}} = \frac{0.467}{0.920} = 0.508$$

The number of O^{2-} ions and the fraction of the anions which they represent can be calculated, starting from (6.6):

$$n_{O^{2-}} = n_{CaO} + n_{MgO} + b_{FeO} - 2n_{SiO_2} - 2n_{Fe_2O_3} = 0.760$$

and

$$N_{O^{2-}} = \frac{n_{O^{2-}}}{n_{SiO_4^{4-}} + n_{Fe_2O_5^{4-}} + n_{O^{2-}}} = 0.905$$

The activity of FeO in the slag will then be:

$$a_{FeO} = 0.508 \times 0.905 = 0.46$$

The activity coefficient of FeO in this slag ($\gamma_{FeO} = a_{FeO}/N_{FeO}$) is slightly smaller than one, which is normal for a basic slag with a low content of silica, as will be seen later (Fig. 92).

The application of Temkin's theory assumes that all the ion species in the slag are known. In a strongly basic slag such as the above example, there is no doubt that simple SiO_4^{4-} anions are formed, but it is quite possible that in an acid slag, ions $Si_2O_7^{6-}$, $Si_3O_{10}^{8-}$, etc., exist in solution. In these circumstances, application of this theory is difficult, although various attempts have been made to do so with some modifications, especially by Herasymenko and Speight.[12, 13]

(B) FLOOD'S THEORY

The theory of Flood, Forland and Grjotheim[14] is based partly on that of Temkin, and considers the equilibrium between the elements dissolved in the metallic phase and its compounds or ions in the slag. For example, the sulphur-oxygen equilibrium between metal and slag is expressed by:

$$(\underline{S})_m + (O^{2-})_s \rightleftarrows (S^{2-})_s + (\underline{O})_m \tag{6.16}$$

and the equilibrium constant of (6.16) will be equal to:

$$K = \frac{a_{(\underline{O})_m} \cdot a_{(S^{2-})_s}}{a_{(\underline{S})_m} \cdot a_{(O^{2-})_s}} \tag{6.17}$$

According to Temkin's definition, the activities of the anions S^{2-} and O^{2-} are equal to their anionic fractions $N_{(S^{2-})}$ and $N_{(O^{2-})}$. The equilibrium constant will therefore be:

$$K = \frac{(\%\ O_m) \cdot f_{(\underline{O})_m} \cdot N_{(S^{2-})_s}}{(\%\ S)_m \cdot f_{(\underline{S})_m} \cdot N_{(O^{-2})_s}} = K' \cdot g(f) \tag{6.18}$$

where $$K' = \frac{(\%\ O)_m \cdot N_{(S^{2-})_s}}{(\%\ S)_m \cdot N_{(O^{2-})_s}}$$

The equilibrium constant K' is equal to K while Henry's Law is followed, i.e. while the activity coefficients $f_{(\underline{O})_m}$ and $f_{(\underline{S})_m}$, the ratio of which is represented by $g(f)$, are one.

In a complex slag, the sulphides and oxides that can exist are those of sodium, calcium, magnesium, iron, manganese, etc., but to simplify, it will be assumed that only two oxides react: CaO and Na_2O for example. (6.16) can therefore be written:

$$(CaO,\ Na_2O)_s + (\underline{S})_m \rightleftarrows (CaS,\ Na_2S)_s + (\underline{O})_m \tag{6.19}$$

and its free energy will be:

$$\Delta G_{(Ox.\ \to\ Sulph.)} = \Delta G^o_{(Ox.\ \to\ Sulph.)} + RT\ \ln K \tag{6.20}$$

K is the equilibrium constant given by (6.17). (6.19) can be expressed in the following steps:

(i) Separation of the oxides participating in the reaction:

$$(CaO,\ Na_2O) \rightleftarrows (CaO) + (Na_2O) \tag{6.21}$$

The free energy of (6.21) is equal to $-G^M_{Ox.}$ and can be expressed as a function of the molar fraction of each oxide in solution and the partial free energies of mixing:

$$-G^M_{Ox.} = -\left(N_{CaO} \cdot \bar{G}^M_{CaO} + N_{Na_2O} \cdot \bar{G}^M_{Na_2O}\right) \tag{6.21a}$$

In this relationship, N_{CaO} and N_{Na_2O} are equal to:

$$N_{CaO} = \frac{n_{CaO}}{n_{CaO} + n_{Na_2O}} \quad \text{and} \quad N_{Na_2O} = \frac{n_{Na_2O}}{n_{CaO} + n_{Na_2O}}$$

The molar fractions are equal to the "electrically equivalent fractions" $N_{Ca^{2+}}$ and N_{Na^+}:

$$N_{Ca^{2+}} = \frac{2n_{Ca^{2+}}}{2n_{Ca^{2+}} + n_{Na^+}} \quad \text{and} \quad N_{Na^+} = \frac{n_{Na^+}}{2n_{Ca^{2+}} + n_{Na^+}}$$

$2n_{Ca^{2+}}$ and n_{Na^+} are the numbers of equivalents of the cations Ca^{2+} and Na^+ in the solution. (6.21a) then can be written:

$$-G^M_{Ox.} = -\left(N_{Ca^{2+}} \cdot \bar{G}^M_{CaO} + N_{Na^+} \cdot \bar{G}^M_{Na_2O}\right) \tag{6.21b}$$

(ii) Independent reaction of each oxide with sulphur to form sulphide:

$$\left.\begin{array}{l}(Na_2O)_s + (\underline{S})_m \rightleftarrows (Na_2S)_s + (\underline{O})_m \\ \\ (CaO)_s + (\underline{S})_m \rightleftarrows (CaS)_s + (O)_m\end{array}\right\} \qquad (6.22)$$

The free energy of transformation of N_{M^+} moles of each oxide to sulphide will be:

$$\Delta G = N_{Ca^{2+}} \cdot \Delta G_{Ca} + N_{Na^+} \cdot \Delta G_{Na} \qquad (6.22a)$$

where $\Delta G_M = \Delta G^o_M + RT \ln K_M$ and K_M is the equilibrium constant of (6.22) when the slag in equilibrium with the metal contains only one species of cation M^+.

(iii) Solution of the sulphides:

$$(CaS) + (Na_2S) \rightleftarrows (CaS, Na_2S) \qquad (6.23)$$

The free energy of mixing of sulphides is equal to:

$$G^M_{Sulph.} = N_{Ca^{2+}} \cdot \bar{G}^M_{CaS} + N_{Na^+} \cdot \bar{G}^M_{Na_2S} \qquad (6.23a)$$

The sum of the free energies of (6.21), (6.22) and (6.23) is equal to the free energy of (6.19), since the initial and final states are the same, i.e.:

$$\Delta G_{(Ox. \rightarrow Sulph.)} = N_{Ca^{2+}} \cdot \Delta G_{Ca} + N_{Na^+} \cdot \Delta G_{Na} + N_{Ca^{2+}} \left(\bar{G}^M_{CaS} - \bar{G}^M_{CaO}\right)$$

$$+ N_{Na^+} \left(\bar{G}^M_{Na_2S} - \bar{G}^M_{Na_2O}\right) \qquad (6.24)$$

It may be assumed, as a first approximation, that the partial free energies of mixing of sulphides and oxides (which are small with respect to the free energies of reaction) are similar to each other and, consequently, the two last terms of (6.24) will vanish. This equation is therefore reduced to:

$$\Delta G_{(Ox. \rightarrow Sulph.)} = N_{Ca^{2+}} \cdot \Delta G_{Ca} + N_{Na^+} \cdot \Delta G_{Na} \qquad (6.25)$$

or

$$\Delta G^o_{(Ox. \rightarrow Sulph.)} + RT \ln K = N_{Ca^{2+}} \cdot (\Delta G^o_{Ca} + RT \ln K_{Ca})$$

$$+ N_{Na^+} \cdot (\Delta G^o_{Na} + RT \ln K_{Na}) \qquad (6.26)$$

Since the standard free energy of transformation of oxide to sulphide is equal to:

$$\Delta G^o_{(Ox. \rightarrow Sulph.)} = N_{Ca^{2+}} \cdot \Delta G^o_{Ca} + N_{Na} + \Delta G^o_{Na}$$

(6.26) is simplified to give:

$$\ln K = N_{Ca^{2+}} \cdot \ln K_{Ca} + N_{Na^+} \cdot \ln K_{Na} \qquad (6.27)$$

The terms $\ln K_{Ca}$ and $\ln K_{Na}$ are constant at a given temperature and can be designated by A_{Ca} and A_{Na} respectively. On the other hand, K can be expressed as a function of K' and g(f), as indicated in (6.18). In this way, the following relationship is found between K' (apparent equilibrium constant) and the electrically equivalent fraction of basic oxide, N_{M^+}:

$$\ln K' = A_{Ca} \cdot N_{Ca^{2+}} + A_{Na} \cdot N_{Na^+} - \ln g(f) \qquad (6.28)$$

For equilibrium between i components participating in the reaction, (6.28) is still valid, and can be generalised in the form:

$$\ln K' = \sum A_i \cdot N_{i^+} - \ln g(f) \qquad (6.29)$$

In the case of complex equilibria such as those of sulphur and oxygen in metal and slag, each basic oxide in the slag (CaO, MgO, FeO, Na_2O, etc.) participates proportionately to the equivalent fraction N_{i^+} of its cation. The proportionality coefficient A_i is a function of the relative affinities of the elements for oxygen and sulphur. (6.29) can be applied for any type of complex equilibrium, and will be used in Chapter 7 to explain the partition of phosphorus and sulphur between metal and slag.

(C) MASSON'S THEORY[28, 29]

Masson's theory can be used to calculate the activity of a basic oxide in a silicate slag. According to Masson, slags are complex solutions containing polymeric silicate anions, the degree of polymerisation being controlled by the character and quantity of basic oxide present. Thus, a highly basic slag will contain the silica mainly as SiO_4^{4-}; a more acid slag will contain the ions SiO_4^{4-}, $Si_2O_7^{6-}$, $Si_nO_{10}^{8-}$, ..., $Si_nO_{3n+1}^{2(n+1)}$, all in equilibrium with each other. The individual equilibria can be written:

$$SiO_4^{4-} + SiO_4^{4-} = Si_2O_7^{6-} + O^{2-} \qquad (6.30a)$$

$$Si_2O_7^{6-} + SiO_4^{4-} = Si_3O_{10}^{8-} + O^{2-} \qquad (6.30b)$$

$$Si_3O_{10}^{8-} + SiO_4^{4-} = Si_4O_{13}^{10-} + O^{2-} \qquad (6.30c)$$

Since these reactions all consist of the addition of a silica tetrahedron to the existing chain, it can be assumed as a first approximation that the free energies are substantially the same as the equilibrium constants k_a, k_b, k_c, ..., and consequently:

$$N_{Si_2O_7}{}^{*} = \frac{kN_{SiO_4}}{N_{O^{2-}}} \cdot N_{SiO_4} \tag{6.31a}$$

$$N_{Si_3O_{10}} = \frac{kN_{Si_2O_7}}{N_{O^{2-}}} \cdot N_{SiO_4} = \left(\frac{kN_{SiO_4}}{N_{O^{2-}}}\right)^2 \cdot N_{SiO_4} \tag{6.31b}$$

$$N_{Si_4O_{13}} = \frac{kN_{Si_3O_{10}}}{N_{O^{2-}}} \cdot N_{SiO_4} = \left(\frac{kN_{SiO_4}}{N_{O^{2-}}}\right)^3 \cdot N_{SiO_4} \tag{6.31c}$$

Because the sum of the series $1 + x + x^2 + \ldots + x^n \ldots = 1/(1 - x)$ when $x < 1$, and inserting $kN_{SiO_4}/N_{O^{2-}} = x$, we obtain:

$$\sum N_{\text{silicate ions}} = N_{SiO_4} + N_{Si_2O_7} + N_{Si_3O_{10}} + \ldots + N_{Si_nO_{3n+1}}$$

$$= \frac{N_{SiO_4}}{1 - kN_{SiO_4}/N_{O^{2-}}} \tag{6.32}$$

When there are only silicate and oxygen ions as anions, Temkin's theory gives:

$$\sum N_{\text{silicate ions}} = 1 - N_{O^{2-}} \tag{6.33}$$

and N_{SiO_4} can be calculated from (6.32) and (6.33) as a function of k and $N_{O^{2-}}$:

$$N_{SiO_4} = \frac{1 - N_{O^{2-}}}{1 - k\left(1 - \frac{1}{N_{O^{2-}}}\right)} \tag{6.34}$$

The fractions of the other silicate ion polymers can be calculated from the equilibria in (6.31a, b, c).

The silica content found by chemical analysis, N_{SiO_2} can be related to $N_{O^{2-}}$ as follows:

$$N_{SiO_2} = \frac{n_{SiO_2}}{n_{SiO_2} + n_{MO(\text{free})} + n_{MO(\text{as silicate})}}$$

---

*For simplicity, the charge associated with each $Si_nO_{3n+1}^{2(n+1)}$ ion is not indicated in the expression for ion fractions.

$$= \frac{N_{SiO_4} + 2N_{Si_2O_7} + 3N_{Si_3O_{10}} + \ldots}{N_{O^{2-}} + 3N_{SiO_4} + 5N_{Si_2O_7} + 7N_{Si_3O_{10}} + \ldots} \tag{6.35}$$

By combining (6.35) with (6.33) and (6.34), and using the knowledge that the sum of the series $1 + 2x + 3x^2 + \ldots = 1/(1 - x)^2$ and $3 + 5x + 7x^2 + 9x^3 + \ldots = 3 - x/(1 - x)^2$, (6.35) can be put in the form:

$$N_{SiO_2} = 1\Bigg/\left[3 - k + \frac{N_{O^{2-}}}{1 - N_{O^{2-}}} + \frac{k(k-1)}{N_{O^{2-}}\Big/\left(1 - N_{O^{2-}}\right) + k}\right] \tag{6.36a}$$

The anion fraction represented by $N_{O^{2-}}$ can be calculated from (6.36) if k is known. With the aid of Temkin's relationship given in (6.15), $\left(N_{O^{2-}} = a_{MO}/N_{M^{2+}}\right)$, the activity of the basic oxide MO can be obtained.

Relationship (6.36a) is valid when the silicate chains are linear, but SiO_2 tetrahedra develop in three dimensions and Whiteway, Smith and Masson(29) have derived a calculation for branching chains which leads to the equation:

$$\frac{1}{N_{SiO_2}} = 2 + \frac{1 - 1\Bigg/\left(1/3 + \dfrac{N_{O^{2-}}}{k'\left(1 - N_{O^{2-}}\right)}\right)}{1 - N_{O^{2-}}} \tag{6.36b}$$

The values calculated for the activities of a number of basic oxides (CaO, FeO, MnO, etc.) in silicate slags, using these equations agree rather well with experimental results, but the theory is not applicable to all slag systems. This is because the constants for the initial polymers, Si_2O_7, Si_3O_{10}, are different from those for the more complex members. In addition, Masson's theory takes account only of linear or branching chains and, with high polymers, it is probable that cyclic forms exist. This theory, like the Temkin and Flood theories, assumes ideal behaviour for anions and cations separately, which goes a long way toward representing the real structure.

(D) TOOP AND SAMIS' THEORY(30)

According to this theory, oxygen may exist in three forms in a liquid solution of SiO_2 and basic oxide, MO: doubly bonded with Si, O^o, (≡Si-O-Si≡), singly bonded and free ions, O^- and O^{2-}, as Fincham and Richardson suggested. In highly siliceous melts, there will be more oxygen bonded with two atoms of Si than in basic melts. The equilibrium between the three forms of oxygen can be written:

$$2(\text{Si-O}^-) \rightleftharpoons (\equiv\text{Si-O-Si}\equiv) + (O^{2-}) \tag{6.37}$$

or

$$2O^- \rightleftharpoons O^o + O^{2-} \tag{6.37a}$$

and there is an apparent equilibrium constant:

$$k = \frac{n_{O^o} \times n_{O^{2-}}}{n^2_{O^-}} \tag{6.37b}$$

where n_{O^o}, n_{O^-} and $n_{O^{2-}}$ are the equilibrium numbers of moles of O^o, O^- and O^{2-} per mole of melt. It is assumed that k does not change with the molar fraction of SiO_2 and depends only on the temperature and on the cations present in the binary or tertiary melt: the value of k will increase from one basic oxide to another, as the free energy of formation of the silicate $MO\text{-}SiO_2$ becomes more negative.[31]

In a solution where the molar fractions of SiO_2 and MO are respectively N and 1 - N, each Si is bonded either with $4O^-$ ($n_{O^-} = 4N$), or with $2O^o$ ($n_{O^o} = 2N$), or with a percentage of each and the charge balance is given by:

$$2n_{O^o} + n_{O^-} = 4N \tag{6.38}$$

In the same way, each cation M^{2+} added to the melt gives either one free oxygen ($n_{O^{2-}} = 1 - N$) or two singly bonded oxygens ($n_{O^-} = 2(1 - N)$) or both and consequently:

$$n_{O^{2-}} + \tfrac{1}{2}n_{O^-} = 1 - N \tag{6.39}$$

In a binary SiO_2-MO solution, for a given oxide MO, i.e. for a given value of k, equations (6.37b) to (6.39) make it possible to calculate $n_{O^{2-}}$, n_{O^-} and n_{O^o} for various molar fractions of SiO_2. The equilibrium constant, k, may be determined for each oxide by comparing the free energy of mixing, $\Delta G_m = \tfrac{1}{2}n_{O^-}$ (RT log k), calculated from the free energy of (6.37a) with the experimental value. As Table 17 shows, k decreases when there is increased reaction between SiO_2 and MO, i.e. when a definite compound forms.

TABLE 17 Value of k for Various Silicate Systems

System	k	Number of compounds in phase diagram	Temperature of melt °C
$Cu_2O - SiO_2$	0.35	0	1100
$FeO - SiO_2$	0.17	1	1600
$ZnO - SiO_2$	0.06	1	1300
$PbO - SiO_2$	0.04	2	1100
$CaO - SiO_2$	0.0017	2	1600

The agreement between calculated and experimental curves,[30] even though k is expressed in terms of concentration instead of activity, indicates that the free energy of mixing in a binary silicate melt may arise entirely from the interaction of oxygen and silicon ions and consequently the activity of O^{2-} (or $n_{O^{2-}}$ calculated above) should therefore be equal to the activity of MO:

$$a_{MO} = a_{O^{2-}} = n_{O^{2-}} \tag{6.40a}$$

The activity of SiO_2 can be calculated from the free energy of mixing:

$$\Delta G_m/RT = (1 - N) \log a_{MO} + N \log a_{SiO_2} \tag{6.40b}$$

In ternary silicate melts, MO - M'O - SiO_2, the application of ionic Gibbs-Duhem equations allows the calculation of $a_{O^{2-}}$, $a_{M^{2+}}$ and $a_{M'^{2+}}$. Introducing the ionic definitions of the activity of MO and M'O:

$$a_{MO} = a_{M^{2+}} \cdot a_{O^{2-}} \tag{6.40c}$$

the activities of the basic oxides may be calculated. As for binary melts, the activity of SiO_2 is calculated from the free energy of mixing.

6.3.2. CELL MODEL OF KAPOOR AND FROHBERG[33]

Kapoor and Frohberg visualise that in a silicate melt the structure consists of cells formed by one oxygen and two cations. For example, in a slag of CaO, FeO and SiO_2 there is a random distribution of the following cells: Ca-O-Ca, Ca-O-Fe, Ca-O-Si, Fe-O-Fe, Fe-O-Si and Si-O-Si. Statistical thermodynamic treatment makes it possible to derive G_m and the activity of each compound from a material balance and the overall energy of the system. The energy is the sum of (a) the formation energies, E_F, of the asymmetric cells Ca-O-Fe, Fe-O-Si, and Ca-O-Si from the symmetric cells Ca-O-Ca, Si-O-Si and Fe-O-Fe according to:

$$\text{Ca-O-Ca} + \text{Si-O-Si} = 2(\text{Ca-O-Si})$$

which relates to the anion-cation and cation-cation interactions, and (b) the interaction energies E_I between cells. E_F and E_I can be determined from the thermodynamic properties of each oxide in the corresponding binary systems CaO-FeO, CaO-SiO_2 and FeO-SiO_2. This model appears to be capable of more general application than those based on ions and there is good agreement between the results of calculation and experiment. However, the calculations are complex enough to require a computer.

6.3.3. MOLECULAR THEORY

The molecular theory, proposed by Schenck[15] in 1934, assumes ideal behaviour by all of the molecules existing in slags. The simple oxides (CaO, MgO, FeO, Al_2O_3,

MnO, Fe_2O_3, SiO_2, etc.) associate to form complex molecules such as $CaAl_2O_4$, $Ca_4P_2O_9$, etc., or remain as non-combined or free compounds. Each oxide will then exist in different forms which are in equilibrium and depend on the relative content of the other oxides. The problem is complex, and to solve it the following method may be used. All of the possible molecules that can be formed are considered. For a ternary slag which contains FeO, CaO, and SiO_2, the molecules that can exist in solution will be mainly FeO, CaO, Ca_2SiO_4, $CaSiO_3$, $FeSiO_3$, Fe_2SiO_4, $Ca_2Si_2O_6$, and $Ca_4Si_2O_8$. To determine the mole fraction of each, eight equations are needed. Three are obtained from the mass balance (one for each oxide). For calcium oxide, the equation of mass balance is expressed in the form:

$$n_{CaO_{free}} + n_{CaSiO_3} + 2n_{Ca_2SiO_4} + 2n_{Ca_2Si_2O_6} + 4n_{Ca_4Si_2O_8} = n^*_{CaO} \quad (6.41a)$$

The total numbers of moles of calcium oxide n^*_{CaO}, silica $n^*_{SiO_2}$, and iron oxide n^*_{FeO} are given by chemical analysis of the slag.

Another five can be deduced from the equilibrium constants of the type:

$$(Ca_4Si_2O_8) \rightleftarrows (Ca_2Si_2O_6) + 2(CaO) \qquad K_1 = N^2_{CaO} \cdot N_{Ca_2Si_2O_6}/N_{Ca_4Si_2O_8} \quad (6.41b)$$

and

$$(Ca_4Si_2O_8) \rightleftarrows 2(Ca_2Si_2O_4) \qquad K_2 = N^2_{Ca_2SiO_4}/N_{Ca_4Si_2O_8} \quad (6.41c)$$

By solving these eight equations, it is possible to determine the mole fraction of the actual species which will exist in the slag and that exhibit ideal behaviour. It is evident that in a metal-slag equilibrium, only the free species are capable of participating in reactions.

Winkler and Chipman[16] applied a similar method to determine the activity of lime and other components of a complex slag in equilibrium with a steel that contained phosphorus (Problem No. 21). This method can be useful in particular cases. Nevertheless, the assumption that slags show ideal behaviour is not always valid. On the other hand, the actual molecules that are formed and the equilibrium constants which relate the mole fractions of these molecules are not often known with any degree of accuracy.

6.3.4. EXPERIMENTAL RESULTS

As a result of the difficulties experienced in establishing a satisfactory theory from which to calculate the activities of the components of liquid slags, it is necessary to determine the values by experiment. The extensive work carried out for this reason has produced results which are of interest to geologists and ceramic scientists, as well as to metallurgists.

Measurements of metal-slag-atmosphere equilibria constitute the most widely used method for slag activity determination (see Chapter 3). This method cannot be applied when the oxide MO in the slag in equilibrium with the element M in the metal is too stable. In this case, as the equilibrium is displaced towards the formation of the oxide, it is not possible to measure the activity of the element in the metal and, consequently, determine that of the oxide in the slag.

Determination of the activities of such oxides in binary and ternary systems is carried out by means of integration of the Gibbs-Duhem equation. The results can be checked along the liquidus line of the diagram under consideration, where the liquid is in equilibrium with a pure solid compound for which the free energy of formation is known. For example, in the ternary diagram $CaO-SiO_2-Al_2O_3$ of Fig. 88, at a point on the liquidus line where there is equilibrium between liquid and solid $2CaO \cdot SiO_2$, we have:

$$2(\underline{CaO})_s + (\underline{SiO_2})_s \rightleftarrows <2CaO \cdot SiO_2> \qquad (6.42)$$

The standard free energy of formation of the solid compound given in Table 18 as a function of temperature is expressed as a function of the activities of the reactants and product referred to the solid by:

$$\Delta G^o = -RT \ln \left[\frac{a_{2CaO \cdot SiO_2}}{a^2_{CaO} \cdot a_{SiO_2}} \right] \qquad (6.43)$$

The activity of the compound is unity, since it forms a separate phase, and the activity of silica at the specified point can be determined by metal-slag equilibrium. From (6.43) the activity of calcium oxide at this point of the liquidus can be calculated.

A simpler method of expressing the experimental results is to draw the curves of the activity of each component as a function of its mole fraction or weight percent. Although this method is feasible in the case of binary and ternary slags, it cannot be applied for complex slags. The difficulty may be lessened by grouping oxides of similar characteristics in a diagram described as "pseudo-ternary". For example, in a complex slag which contains SiO_2, P_2O_5, CaO, MgO, MnO, and FeO, the pseudo-ternary diagram (SiO_2, P_2O_5)-(CaO, MgO, MnO)-FeO, can be constructed.

The binary and ternary diagrams most widely used in pyrometallurgy are examined below.

(A) BINARY DIAGRAMS

The most interesting binary systems in iron and steelmaking are those formed by silica and a basic oxide. These can be divided into two groups, according to whether they indicate a miscibility gap in the liquid state. As is shown in Fig. 86(a), at the silica-rich end, the systems SiO_2-alkaline oxides and SiO_2-BaO indicate complete solubility in the liquid state and complete insolubility in the solid state. In the systems SiO_2-MgO, SiO_2-FeO, etc., there is a zone of immiscibility in the liquid state, and two liquids are formed at temperatures above the monotectic for certain concentrations of silica. Immiscibility is complete in the solid state.

A binary system with two phases is monovariant and, at a given temperature, the activities of the components of a liquid in equilibrium with a solid have constant values over the whole of the domain where the two phases coexist. In activity diagrams, this is represented by a horizontal straight line in the two-phase domain (Fig. 86(b), (c) and (d)). In a domain where there is only one phase, the variance is two, and at a given temperature the activity of the components will vary continuously with the composition of the solution.

TABLE 18 Standard Free Energy of Formation of Certain Compounds[22]
(C = CaO, S = SiO_2, M = MgO, A = Al_2O_3)

Compound		ΔG^o of Formation (1700 - 1900°C)	$\frac{CaO + MgO}{SiO_2}$	T_f (°C)
CS	Pseudo-wollastonite	-19,900 - 0·7 T	0·93	1530
CS	Liquid	- 5,200 - 8·8 T	0·93	-
C_2S	Dicalcium silicate	-24,400 - 5·8 T	1·87	1700
C_3A		- 5,200 - 7·0 T	-	-
CA		- 4,600 - 4·2 T	-	-
CA	Liquid	+14,200 - 14·2 T	-	-
CA_2		- 4,000 - 6·1 T	-	-
S_2A_3	Mullite	- 1,040 - 2·5 T	-	-
CAS_2	Anorthite	-18,400 - 7·9 T	0·47	1540
CAS_2	Liquid	- 1,000 - 17·6 T	-	-
C_2AS	Gehlenite	-27,800 - 9·3 T	1·87	1580
MS	Clinoenstatite	- 9,100 + 1·0 T	-	-
MS	Liquid	+ 5,700 - 7·1 T	-	-
M_2S	Forsterite	-13,600 - 0·8 T	-	-
MA	Spinel	- 4,500 - 1·5 T	-	-
CMS_2	Liquid	-19,200 - 12·4 T	-	-
C_2MS_2	Akermanite	-44,500 + 0·4 T	1·27	1370
C_2MS_2	Liquid	-17,600 - 15·2 T	1·27	1370

In the SiO_2-CaO diagram (Fig. 86(b)), the activity of silica in the liquid in equilibrium with cristobalite at 1600°C, is equal to one, whereas that of the lime is very low. The activity of silica in the liquid decreases rapidly as CaO is added, until it reaches the domain of equilibrium between the liquid and solid, $2CaO \cdot SiO_2$, where it is of the order of 10^{-2}. The activity of CaO in this domain can be calculated in the manner previously described.

In the SiO_2-MnO diagram (Fig. 86(c)), the compounds $MnO \cdot SiO_2$ and $2MnO \cdot SiO_2$ decompose before they reach their melting point. As in the previous case, the activities remain constant as long as equilibrium exists between the two phases, but change rapidly with the composition of the liquid in the single phase domain.

The only compound in the system FeO-SiO_2 is fayalite, $2FeO \cdot SiO_2$, which has a low melting point. The activity of FeO is ideal in the liquid phase (Fig. 86(d)).

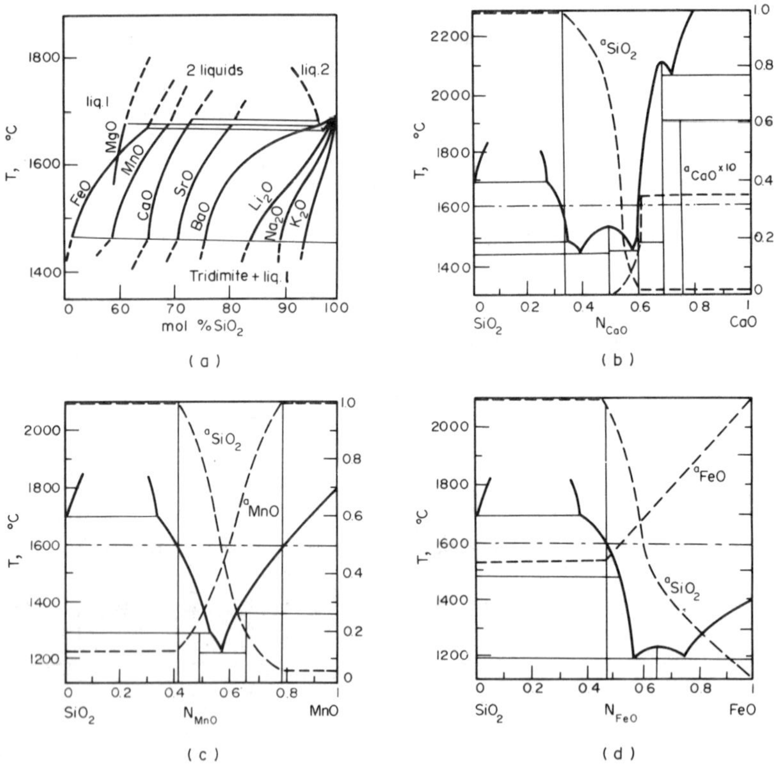

Fig. 86. Binary phase diagrams, SiO_2 - basic oxides.
Activities of the components at 1600°C.(17, 18, 19)

(B) TERNARY DIAGRAMS

The ternary systems SiO_2-CaO-Al_2O_3 and SiO_2-CaO-FeO are of particular interest. The first includes iron blast furnace slags, and the second the slags associated with the refining processes.

(i) CaO-SiO_2-Al_2O_3

The phase diagram for the CaO-SiO_2-Al_2O_3 system (Fig. 87) indicates the existence of several binary compounds, some stable at the melting point (CaO . SiO_2, 2CaO . SiO_2, 5CaO . $3Al_2O_3$), and others unstable (3CaO . $2SiO_2$, 3CaO . SiO_2, 3CaO . Al_2O_3 and $3Al_2O_3$. $2SiO_2$), as well as two ternary compounds which are stable at their melting points (CaO . Al_2O_3 . $2SiO_2$ and 2CaO . Al_2O_3 . SiO_2).

Two zones are of special importance in ironmaking (shaded in Fig. 87); the first represents the composition of basic blast furnace slags, and the second the composition of acid slags. These have been selected for having a very low melting point, generally lower than 1400°C.

The isothermal section of this diagram at 1600°C (Fig. 88) shows a liquid domain and several domains where liquid and solid phases coexist in equilibrium. The solid phases are either a pure oxide or a binary oxide compound. Rein and Chipman[21, 22] determined the activity of silica in CaO-SiO_2-Al_2O_3 slags by equilibrium with a metallic phase, of carbon-saturated iron with silicon in solution (Fig. 88(a)). By integration of the Gibbs-Duhem law (Schuhmann's method[23]), the activities of lime and alumina were calculated in all of the liquid domains (Fig. 88(b)). The values so obtained can be checked along the liquidus lines where two phases are in equilibrium.

(ii) CaO-FeO-SiO_2

This system is the basis of a large number of industrial slags, including those involved in acid and basic steelmaking, nickel and copper matte smelting, and the production of black or blister copper, lead, and tin. Minor quantities of Al_2O_3, MgO, and other flux impurities may make up 10% of the slag weight. The combinations of oxidising power, acidity or basicity, and free flowing temperature most suited to production of the metal involved, are found in the areas outlined in Fig. 89.

There are several binary compounds which are stable at their melting points (CaO . SiO_2, 2CaO . SiO_2, 2FeO . SiO_2), two which are unstable (3CaO . $2SiO_2$, 3CaO . SiO_2) and two ternary compounds (CaO . 2FeO . $2SiO_2$ and CaO . FeO . SiO_2).

In Fig. 90[22] the activity values for FeO in the pseudo ternary FeO-SiO_2-(CaO + MgO) system are indicated. They have been determined from equilibria between liquid iron that contains oxygen and the slag. By integrating the Gibbs-Duhem equation, it is possible to calculate the activities of silica and lime[25] (Fig. 91).

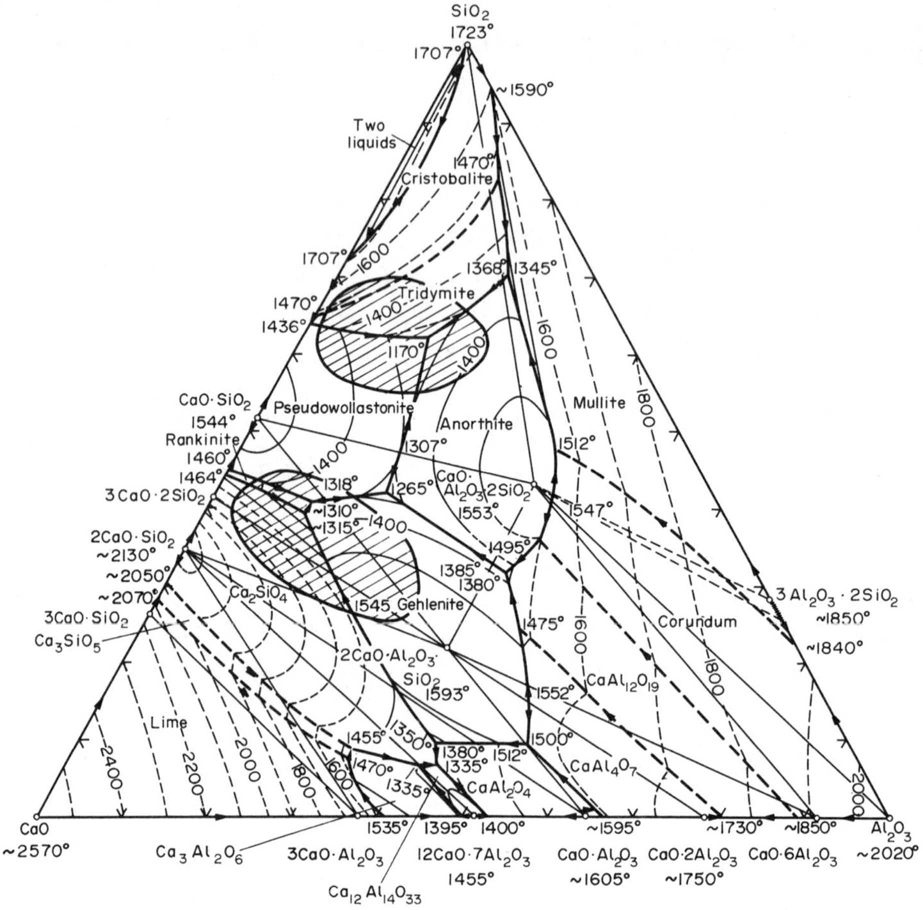

Cristobalite Tridymite	SiO_2
Pseudowollastonite	$CaO \cdot SiO_2$
Rankinite	$3CaO \cdot 2SiO_2$
Lime	CaO
Corundum	Al_2O_3
Mullite	$3Al_2O_3 \cdot 2SiO_2$
Anorthite	$CaO \cdot Al_2O_3 \cdot 2SiO_2$
Gehlenite	$2CaO \cdot Al_2O_3 \cdot SiO_2$

Fig. 87. $CaO-Al_2O_3-SiO_2$ phase diagram.[20]

The activity of FeO in the pseudobinary sections is given in Fig. 92. As has been seen, the behaviour of FeO is ideal in the liquid phase of the $FeO-SiO_2$ system (curve a of Fig. 92). In the CaO-FeO system (curve b), the FeO has an activity coefficient of less than one in the liquid phase. Fetters and Chipman[27] explain this behaviour by taking into consideration that calcium

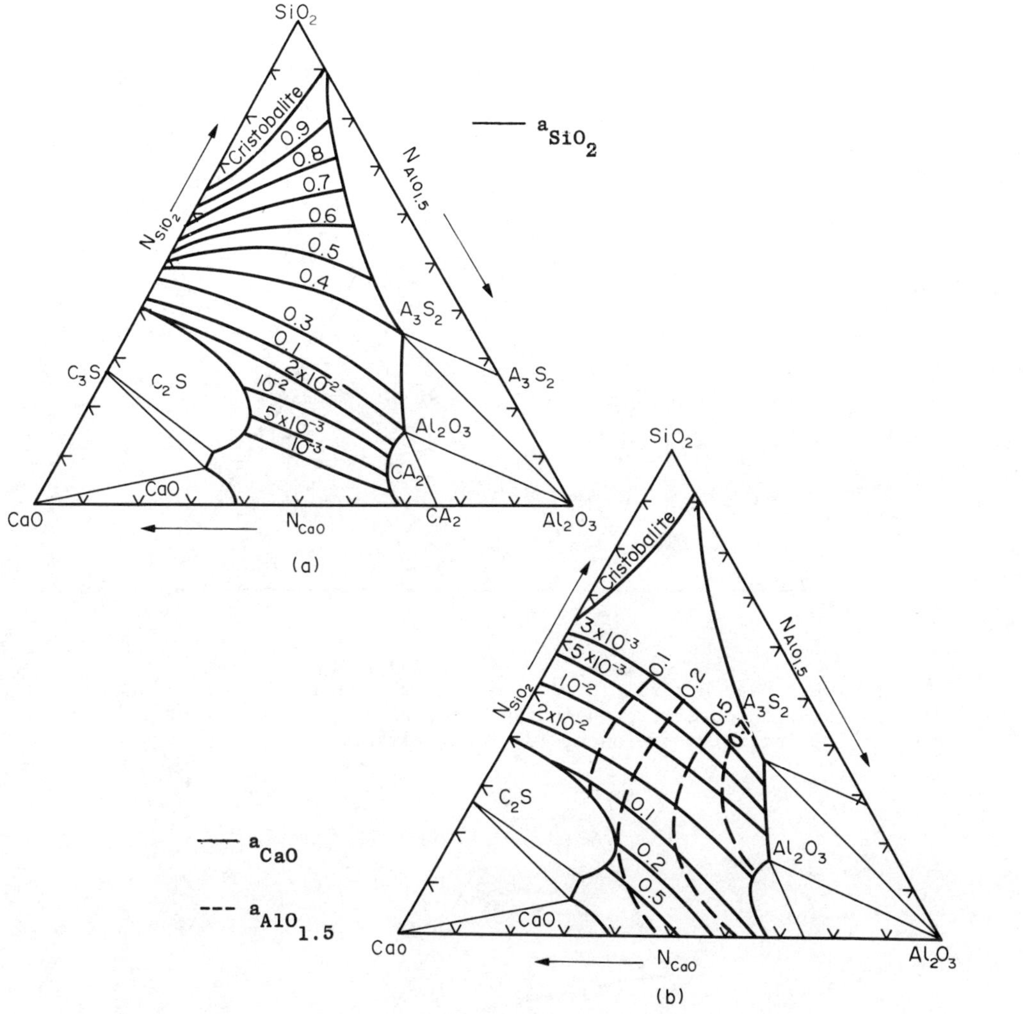

Fig. 88. (a) Activity values for SiO_2 in the $CaO\text{-}SiO_2\text{-}Al_2O_3$ system at 1600°C[22]
(b) Activity values for CaO and Al_2O_3 in the $CaO\text{-}SiO_2Al_2O_3$ system at 1600°C.

ferrite ($CaFe_2O_4$ or CaO . Fe_2O_3) in which iron has an oxidation number of 3^+, is formed in basic slags. Taking into account the formation of this complex molecule, the behaviour of FeO would be ideal according to the molecular theory. The deviation from the ideal condition in the pseudo-binaries CaO . SiO_2-FeO (curve c), 2CaO . SiO_2-FeO (curve d), and 3CaO . SiO_2-FeO (curve e), is positive, and maximum for the pseudobinary 2CaO . SiO_2-FeO. The association between molecules of CaO and SiO_2 is so strong that those of iron oxide cannot form associated molecules, and they are repulsed, which confers on the iron oxide a higher activity than its mole fraction. Nevertheless, Taylor and Chipman[26]

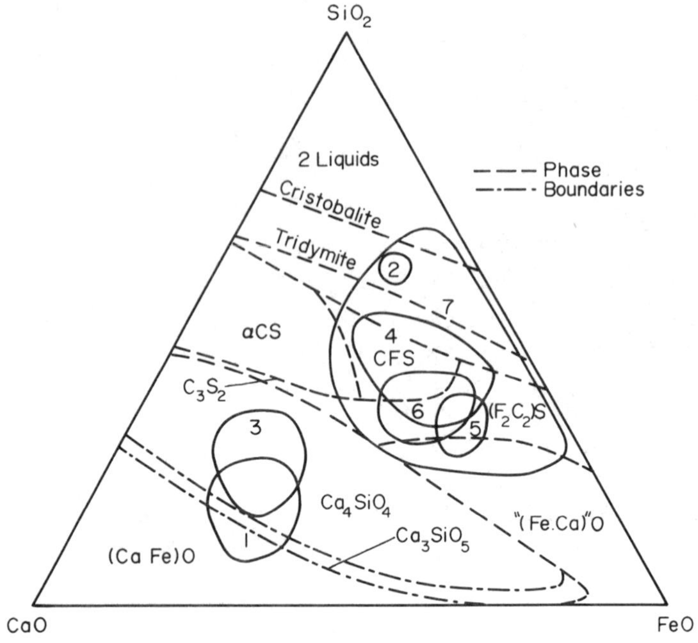

Fig. 89. SiO_2-FeO-CaO phase diagram.(24)
(1) Basic open hearth steel furnace; (2) Acid open hearth steel furnace; (3) Basic oxygen converter; (4) Copper reverberatory; (5) Copper oxide blast furnace; (6) Lead blast furnace; (7) Tin smelting.

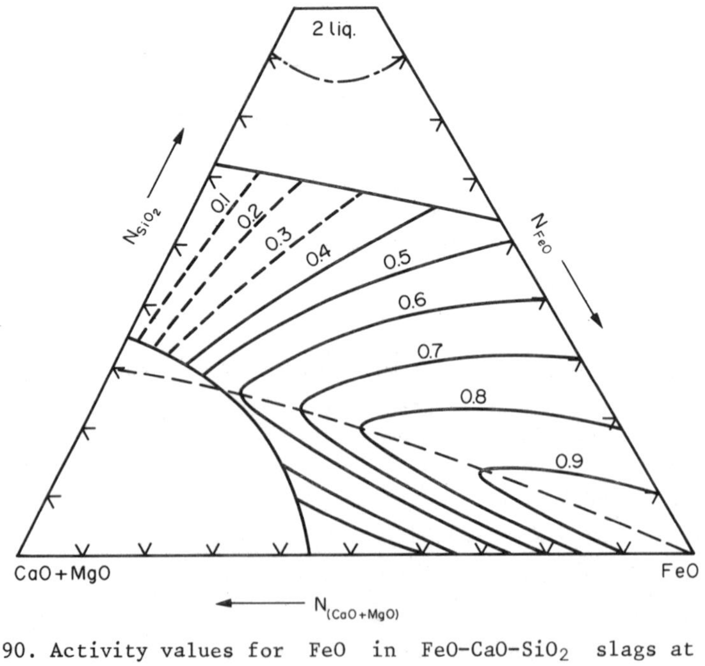

Fig. 90. Activity values for FeO in FeO-CaO-SiO_2 slags at 1600°C.(22)

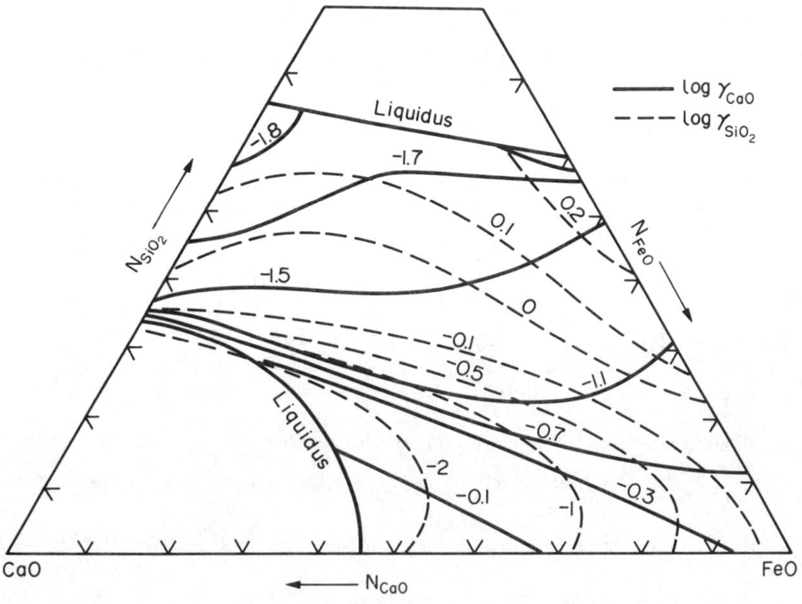

Fig. 91. Activity coefficient values for SiO_2 and CaO in SiO_2-FeO-CaO slags at 1600°C.(25)

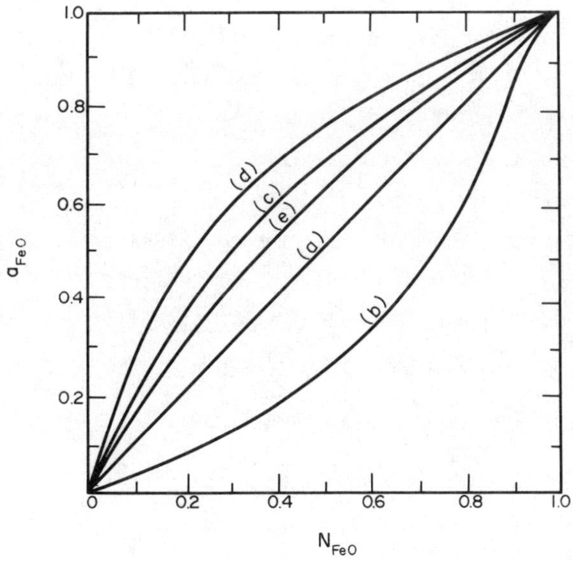

Fig. 92. Activity values for FeO in the binary or pseudo-binary systems in CaO-SiO_2-FeO slags at 1600°C.(27)
(a) SiO_2-FeO; (b) CaO-FeO; (c) CaO . SiO_2-FeO;
(d) 2CaO . SiO_2-FeO; (e) 3CaO . SiO_2-FeO.

have shown that if, instead of assuming the formation of a molecule of orthosilicate $2CaO\text{-}SiO_2$, the double molecule is postulated $(2CaO \cdot SiO_2)_2$, the behaviour of the iron oxide in the pseudobinary $2CaO \cdot SiO_2\text{-}FeO$ is ideal. The increased activity of FeO is now explained on the basis of the ionic theory, and the difference in cation-oxygen attraction between FeO and CaO.

REFERENCES

1. L. Pauling, *The Nature of the Chemical Bond*, 2nd Ed., Cornell University Press, N.Y. (1948).
2. F.D. Richardson, *J. Soc. Chem. Ind.* 71, 50 (1952).
3. B.G. Baldwin, *J. Iron & Steel Inst.* 186, 388 (1957).
4. J.S. Machin, T.B. Yee and D.L. Hanna, *J. Am. Ceram. Soc.* 51, 569 (1954).
5. J.O. M'Bockris and D.C. Lowe, *Proc. Royal Soc.* A226, 223-235 (1954).
6. J.S. Machin and T.B. Yee, *J. Am. Ceram. Soc.* 31, 203 (1948).
7. J.S. Machin and D.L. Hanna, *J. Am. Ceram. Soc.* 28, 310-316 (1954).
8. J.S. Machin, T.B. Yee and D.L. Hanna, *J. Am. Ceram. Soc.* 35, 323 (1952).
9. P. Kozakevitch, *Rev. Metallurgie*, 51, 569-587 (1954).
10. P. Kozakevitch, *Physical Chemistry of Process Metallurgy*, pp. 97, 116, Interscience Publ., N.Y. (1961).
11. M. Temkin, *Acta Physicochimika, URSS*, 20, 211 (1945).
12. P. Herasymenko and G.E. Speight, *J. Iron and Steel Inst.* 166, 169-183 (1950).
13. Ibid., p. 289.
14. H. Flood and K. Grjotheim, *J. Iron and Steel Inst.* 171, 64-70 (1952).
15. H. Schenck, *Physico-Chemistry of Steel Making*, BISRA Transl., p. 455, London (1945).
16. T.B. Winkler and J. Chipman, *Trans. A.I.M.E.* 167, 111-133 (1946).
17. N.L. Bowen and J.F. Schairer, *Am. J. Sci.* 5th Ser., 24, 200 (1932).
18. F.L. Glaiser, *Am. J. Sci.* 256, No. 6, 405 (1958).
19. B. Phillips and A. Muan, *J. Am. Ceram. Soc.* 42, No. 9, 414 (1959).
20. E.F. Osborn and A. Muan, *Phase Equilibrium Diagrams of Oxide Systems.* Plate 1, Am. Ceram. Soc. (1960).
21. R.H. Rein and J. Chipman, *Trans. A.I.M.E.* 227, 1193 (1963).
22. R.H. Rein and J. Chipman, *Trans. A.I.M.E.* 233, 415 (1965).
23. R. Schumann, *Acta Met.* 3 (May 1955).

24. E.F. Osborn and A. Muan, Op. cit., Plate 7 (1960).

25. J.F. Elliott, *Trans. A.I.M.E.* 203, 485 (1955).

26. C.R. Taylor and J. Chipman, *Trans. A.I.M.E.* 154, 228-247 (1943).

27. K.L. Fetters and J. Chipman, *Trans. A.I.M.E.* 145, 95-112 (1941).

28. C.R. Masson, *Proc. Roy. Soc.* A287, 201-221 (1965).

29. S.G. Whiteway, I.B. Smith and C.R. Masson, *Can. J. Chem.* 48, 33-45 (1970).

30. G.W. Toop and C.S. Samis, *Trans. Met. Soc. A.I.M.E.* 224, 878 (1962).

31. C.J.B. Fincham and F.D. Richardson, *Proc. Roy. Soc.* A223, 40 (1954).

32. G.N. Lewis and M. Randall, *Thermodynamics*, p. 326, McGraw-Hill Book Co., New York (1923).

33. M.L. Kapoor and M.G. Frohberg, *Chem. Met. of Iron & Steel*, Symposium, Iron & Steel Inst., Sheffield (1971).

34. J.D. Mackenzie, *Adv. Inorg. Chem., Radiochemistry*, 4, 293 (1962).

GENERAL READING

Bodsworth, C., *Physical Chem. of Iron and Steel Making*, Longmans, London (1963).

Hopkins, D.W., *Physical Chemistry and Metal Extraction*, I. Garnet Miller Ltd., London (1954).

Kozakevitch, P., *Physical Chemistry of Process Metallurgy*, Interscience Publ., N.Y. (1961).

Levin, E.M., C.R. Robbins and H.F. McMurdie, *Phase Diagrams for Ceramists*, Am. Ceramic Soc. (1964).

Mudd Series, S.W., *Basic Open Hearth Steelmaking*, A.I.M.E. Soc., N.Y. (1964).

Richardson, F.D., *Physical Chemistry of Melts in Metallurgy*, Vols. 1 and 2, Academic Press, London (1974).

St. Pierre, G.R., (Ed.) *Physical Chemistry of Proc. Met.*, Part I, Metallurgical Soc. Conferences of A.I.M.E., Interscience Publ., N.Y. (1959).

Ward, R.G., *An Introduction to the Phys. Chem. of Iron and Steel Making*, E. Arnold Ltd., London (1962).

Chapter 7

IRON AND STEELMAKING

INTRODUCTION

The extraction of iron from ore and its conversion into mainly iron-carbon alloys constitutes the largest use of metallurgical extraction and refining processes. This derives from a combination of abundant mineral supply, relatively low processing costs, and the wide range of useful forms and mechanical properties exhibited by the products. These extend from the cast products containing over 1.7% carbon to the heavily worked packaging sheet in which the carbon content is below 0.05%. The versatility can be increased even further by the addition of such alloying elements as nickel, chromium, manganese and silicon.

More than 95% of the ore treated is reduced to metal in the blast furnace, but the increasing cost and diminishing availability of coal suitable for coke production has stimulated the examination of alternative processes. There has been greater recognition of the disadvantages of blast furnace iron in terms of the substantial quantities of carbon, silicon, and phosphorus which have to be removed in any refining process.

The use of electric power for ore smelting has been established in Norway and Sweden for a long period, and is obviously attractive in any other country where there is an abundance of hydro-electric power, especially if it is accompanied by the absence of indigenous coal supplies. Natural gas has also been utilised for "direct reduction"* as in the H y L process in Mexico. The Wiberg process uses electric power and coke for production of the reducing gases. The Krupp-Renn, SLRN and several other processes use low grade coal as a reducing agent in kilns and other vessels. The product of these "direct reduction" processes may be fed into an electric furnace for the production of steel, or used to increase the throughput of a conventional blast furnace. "Metallised" iron ore pellets, in which about 92% of the oxide has been reduced to metal at the ore field, offer considerable economies in transport and subsequent processing.

---

*In this context "direct reduction" describes the reduction of iron oxide to metal without melting. It must be distinguished from the reaction between carbon and ferrous oxide in the lower part of the iron blast furnace which is described as "direct reduction" in order to draw attention to the difference from the "indirect reduction" of iron oxides by CO in the shaft of the furnace.

In this chapter, after a brief description of the production of iron in the blast furnace and the processes available for the production of steel, the principal emphasis is on the equilibria between metal slag and atmosphere, which control the composition of the blast furnace iron and the steel produced. Figure 93a has been included to indicate the relative importance of the various routes for the conversion of iron ore to steel.

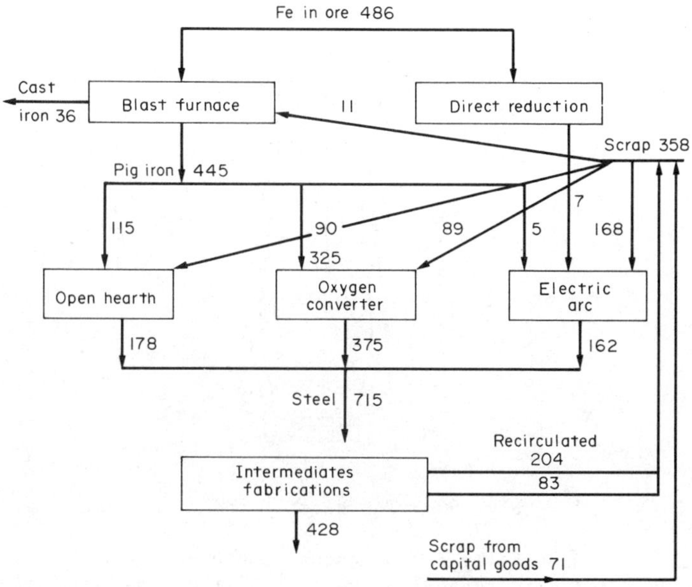

Fig. 93a. World balance in iron and steel (in Mt Fe).[53]

7.1. THE IRON BLAST FURNACE, STEELMAKING PROCESSES

Using this method, steel is produced from iron ore in two stages. In the blast furnace, iron oxide is reduced by carbon monoxide and carbon to give a metal saturated with carbon and containing significant percentages of silicon, manganese and phosphorus. The gangue is fluxed by the limestone added in the charge, and most of the sulphur in the feed material is removed in the slag. In the steel furnace, most of the silicon, sulphur and phosphorus are eliminated and the carbon and manganese are controlled to specified levels. Alloying elements which confer special properties, are added at this stage.

7.1.1. THE IRON BLAST FURNACE

The structure, which may be over 50 m high and up to 14 m diameter in the hearth, must be of such a form that up to 10,000 tonnes of pig iron, 3000 tonnes of slag, and 1.5×10^6 m^3 of exit gas can be disposed of as a result of the processing of about 16,000 tonnes of ore, 5000 tonnes of coke, 2000 tonnes of limestone, and 10^6 m^3 of oxygenated blast in 24 hours. The upper portion or shaft of the furnace increases in diameter downward from the charging level until it meets the upper edge of the bosh. This lower portion slopes in at a slightly sharper angle and joins the lowest section - the cylindrical hearth - in which slag and metal collect (Fig. 93b). At the top of the furnace, there are offtakes for gas and the charging system, whereby the ore, limestone and coke transported by a belt or buckets, can pass into the shaft without the emission of gas. The air blast, which may be heated to about 1500 K and at a pressure of up to 3 atm, enters through up to 40 tuyeres around the bottom end of the bosh. Oxygen, coal, oil, or hydrocarbon gas may be added to the blast, and the humidity may be controlled. Tap-holes for iron and slag are provided at different levels in the hearth.

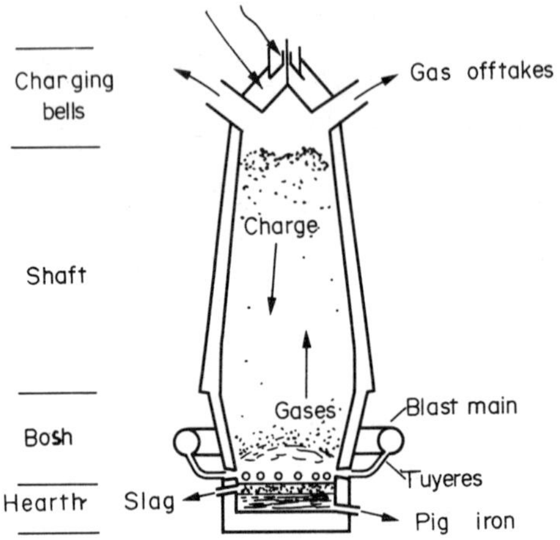

Fig. 93b. The blast furnace

The blast furnace is essentially a continuous counter-current reactor in which the descending solids react with ascending gas derived from the combustion of carbon at the tuyeres. Some iron oxide is involved in direct reaction with carbon, and the gangue and coke ash are fluxed by limestone to form the slag of silica, alumina, and lime, containing less than 1% Fe. Silica, manganese oxide, and phosphates are reduced to a greater or less extent, and silicon, manganese, and phosphorus dissolve in the iron. Sulphur passes mainly to the slag as calcium and manganese sulphides, and the iron is saturated with carbon at the temperature in the hearth.

The combustion of coke at the tuyere level raises the temperature to between 1800 and 2100°C, and melts the metal and slag. The gas, consisting mainly of CO and N_2, with some H_2 from moisture in the blast and any hydrocarbon added with it, ascends through the charge, with the N_2 acting as an important heat transfer medium, and reduce a large part of the wustite with the production of CO_2. This is quickly reduced by coke according to the reaction:

$$CO_2 + C = 2CO$$

the kinetics of which are rapid only above 1250 K. Further reaction between CO and higher oxides of iron will produce CO_2 which accumulates in the ascending gases, and the content is augmented by the decomposition of descending $CaCO_2$ at about 950°C. The exit gas may contain about 22% CO_2, with about 22% CO and 48% N_2, and leaves the furnace at about 500-600 K.

The changes in the descending charge can be considered as taking place in four zones, starting from the top:

Carbonates and hydrates are decomposed and the higher oxides of iron are reduced to wustite ($FeO_{1.12}$) in this "conditioning zone" in which the gas is cooled by contact with fresh charge.

The next zone extends almost to the top of the bosh and constitutes a thermal and chemical reserve, in that there is little change in the temperature or composition of gas or solid until the reduction of wustite commences.

In the bosh, wustite is reduced to metal by CO and C; slag is formed, and Si, Mn, and P are reduced from compounds and dissolved by the iron. The gas and solid temperatures rise sharply from the top to the bottom of this zone.

In the hearth, the sulphur in the iron, derived mainly from FeS in the coke, is substantially removed by reaction between metal droplets and the slag layer. The carbon content of the metal is raised to saturation by contact with the mass of unburnt coke which extends to the bottom of the hearth.

7.1.2. BALANCE OF C IN THE BLAST FURNACE

The coke rate of a blast furnace can be calculated from a mass balance of the reacting compounds. This can be represented by the operating line described in section 4.2.3. and already utilised in section 5.3.1. to explain the behaviour of a counter-current solid/gas reaction in the Direct Reduction processes.

The calculation will be carried out for the ideal case, where a reserve of wustite exists and only coke is used for heating, reducing oxides and carburising the iron, i.e. without liquid or solid fuel additions at the tuyeres. It is obvious that the removal of the wustite reserve will increase coke consumption, but the addition of liquid or solid fuel will decrease it.

The behaviour of iron oxide in the blast furnace is more complex than in the Wiberg process, mainly for the following reasons: there are several sources of oxygen, in addition to the iron oxides; the principal sources are air blown through the tuyeres and other oxides that are reduced, such as SiO_2, MnO, P_2O_5, etc.; simultaneous with the oxidation of CO to CO_2 by iron oxide reduction, regeneration of the gas takes place by direct reduction of the oxides, or by the Boudouard reaction:

$$\langle C \rangle + [CO_2] \rightleftarrows 2[CO]$$

In the Wiberg process, this reaction occurs in a separate vessel (carburettor). It is then necessary to consider:

Reactions utilising CO such as the indirect reduction of hematite, magnetite and wustite.

Reactions forming CO such as direct reduction of wustite with carbon, reduction of other oxides, regeneration of CO from CO_2 by coke, and oxidation of carbon to CO by the air from the blast.

In all of them, oxygen atoms are interchanged, and the distribution can be used as a measure of the extent of both reactions. This can be represented by Fig. 93c in which a system of rectangular coordinates, OX and OY expresses the amounts of interchanged oxygen in such a form that:

> Oxygen removed as CO and CO_2, or rather the ratio O/C, is indicated on the axis OX. The left-hand part of the diagram ($0<O/C<1$) represents reactions generating CO; the right-hand part ($1<O/C<2$) represents reactions utilising CO (indirect reduction of iron oxides).
>
> The sources of oxygen are represented on the axis OY; the positive ordinates indicate the number of atoms from iron oxides, expressed as O/Fe (instead of F[O-1] of Fig. 68c). The negative ordinates indicate the oxygen entering from other sources.

When a wustite reserve zone exists in the blast furnace, the gas emerges from it with a composition N_1^* and the operating line then passes through W. It can be shown that this line passes through another fixed point, P,† the coordinates of which can be determined by mass and thermal balances. The operating line obtained in Fig. 93c makes possible the determination of:

> The percentage of direct reduction (i.e. of the iron oxides by carbon). This is equal to the ratio BE/AE.
>
> The ratio $CO/(CO + CO_2)$ in the exit gas represented by HR/HA in Fig. 93c.
>
> The amount of carbon, M, necessary to produce one tonne of pig iron.

The slope of the operating line, $\mu = (O/Fe)/(O/C) = C/Fe$, gives the amount of carbon necessary for reduction of the oxides and heating the furnace (in atoms of carbon per atom of iron produced). If the C dissolved in the pig iron, γ(atoms C/atoms Fe) is added, the coke rate, M, will have the value:

$$M = 1000 \frac{12}{55.8} (\mu + \gamma) \text{ (kg C/tonne iron)}$$

The consumption of carbon in a modern blast furnace (coke in the charge and carbon injected through the tuyeres as oil, gas or pulverised coal) is between 430 and 500 kg $tonne^{-1}$ of pig iron. It is as low as this because of:

- lowering of the thermal demand by the use of prereduced charge, making low Si iron, raising the blast temperature and lowering the heat losses by creating a temperature gradient between the centre of the furnace and the walls, mainly by increasing the permeability of the charge in the centre. This corresponds to raising the point P in Fig. 93c;

- increasing the reducibility of the charge by substituting pellets for sized ore. The wustite can then be reduced by CO at lower temperatures and the efficiency of utilisation of the gas is increased. This corresponds to shifting the points W and R of Fig. 93c to the right, so that the slope of the operating line is less.

---

†A. Rist and N. Meysson, *Rev. de Metal.* 61, No. 2, 121-145 (1964).

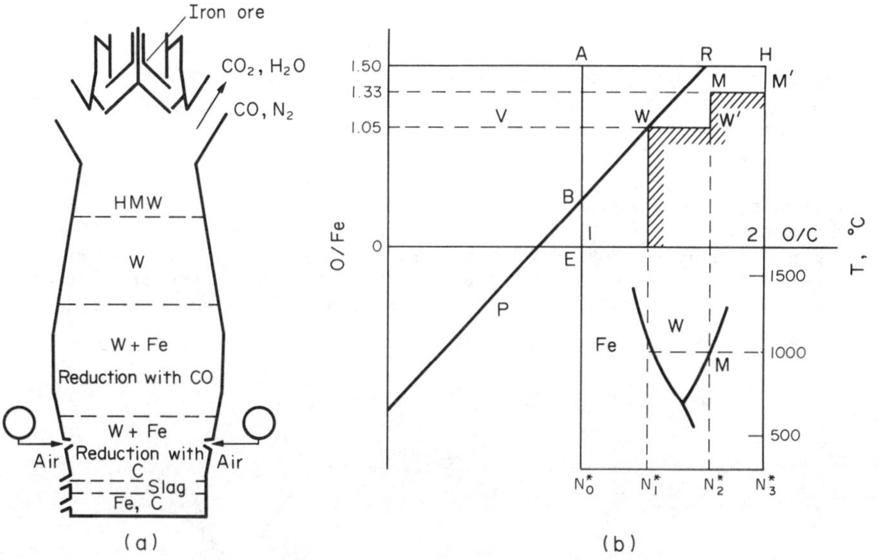

Fig. 93c. Diagram and operating line of the blast furnace

7.1.3. STEELMAKING PROCESSES

The production of steel from pig iron is essentially an oxidising process in which carbon, silicon, manganese, and phosphorus are removed by oxidation. At this stage, there is an excess of oxygen in the molten metal relative to that which is desirable in the solid, and de-oxidising elements having a high affinity for oxygen such as silicon, manganese and aluminium are added to form SiO_2, MnO and Al_2O_3, which are dissolved in the slag. To some extent throughout the process, and especially during this de-oxidising period, sulphur is removed by transfer to a slag rich in CaO as sulphides of calcium and manganese.

Two types of steelmaking process can be distinguished by the type of slag used, either acid or basic. The acid process can only be used to make steel from blast furnace iron low in phosphorus and sulphur, since these elements are not removed by the slag during the refining process. The slag is essentially an iron-manganese silicate, and the refractories in the smelting furnace must be high grade silica. If the blast furnace iron contains more than about 0.04% sulphur and phosphorus, it is necessary to use a basic slag in which there is a high proportion of CaO and the furnace refractories are of dolomite or magnesite.

Two types of refining process may be distinguished by the mode of transfer of oxygen to the metal bath:

In the open hearth and electric arc processes, oxygen is transferred to a shallow metal bath, of large surface area, from the atmosphere and from added iron oxide by contact with slag floating on the surface.

However, in order to reduce the time of refining, processes derived from the Open Hearth, such as Ajax, Dual and Tandem utilise oxygen blown into the charge by lances passing through the roof. In the same way, oxidation in the electric arc furnace is accelerated by the use of oxygen lances or refining can be carried out in the ladle.

In the converter or pneumatic processes, the metal is oxidised directly, either by air blown through it from below, as in the Bessemer and Thomas processes, by oxygen from bottom tuyeres, as in the Q-BOP process, or by oxygen blown from a lance above the bath, as in the LD, Kaldo, Rotor, LD-AC and OLP processes.

The slag-transfer processes require external heat in order to maintain the system in molten reactive condition, but the pneumatic processes are autogenous, in that the heat generated by oxidation of the impurities is sufficient to melt the solid constituents of the charge and to raise the temperature to the required reaction and casting temperature. The changes in the methods or producing steel in the last 100 years, with a projection to A.D. 2000 are given in Fig. 94.[55]

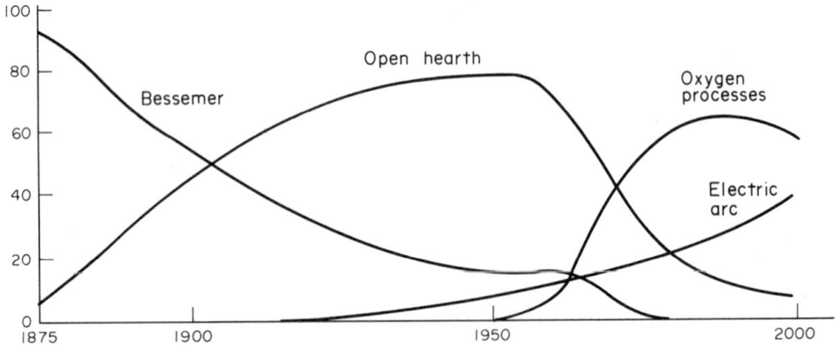

Fig. 94. Percentage of world steel by main processes 1875-1980, with projection to 2000[55]

(A) SLAG TRANSFER PROCESSES

(i) The Open Hearth Furnace

The open hearth furnace has been the principal method of producing steel for over 100 years, but it is being replaced rapidly by pneumatic converters, and by the increased use of electric furnaces. It is likely to remain as a minority method of production, and study of this process has provided data which have assisted in the development of steelmaking generally.

With the decline in availability of blast furnace iron sufficiently pure for use in the acid process, the refractories in the smelting chamber have been changed from silica to dolomite and magnesite. Generous provision of doors all along one side of the furnace make it possible to charge the large amount of scrap which this furnace can accommodate in a relatively short time. Current practice with about 50% scrap and 50% hot metal has developed from the use of scrap and cold pig iron. Heating is effected by combustion of gas or oil through burners operated alternately at the ends of the smelting chamber, with regenerators for preheat of combustion air and gas.

Oxidation of the metallic phase is effected by the addition of iron ore to the slag and by the transfer of oxygen from the atmosphere through the slag. Partition of oxygen between slag and metal occurs as a result of transfer across the interface. The area of this interface is increased and maintenance of suitable concentrations of reactants and products in the adjacent slag and metal is more effective as a result of agitation by CO, resulting from oxidation of carbon in the metal. This oxygen transfer is illustrated in Fig. 95b.

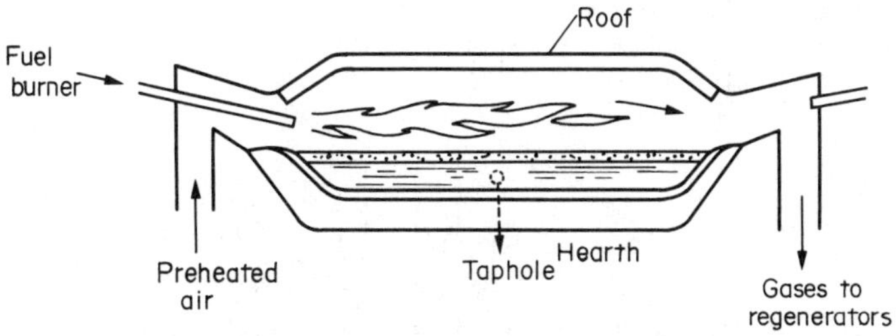

Fig. 95a. The open hearth furnace

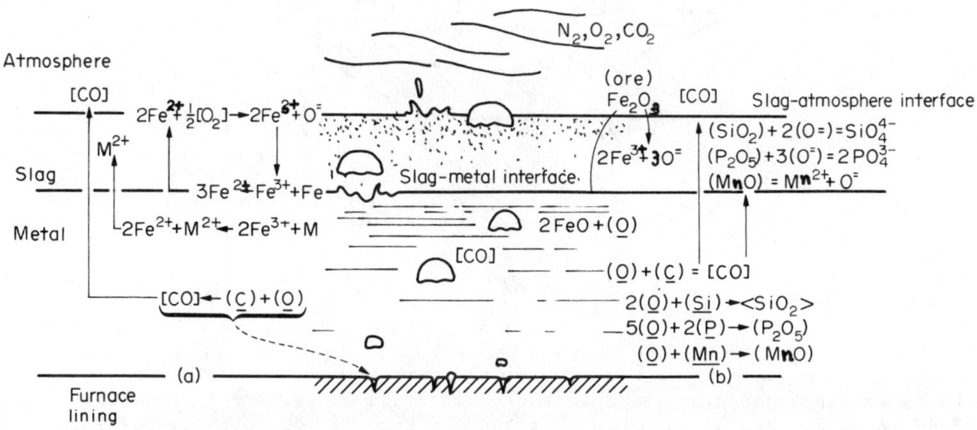

Fig. 95b. Mechanism of oxygen transfer in the slag-transfer processes: (a) from the atmosphere; (b) by means of iron ore

This process is capable of utilising large quantities of scrap, the nitrogen content of the product is lower than in metal made by converters using air and there is close control of composition during the smelting period of 4-6 hours. Removal of sulphur is only about 50% because the state of oxidation of the slag is always relatively high, on account of contact with air and products of combustion.

(ii) Electric Arc Furnace

The use of this type of furnace is increasing rapidly, especially for the production of steel from 100% scrap charges. The high slag temperatures which accompany the use of arcs between the electrodes and the metal bath as the method of heating, will result in very rapid refining. Oxygen is added to the slag as iron oxide, but the use of oxygen gas, blown directly into the metal is a considerable advantage in accelerating the process. Carbon, silicon, phosphorus, and manganese are removed by operating with a hot basic slag rich in oxygen and lime. This slag is removed by pouring; a new slag containing lime is made up, ferrosilicon and coke are added to the bath, and the furnace is closed. Under these conditions, the oxygen contents of the slag and then the metal are reduced to very low levels, and sulphur is transferred from metal to slag with a high degree of efficiency. Steels with phosphorus and sulphur contents of the order of 0.01% can be produced by this combination, and the electric arc furnace is particularly suitable for the manufacture of high quality carbon and alloy steels. More recently there is a tendency to use the arc furnace only for melting and to carry out the refining in a converter, or in the ladle.

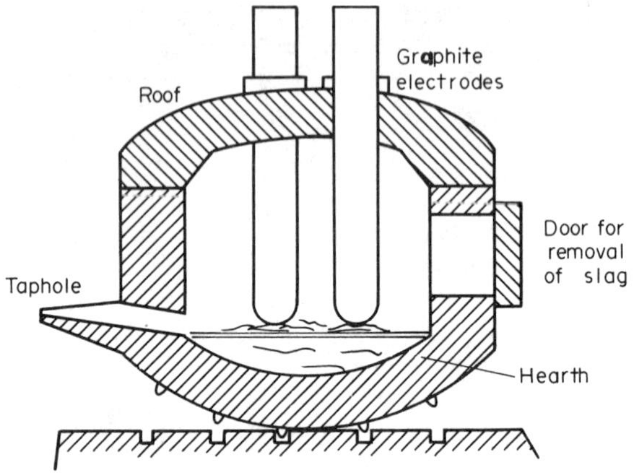

Fig. 96. Electric arc furnace

(B) PNEUMATIC PROCESSES

(i) Processes using Air

These were represented by the acid and basic Bessemer processes, the latter being widely known as the Thomas process, after the name of the investigator who pioneered the use of a basic dolomite lining in the converter designed by Bessemer.

The charge of molten pig iron, scrap, and flux was blown with air admitted through tuyeres in the bottom of the vessel. Because of the intimate contact between air and metal and the extremely high degree of agitation, reaction was rapid, and a charge of 25 tonnes was converted in about 15 minutes. The heat required to melt the 10% of scrap included in the charge and the flux, and to raise the metal to about 1600°C, was obtained mainly by oxidation of the silicon (acid process) and phosphorus (basic process) in the special grades of blast furnace iron used for these processes. Table 19 gives the temperature increments resulting from the oxidation of iron and several of the other elements which may be present. Up to 10% of the iron in the charge was lost as oxide in slag and fume, and the steel might contain 0.014% N_2, compared with 0.004% in open hearth steel. This nitrogen-rich steel was not suitable for the manufacture of flat rolled products and the process was abandoned after the development of the L.D. and other processes using oxygen.

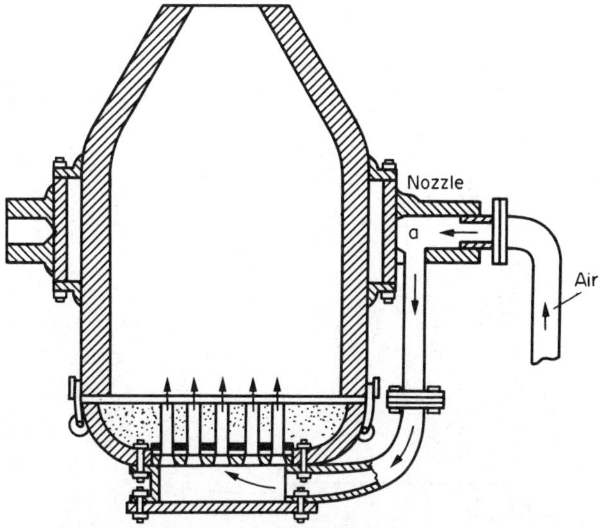

Fig. 97. Bessemer converter (from *Iron and Steel* by H.M. Boylston)

TABLE 19 Rise in Temperature Due to the Oxidisation of 0.01% Weight of an Element

Element	Fe	Mn	Cr	P	Si	C	C
Product	FeO	MnO	Cr_2O_3	P_2O_5	SiO_2	CO	CO_2
ΔT (^oC)	0.5	0.7	1.5	2.3	3.1	1.2	3.8

(ii) Processes Using Oxygen

These processes, which have been developed on an industrial scale since 1952, are capable of producing steel on a larger scale and at a very much higher rate than earlier versions. Because of the rapid destruction of converter bottoms when pure oxygen was used as the blast, modifications using oxygenated air, O_2-CO_2 and O_2-H_2O (steam) were introduced. All were capable of producing low-N_2 steel at a high rate, but the product of the last tended to be high in hydrogen. None was adopted on a large scale. The Q-BOP process has very recently been adopted on a substantial scale in industry. This utilises an oxygen blast introduced through the bottom of a Bessemer-pattern converter, but rapid destruction of the bottom is prevented by the use of a concentric shield of hydrocarbon gas around each tuyere. No data is yet available on sustained industrial application of this process, but the scale of plant construction is indicative of considerable confidence.

(a) The LD, LD-AC and OLP processes (top blowing). The first of these processes was the L.D. (Linz-Donawitz). Oxygen from the water-cooled lance was blown down vertically as a supersonic jet which penetrated the slag and impinged directly on the liquid metal in a vessel lined with dolomite.

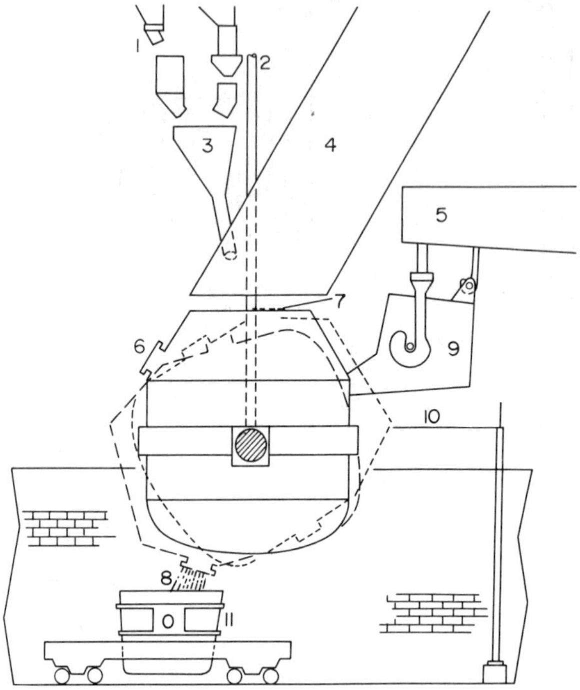

Fig. 98a. Important parts of an LD plant (after Jackson):[39] (1) flux material hoppers; (2) oxygen lance; (3) flux charging chute for limestone, scale, etc.; (4) exhaust hood; (5) hot-metal crane; (6) refining position; (7) charging position; (8) tapping position; (9) hot-metal ladle; (10) charging floor; (11) teeming ladle.

Refining was completed in a very short time under the influence of the high degree of agitation, the high oxygen concentration in all three phases of metal, slag and gas, and the high temperature in the crater formed by the gas at the point of contact with metal and slag. Later developments have included adjustment of the height of the lance during the process, so as to adjust the state of oxidation of metal or slag as required. The process is now known more generally as the B.O.F. (Basic Oxygen Furnace) Process, and converters capable of holding charges of 250 tonnes of liquid iron and 100 tonnes of scrap are in regular use. The blowing time is about 20 minutes, and tap-to-tap about 40 minutes.

Iron containing more than about 0.5% P can be treated in the normal process using two slags, or the modifications described by the initials LD-AC and O.L.P. (Oxygen Lance Powder) can be used. In these, either limestone for the second slag may be added directly in lumps to the metallic bath, or lime powder may be blown in, using oxygen in a special lance fitted with a venturi orifice.

(b) Bottom blowing. Since about 1970, bottom blown converters, using oxygen, have been used for steelmaking. The early high rate of destruction of tuyeres, caused by the high temperatures resulting from the use of pure oxygen, has been eliminated by the use of concentric nozzles, in which the oxygen jet is surrounded by a stream of hydrocarbon gas, fuel oil, CO_2 or Argon, which cools the refractory by cracking or otherwise absorbing heat. The original Maxheutte version, using hydrocarbon gas was entitled O.B.M. (Oxygen Blown Maxheutte), but Q-BOP is used more generally in current literature, with L.W.S. (Creusot-Loire-Wendel-Sidelor and Sprunck for the versions using fuel oil). The A.O.D. (Argon-Oxygen-Decarburisation) converters use a mixture of O_2 and N_2 or Ar and O_2, varying the O_2/N_2 or O_2/Ar ratios at different stages of the process. They are generally used for the refining of low-carbon stainless steels, melted in an electric arc furnace.

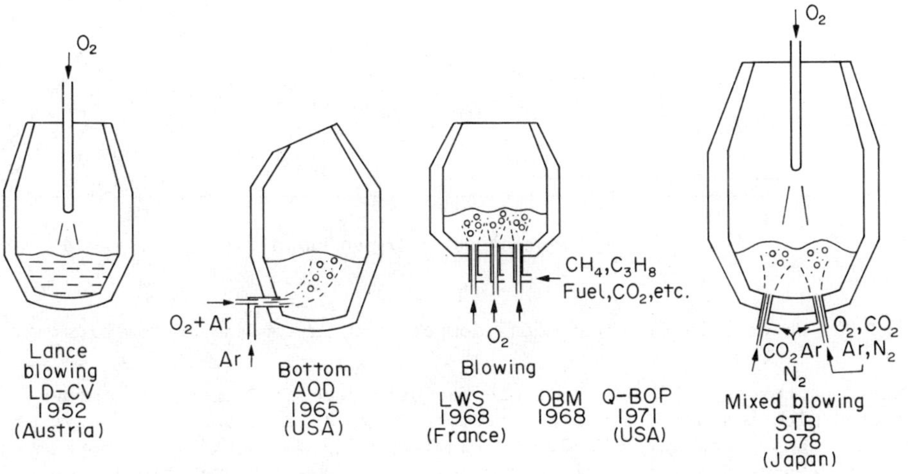

Fig. 98b. Diagrammatic representations of converter blowing using oxygen.[57]

(c) <u>Comparison between lance and bottom blowing.</u>[56] In the lance processes, oxidation is carried out through the slag, which is very highly oxidised. With bottom blowing, the metal is oxidised directly, the bath is more homogeneous and the slag/metal equilibrium more complete, with the following consequences:

In the bottom blowing process, removal of P, which needs an oxidising slag (Section 7.2.5), starts after complete oxidation of C and Si, whereas, in the top-blowing processes removal of C and P proceed simultaneously (Fig. 99b). The loss of iron in the slag is less with bottom blowing (Fig. 99a). The yield of iron is better (98% compared with 97% in the LD) and the final Mn content is 0.2% higher.

Less lime and oxygen are used, 40 kg CaO/tonne instead of 50 and 45 m^3 of O_2 instead of 50.

The heat balance is to the advantage of the LD, because part of the CO generated is burned to CO_2. So it is possible to melt more scrap, (350 kg/tonne^{-1} of pig iron, compared with 250-300 with bottom blowing.

The consumption of refractories in the sidewalls is very similar in both kinds of converter, but with bottom-blowing the bottoms have to be renewed after about 500 casts, compared with 1200 casts for top blowing.

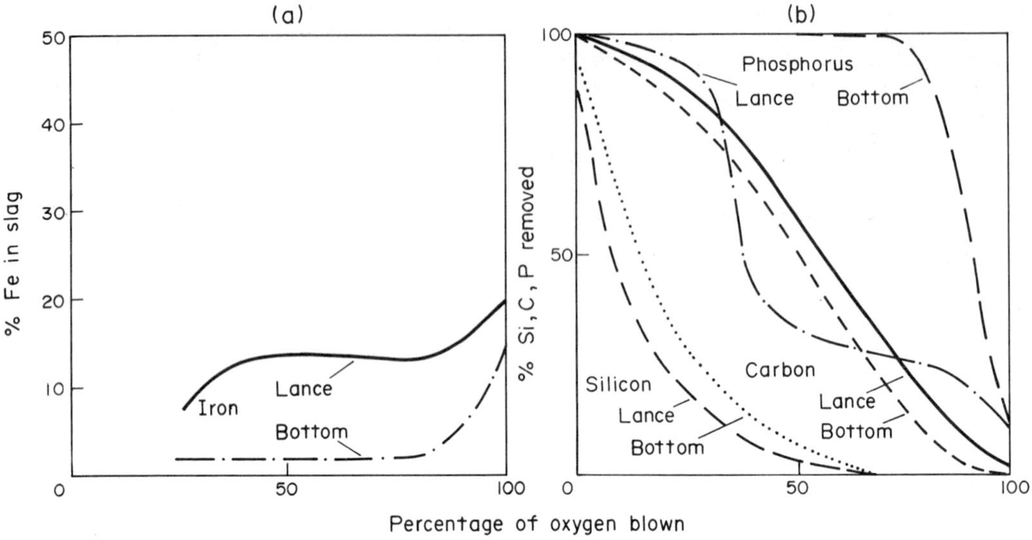

Fig. 99. Changes in slag and metal composition in top and bottom blowing.[56]

(d) Mixed processes.[57, 58] In these recently developed processes, converters are fitted with means for top and bottom blowing, a lance and either tuyeres or a porous refractory insert. The S.T.B. (Sumitomo Top and Bottom) has tuyeres and the L.B.E. (Lance Bubbling Equilibrium) uses a porous plug for blowing O_2, N_2, Ar or CO_2. This combination is particularly useful for the production of low carbon and stainless steels.

(iii) Refining in the Ladle[59]

This method of treatment is being increasingly used to complete the refining of steel produced from either electric arc furnaces or converters and so to increase the productivity of both. There are open and closed ladles, some with electrodes through the roof. They may be equipped with means of stirring and removal of oxygen, sulphur, carbon, phosphorus and dissolved gases, as well as for addition of alloying elements. Figure 136 illustrates some of the processes used for adjusting the composition and removing gases. The return of P to the steel must be prevented by separating slag and metal before deoxidising (finishing) and this can be done by stopping the flow of slag after tapping the metal. Removal of gas and deoxidation are effected by applying a vacuum, or blowing an inert gas through the metal, with or without a vacuum, that is V.O.D. (Vacuum-Oxygen-Decarburisation) or AOD, already mentioned.

7.2. EQUILIBRIA IN IRON AND STEEL PRODUCTION

The study of the variance indicates that, in general, metal of a desired composition can be produced by specifying the temperature and slag composition.

7.2.1. PARTITION BETWEEN SLAG AND METAL

OXYGEN POTENTIAL

The equilibrium between a solute in a metal, gaseous oxygen, and the oxide in a slag, can be expressed as:

$$2(\underline{M})_m + O_2 = 2(\underline{MO})_s \qquad (7.1a)$$

and the free energy change is:

$$\Delta G^o_{(M \to MO)} = -RT \ln \frac{a^2_{(MO)_s}}{a^2_{(\underline{M})_m} \cdot P_{O_2}} \qquad (7.1b)$$

when $\quad a_{(MO)_s} \Big/ a_{(M)_m} = 1; \qquad \Delta G^o_{M \to MO} = RT \ln P_{O_2}$

$RT \ln P_{O_2}$ is termed the "oxygen potential" of the system. This will have the same value for all of the metal-oxygen equilibria involved in a given system.

The oxygen potential in the hearth of the blast furnace is fixed by the C-CO equilibrium, where $a_C = 1$ and $P_{CO} = 0.35$ P, at the prevailing temperature, assumed in this case to be 1800 K. The value of the oxygen potential when $a_C = 1$ and $P_{CO} = 1$ is given by point A on Fig. 100 (ΔG^o for $C + O_2 = 2CO$ at 1800 K) and by A' for the condition when P = 2 atm and $P_{CO} = 0.7$ atm. Since all other equilibria in the hearth must be operating at the same oxygen potential, the plots of $M \to MO$ or MO_2 must all pass through the same point, and the ratio a_{MO}/a_M will not be equal to one. If the plot of ΔG^o for $M \to MO$ passes above A in the standard diagram, adjustment of a_{MO}/a_M to cause it to pass through A will cause it to rotate downward, and to the right. The ratio $a_{(MO)}/a_{(M)}$ will then be less than one, and the activity of the oxide in the slag will be very low. For those plots which pass below A at 1800 K, $a_{(MO)}/a_{(M)}$ will be greater than one, and the oxide will be only slightly reduced to give a relatively low value of $a_{(M)}$ (Fig. 101).

Considering the oxides likely to be present in the slag, P_2O_5 and FeO will be completely reduced; MnO, Cr_2O_3 and SiO_2 will be partly reduced, and Al_2O_3, CaO and MgO will not be reduced to any extent.

The oxygen potential of the blast entering through the tuyers will be that of air at about 800°C (Point E, Fig. 100). Within a very short distance, 1 to 2 metres, the temperature is raised to about 2000°C, and the oxygen potential is lowered to that of CO at that temperature (A). As this gas cools to below 1000°C, any CO_2 formed by reaction with FeO is rapidly removed by reaction with C, and the oxygen potential is close to that given by $2C + O_2 = 2CO$. Below this temperature, the oxidation of CO to CO_2 tends to be more rapid than reduction back to CO, and the oxygen potential is given by the Fe-FeO equilibrium.

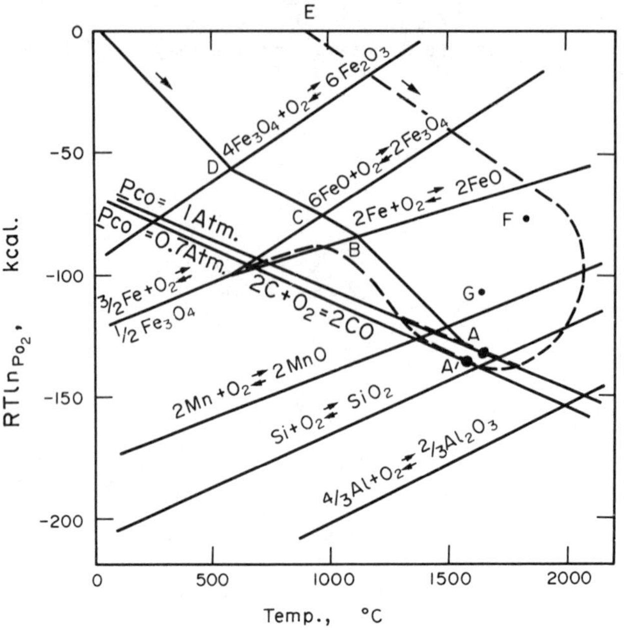

Fig. 100. Oxygen potential in the iron blast furnace. ——— Solids; – – – Gases

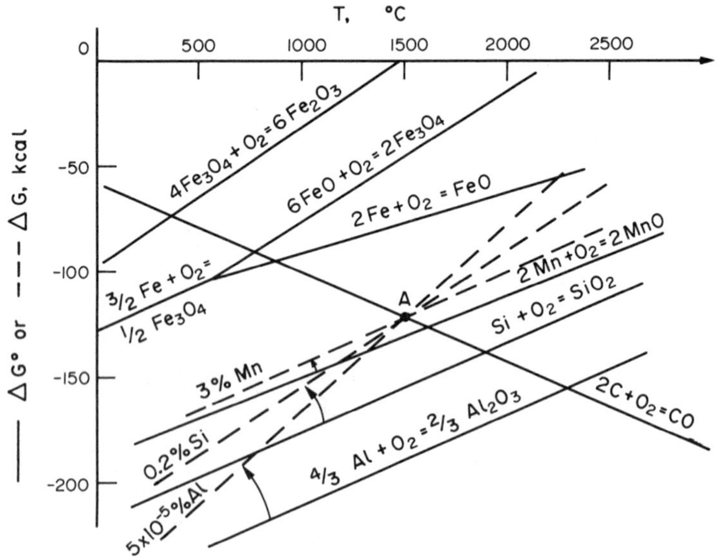

Fig. 101. Metal-oxide equilibria in the iron blast furnace hearth.
——— ΔG°; – – – ΔG

The descending charge is dried and heated by ascending gas, and reduction of Fe_2O_3 to Fe_3O_4 begins at about 500°C (D). Reduction to FeO (C) and Fe (B) follow as the temperature is raised to higher levels and there is contact with gas rich in CO.

The partition of oxygen and other elements is the fundamental basis of the refining of pig iron to produce steel. In the early stages of the process, the state of oxidation can be represented by F in Fig. 100. The metal is in contact with a $CaO-FeO-SiO_2$ slag which also contains P_2O_5 and MgO, and a high value of $a_{(O)_{Fe}}$ is maintained either by blowing gaseous oxygen into the metal or adding oxygen as Fe_2O_3 to the slag. As the oxidisable elements are removed, the excess of oxygen in both slag and metal increases, and it is necessary to lower $a_{(O)_{Fe}}$ by addition of elements having a high affinity for oxygen. These change the conditions to those represented by G in Fig. 100.

The graphical method used previously gives only the ratio of $a_{(MO)}$ to $a_{(M)}$, and the actual partition of M between metal and slag can be calculated only if the activity coefficients of MO and M are known. For dilute solutions in metal, the activity coefficient of any one solute, i or j, depends on the concentrations of all of the solutes present. This is because the solutes interact with each other so as to raise or lower the value of $\gamma_{(M)}$. The activity coefficients can be expressed in terms of the mole fraction, or the weight percentage, in the following expressions (see Chapter 2):

$$\ln \gamma_i = \ln \gamma_i^\infty + \sum_j \varepsilon_i^j N_j \qquad (7.2)$$

and

$$\ln f_i = \sum_j e_i^j \cdot (\% \ j) \qquad (7.3)$$

Values for the interaction parameters ε_i^j and e_i^j between pairs of solutes in iron are given in Tables 20 and 21.

Since there is not a similar method of deriving activity coefficients for the constituents of slags, it is necessary to examine each steelmaking equilibrium experimentally.

7.2.2. EQUILIBRIUM IN THE BLAST FURNACE

(A) CARBURISATION OF PIG IRON

Wustite is completely reduced in the bosh of the blast furnace, and the iron so produced is saturated with carbon by contact with coke in the hearth. The saturated solubility of carbon in iron can be expressed by:

$$\log N_{C_{sat.}} = -\frac{560}{T} - 0.375 \qquad (7.4)$$

or

$$\% \ C_{sat.} = +\ 1.34 + 2.54 \times 10^{-3} t \quad (t\ ^\circ C) \qquad (7.5)$$

TABLE 20 Interaction Parameters ε_i^j of the Solute in Liquid Iron (from Bodsworth[49])

i \ j	Al	C	Cr	Co	Cu	Mn	Mo	Ni	N	O	P	S	Si	Ti	V
Al	6·0	6·7		2·1	1·5	2·8		1·8	0·28	-104	6·5	6·1	6·9		
C	6·7	11·4	-4·7	2·9	4·2	-0·5	-3·5	2·9	6·4	-22·2	6·0	12·0	8·3	-12·8	-8·0
Cr		-4·7	0						-9·8	-13·7		-4·1			
Co	2·1	2·9		0					2·2	1·7	0·9	0·3	1·5		
Cu	1·5	4·2			-6·2				2·4	-2·5	-1·5	-3·1	0		
Mn	2·8	-0·5				0			-4·8	0	-2·8	-5·7	0		
Mo		-3·5							-4·4	1·4					
Ni	1·8	2·9						-0·01	2·5	1·4	0·6	0	1·2		
N	0·28	6·4	-9·8	2·2	2·4	-4·8	-4·4	2·5	0	0	6·4	1·7	5·4		-20
O	-104	-22·2	-13·7	1·7	-2·5	0	1·4	1·4	0	-13·4	-4·0	0	-17·3	-100	-55
P	6·5	6·0		0·9	-1·5	-2·8		0·6	6·4	-4·0	0	5·4	11·0		
S	6·1	12·0	-4·1	0·3	-3·1	-5·7		0	1·7	0	5·4	-3·7	7·5		
Si	6·9	8·3		1·5	0	0		1·2	5·4	-17·3	11·0	7·5	37·0		
Ti		-12·8								-100					
V		-8·0							-20	-55					

TABLE 21 Interaction Parameters $e_i^j \times 10^2$ of the Solute j over i in Liquid Iron[49]

Interaction Parameters $e_i^J \times 10^2$ of the Solute j over i in Liquid Iron.[49]

i \ j	Al	C	Cr	Cu	Mn	Mo	Ni	N	O	P	S	Si	Ti	V
Al	5·4	13		0·6	1·2		0·8	0·5	-200	5	4·2	5·8		
C	5·8	23	-2·2	1·6	-0·2	-0 9	1·2	11	-34	4·7	9·0	7·2	-7	-3·8
Cr		-9·5	0					-17·0	-20·8		-3·1			
Cu	1·4	8·5		-2·3				4·2	-3·8	-1·2	-2·3	0		
Mn	2·7	-0·9			0			-8·3	0	-2·2	-4·3	0		
Mo		-7·2						-7·6	2·1					
Ni	1·6	5·9					0	4·3	2·2	0·5	0	1·1		
N	0·25	13	-4·5	0·9	-2·1	-1.1	1·1	0	0	5	1·3	4·7		-10
O	-120	-45	-6·4	-1·0	0	0.35	0·6	0	-20	-3·2	0	-15·0	-51·0	-26
P	5·6	12		-0·6	-1·2		0·3	11·1	-6·2	0	4·1	9·4		
S	5·5	24	-1·9	-1·2	-2·6		0	3	0	4·2	-2·8	6·5		
Si	5·6	17		0	0		0·5	9·4	-26·3	8·5	5·7	32		
Ti		-25·8							-152				4·8	
V		-16						-35	-84					2·2

derived from the liquidus line of the Fe-C phase diagram (Fig. 102). The effect of a third element on the solubility of carbon in pig iron can be seen in Fig. 103. Because the activity of carbon in the carbon-saturated iron in the hearth is one, and $a_C = \gamma_C \cdot N_C$, then $\ln \gamma_{C_{sat.}} = -\ln N_{C_{sat.}}$. The molar fraction of carbon at saturation is modified by the presence of a second solute in the pig iron in the manner given below:

$$\frac{d \ln N_{C_{sat.}}}{dN_i} = \frac{dN_{C_{sat.}}}{N_{C_{sat.}} \cdot dN_i} = -\frac{d \ln \gamma_{C_{sat.}}}{dN_i} = -\varepsilon^i_{C_{sat.}} \tag{7.6a}$$

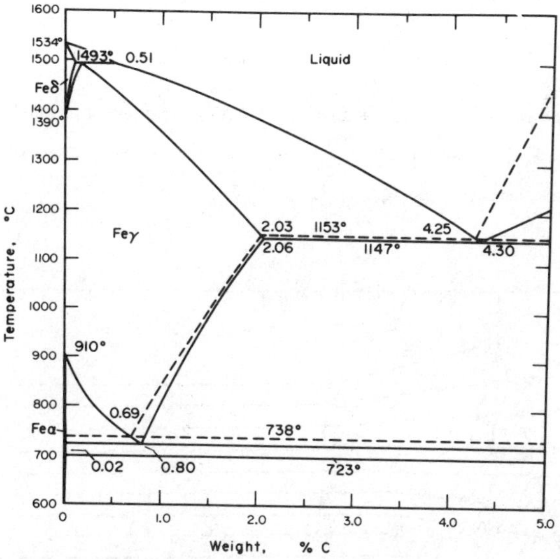

Fig. 102. The iron-carbon phase diagram (after Hansen)

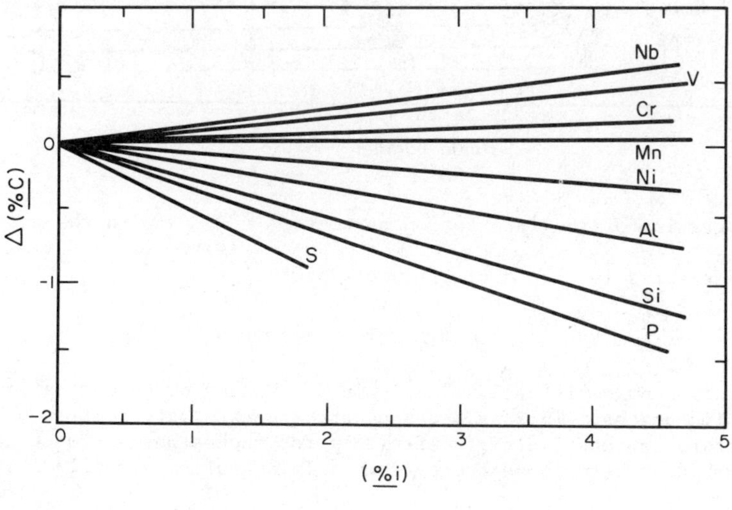

Fig. 103. Influence of a second solute on the solubility of carbon in iron.[2]

Schenck[3] studied the variation of $\varepsilon^{i}_{C_{sat.}}$ with the atomic number of the elements, obtaining straight lines almost parallel to each other (Fig. 104). From this plot, the following conclusions can be drawn:

Noble gases and elements in the same column as iron (Fe, Ru, Os) have no effect on the solubility of carbon in iron. Elements situated to the left of iron in Mendeleev's table raise the solubility of carbon ($\varepsilon^{j}_{C} < 0$), whereas those to the right decrease the solubility ($\varepsilon^{j}_{C} > 0$). Carbon has a self-interaction parameter different from zero:

$$\varepsilon^{C}_{C_{sat.}} = \frac{d \ln \gamma_C}{d N_{C_{sat.}}} = 3.4 \qquad (7.6b)$$

This parameter indicates the variation of $\ln \gamma_C$ when the molar fraction of carbon varies slightly round its saturation value, and can be deduced from the $a_C = f(N_C)$ curve of Fig. 105(a) for $N_C = N_{C_{sat.}}$.

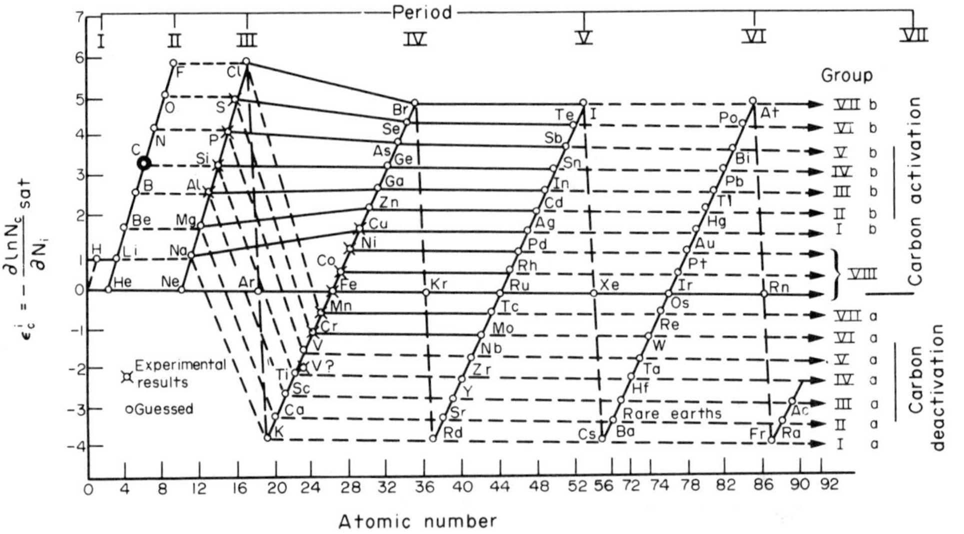

Fig. 104. Variation of the interaction parameter $\varepsilon^{i}_{C_{sat.}}$ with the atomic number of solutes in liquid pig iron at 1550°C.[3]

(B) MANGANESE EQUILIBRIUM

Fe-Mn solutions behave virtually ideally in all ranges of concentration in the liquid and solid states. This is because both elements have similar atomic radii and can therefore be substituted indefinitely for each other in solution, without causing disturbances between adjacent atoms. In an infinitely dilute solution of Mn in Fe, γ^{∞}_{Mn} is equal to one, and the free energy of solution, according to:

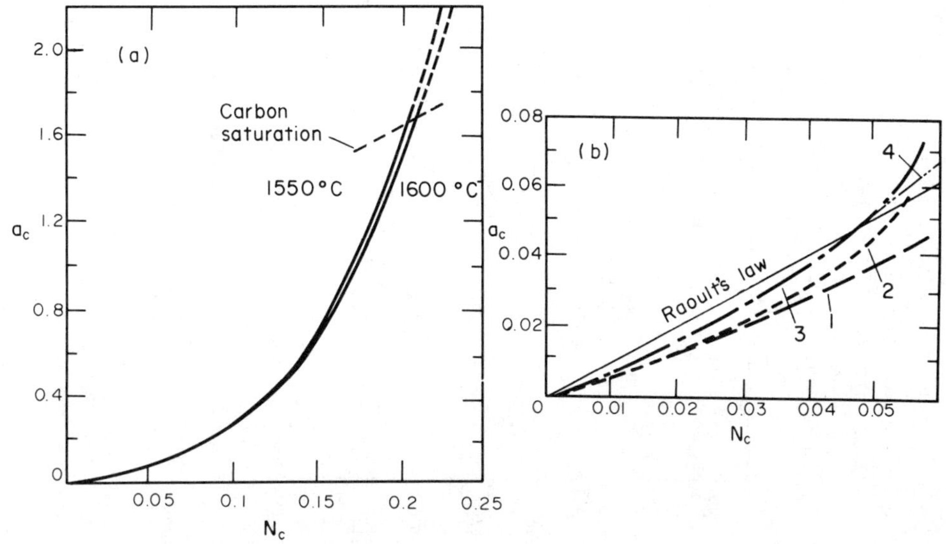

Fig. 105. Activity of carbon dissolved in iron as a function of its molar fraction: (a) according to J.F. Elliott[4] reference state: $f_C = 1$ when $N_C \to 0$; (b) for dilute solutions: 1 - S. Marshall and J. Chipman[5]; 2 - H. Schenck and U. Feldmann[6]; 3 - F.D. Richardson and W. Dennis[7]; 4 - A. Rist and J. Chipman[8].

$$(\mathrm{Mn}) \rightleftarrows (\underline{\mathrm{Mn}})_{1\%} \tag{7.7}$$

is then:

$$\Delta G^{o} = G^{o}_{(\underline{\mathrm{Mn}})_{1\%}} - G^{o}_{(\mathrm{Mn})} = RT \ln \gamma^{\infty}_{\mathrm{Mn}} \frac{M_{\mathrm{Fe}}}{100\, M_{\mathrm{Mn}}} = -9.1\, T$$

Because MnO is a basic oxide, its activity coefficient in an acid slag is low (the MnO reacts with silica), increasing rapidly as the basicity of slag increases (Figs. 106 and 107). In a slag which contains principally CaO, Al_2O_3 and SiO_2, it is possible to note the amphoteric role of alumina: If Al_2O_3 is added to an acid SiO_2-CaO-MnO slag, it neutralises part of the silica, increasing the activity coefficient of MnO. On the contrary, in a slag of high basicity, Al_2O_3 reacts with CaO and the activity coefficient of MnO decreases slightly (Fig. 107).

The manganese equilibrium between carbon-saturated pig iron and a slag composed mainly of CaO, SiO_2 and Al_2O_3 can be calculated starting from the constant K_{Mn} of reaction:

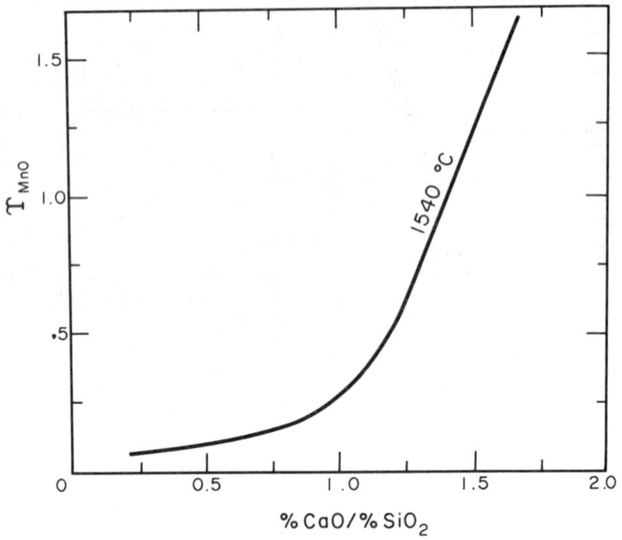

Fig. 106. Variation of the activity coefficient of manganese oxide with slag basicity.[49]

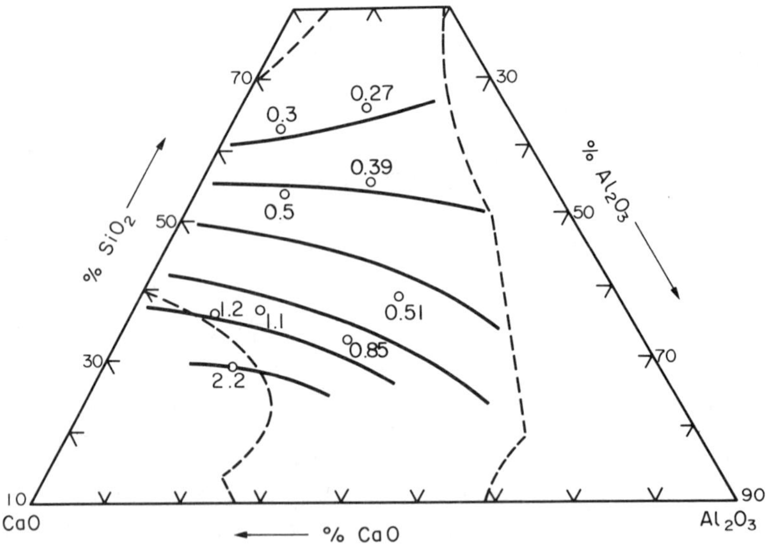

Fig. 107. Variation of the activity coefficient of manganese oxide with the composition of the slag at 1640°C.[9]

$$(MnO) + \langle C \rangle \rightleftarrows (\underline{Mn})_m + [CO] \tag{7.8}$$

which can be expressed as a function of the temperature and composition by:

$$\log K_{Mn} = -\frac{12{,}000}{T} + 9.56 \tag{7.8a}$$

$$K_{Mn} = \frac{P_{CO} \cdot f_{(Mn)_m} \cdot (\%\ Mn)_m}{a_{(C)} \cdot \gamma_{(MnO)_s} \cdot N_{(MnO)_s}} \tag{7.8b}$$

Since the CO pressure in the blast furnace hearth is near to 1 atm and the carbon activity in the saturated pig iron is one, the latter relation can be simplified. In a low-alloy pig iron, $f_{(Mn)}$ depends only on the amount (almost constant) of dissolved carbon. The partition coefficient of manganese (ratio of the manganese dissolved in the pig iron to the manganese in the slag $(\%\ \underline{Mn})/N_{(MnO)}$ or else $(\%\ \underline{Mn})/(\%\ MnO)$), is then proportional to $\gamma_{(MnO)}$:

$$\frac{(\%\ \underline{Mn})_m}{(\%\ MnO)_s} = K'_{Mn} \cdot \gamma_{(MnO)} \tag{7.9}$$

The activity coefficient $f_{(Mn)_m}$ as well as the conversion factor for molar fraction to weight percentage, which varies little, with the composition of the slag, are included in the value of the constant K'_{Mn}, and this constant depends only on the temperature. On the other hand, $\gamma_{(MnO)}$ depends on the basicity of the slag (Figs. 106 and 107) as well as on the temperature. For this reason, the manganese partition coefficient, $(\%\ \underline{Mn})_m/(\%\ MnO)$, also depends on the temperature and basicity of the slag, as is seen in Fig. 108, for 10 and 20% Al_2O_3 and different temperatures. Since the partition coefficient increases rapidly with the temperature and basicity of the slag, it is necessary to have a high reduction temperature in the bosh and a very basic slag, in order to produce high manganese pig iron (Spiegel) from an iron ore containing manganese oxide.

(C) SILICON EQUILIBRIUM

Unlike manganese, silicon has a strong negative deviation from ideality in the Fe-Si binary system (see Problem No. 3), and the free energy of solution of silicon in iron, according to:

$$(Si) \rightleftarrows (\underline{Si})_{1\%} \tag{7.10}$$

is always negative at the normal working temperatures in the blast furnace:

$$\Delta G^o_{Si} = G^o_{(\underline{Si})_{1\%}} - G^o_{(Si)} = RT \ln \gamma^{\infty}_{Si} \frac{M_{Fe}}{100\ M_{Si}} = -28{,}000 - 5.54\ T \tag{7.10a}$$

The activity of silicon in iron was determined by Rein and Chipman[10] (Fig. 109). As can be seen, at 1600 C, γ^{∞}_{Si} is very low, of the order of 10^{-3}.

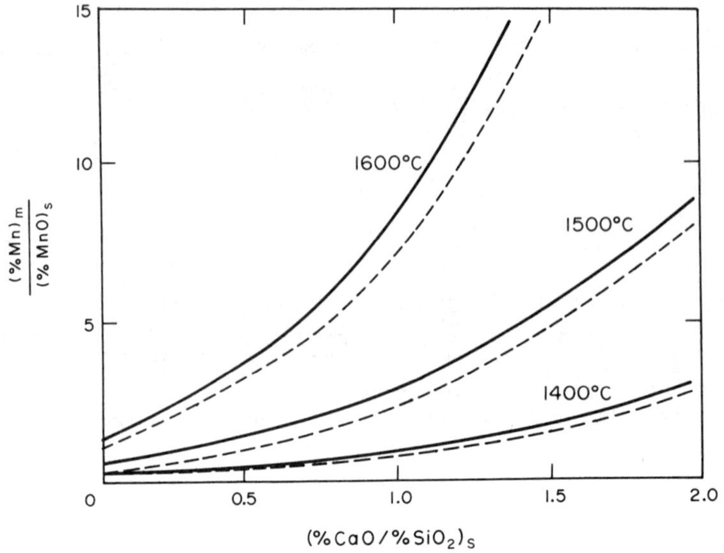

Fig. 108. Manganese partition coefficient between pig iron and a CaO-Al_2O_3-SiO_2 slag.[9] —— 20% Al_2O_3; – – – 10% Al_2O_3

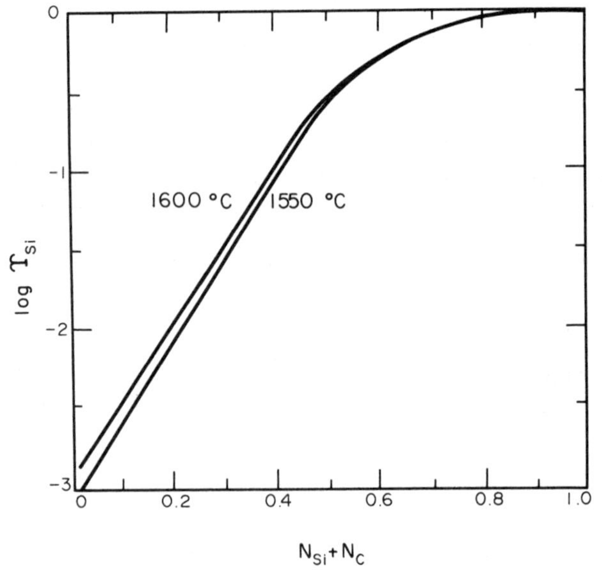

Fig. 109. Activity coefficient of silicon in iron as a function of the mole fractions of silicon and carbon.[10]

The activity of silica in SiO_2-CaO and CaO-SiO_2-Al_2O_3 slags has been studied in the previous chapter (Figs. 86 and 88). It should be noted that the activity of silica decreases rapidly on increasing the basicity of the slag. In the same way as for manganese, the partition coefficient for silicon between iron and slag can be deduced from thermodynamic data. The equilibrium constant of:

$$(SiO_2)_s + 2\langle C\rangle \rightleftarrows 2[CO] + (\underline{Si})_m \tag{7.11}$$

is expressed as a function of the temperature and concentrations of the components of the system from:

$$\log K_{Si} = -\frac{31{,}600}{T} - 0.908 \log T + 21.4 \tag{7.11a}$$

with

$$K_{Si} = \frac{P_{CO} \cdot f_{(Si)_m} \cdot (\%\ Si)_m}{a_C \cdot \gamma_{(SiO_2)} \cdot N_{(SiO_2)}} \tag{7.11b}$$

The partition coefficient $(\%\ \underline{Si})_m/(\%\ SiO_2)_s$ is therefore a function of the activity coefficients $f_{(Si)_m}$ and $\gamma_{(SiO_2)_s}$:

$$(\%\ Si)_m/(\%\ SiO_2)_s = K'_{Si}\ \frac{\gamma_{(SiO_2)_s}}{f_{(Si)_m}} \tag{7.12}$$

The constant K'_{Si} is the product of K_{Si} and the conversion factor for $N_{(SiO_2)}$ to $(\%\ SiO_2)$, and is only a function of temperature, whereas the activity coefficients $\gamma_{(SiO_2)_s}$ and $f_{(Si)_m}$ depend on the temperature, the mole fraction of silica in the slag and of silicon in the metal respectively. The curves of Fig. 110 show the values of the distribution coefficient $(\%\ Si)_m/(\%\ SiO_2)_s$ versus the basicity at different temperatures and for slags with 10 and 20% of alumina, in equilibrium with carbon-saturated iron. Figure 111 is another form of presentation of these results, indicating the weight-percent of Si in carbon-saturated iron in equilibrium with SiO_2-Al_2O_3-CaO slags.[12]

From these figures it can be seen that it is necessary to work at high temperatures and with a high silica slag in order to produce a high silicon pig iron. The first condition is the same as for the production of high-manganese pig iron, but the influence of temperature is greater for the equilibrium SiO_2-Si than for MnO-Mn because the variation of K_{Si} with T is greater than that for K_{Mn}. The slag condition for the production of high Mn in pig iron is evidently very different from that for high Si. A high-Mn pig iron requires a basic slag, whereas to produce a high-Si pig iron it is necessary to use an acid slag.

(D) CHROMIUM EQUILIBRIUM

Iron-chromium solutions exhibit almost ideal behaviour, again because of the similar atom radii. The free energy of solution of chromium in iron, according to the

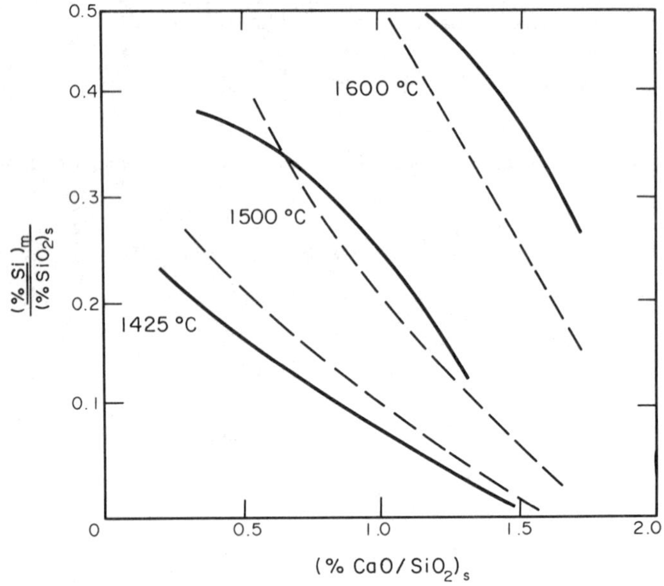

Fig. 110. Distribution of silicon between carbon-saturated iron and a $CaO-SiO_2-Al_2O_3$ slag as a function of the basicity.(11) ——— 20% Al_2O_3; – – – 10% Al_2O_3

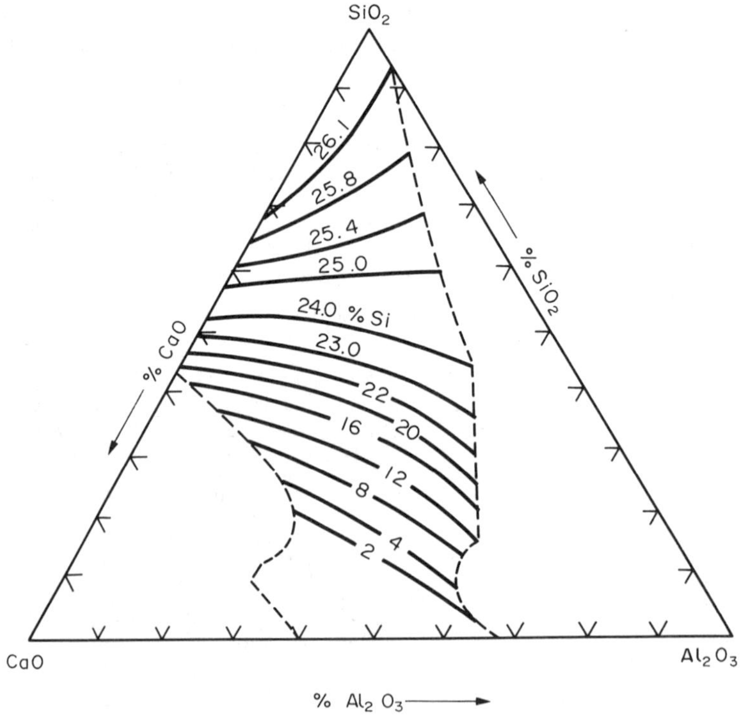

Fig. 111. Weight % of silicon in a carbon-saturated pig iron in equilibrium with a $SiO_2-CaO-Al_2O_3$ slag at 1600°C.(12)

reaction:

$$(Cr) \rightleftarrows (\underline{Cr})_{1\%} \qquad (7.13)$$

is then equal to:

$$\Delta G^{o} = G^{o}_{(\underline{Cr})_{1\%}} - G^{o}_{(Cr)} = RT \log \frac{0.56}{52} = -\ 9.0\ T\ (cal) \qquad (7.13a)$$

The chromium oxides, which are very refractory, are only slightly soluble in slags, from which it can be deduced that the activity coefficients of these oxides are much higher than one. Although the oxidation level of chromium in slags is not known as a certainty, it is assumed that Cr exists as a chromous oxide CrO in reducing and acid slags, whereas chromic oxide Cr_2O_3 predominates in oxidised and basic slags in the form of FeO . Cr_2O_3 and CaO . Cr_2O_3.

The equilibrium constant of the reaction:

$$<Cr_2O_3> + 3<C> \rightleftarrows 2(Cr) + 3[CO] \qquad (7.14)$$

is expressed as a function of the temperature by:

$$\log K_{Cr} = -\frac{42,000}{T} + 31.75 \qquad (7.14a)$$

At 1500°C, this constant has a value of 10^8. As the activity of chromic oxide in the slag is higher than its molar fraction ($\gamma_{Cr_2O_3} > 1$), most of the chromic oxide in the blast furnace is reduced to metal which dissolves in iron.

7.2.3. EQUILIBRIA IN STEEL REFINING

The refining processes which are used to convert pig iron into steel consist, in the first stage, of controlled oxidation to eliminate most of the undesirable impurities which are present in solution. In the second stage, the oxygen dissolved in the bath is eliminated by means of an appropriate reducer which forms a stable oxide insoluble in steel. In both cases there is an equilibrium of the type:

$$(\underline{M})_m + (\underline{O})_m \rightleftarrows (\underline{MO})_s \qquad (7.15)$$

In order to determine the partition coefficients for these elements between the two phases in equilibrium, it is necessary to know the activity of oxygen in the molten metallic bath.

(A) THE Fe-O SYSTEM

The maximum solubility $(\%\ O)_{sat.}$ of oxygen in liquid iron in equilibrium with pure FeO increases with the temperature, as shown in the Fe-O diagram (Fig. 112)[13] and can be expressed by the relation:

$$\log (\%\ O)_{sat.} = -\frac{6320}{T} + 2.734 \qquad (7.16)$$

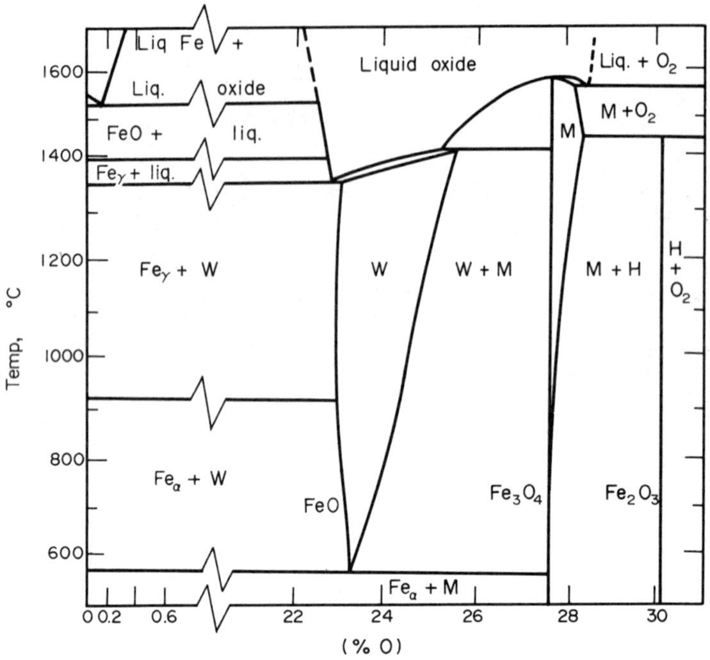

Fig. 112. Phase diagram Fe-O: The solubility of oxygen in liquid iron increases with the temperature.[13]

The free energy of solution of oxygen in iron, according to the reaction:

$$\tfrac{1}{2}[O_2] \rightleftarrows (\underline{O})_{1\%} \tag{7.17}$$

is always negative:

$$\Delta G^o = G^o_{(\underline{O})\%} - \frac{1}{2} G^o_{[O_2]} = -\ 28000 - 0.69T \tag{7.17a}$$

The activity coefficient $\gamma^{\infty}_{(O)}$ which can be deduced from the difference of free energy:

$$\Delta G^o = RT \ln \gamma^{\infty}_{(O)} \cdot \frac{M_{Fe}}{100\ M_O} \tag{7.17b}$$

is then always less than one. This can be explained by the attraction between iron and oxygen.

Oxygen dissolved in iron does not follow Henry's Law; the self-interaction parameter is not zero ($e_O^O = -\ 0.2$). The modifying effect of a second solute on the activity coefficient of the oxygen can be calculated by using equations and tables given earlier ((7.2), (7.3), Tables 20 and 21). For other ranges of concentration of solutes, Fig. 113 gives the variation of $\log f_0^i$ as a function of concentration. Elements which have a larger affinity than iron for oxygen (C, Si, V, Cr, etc.) decrease its activity coefficient.

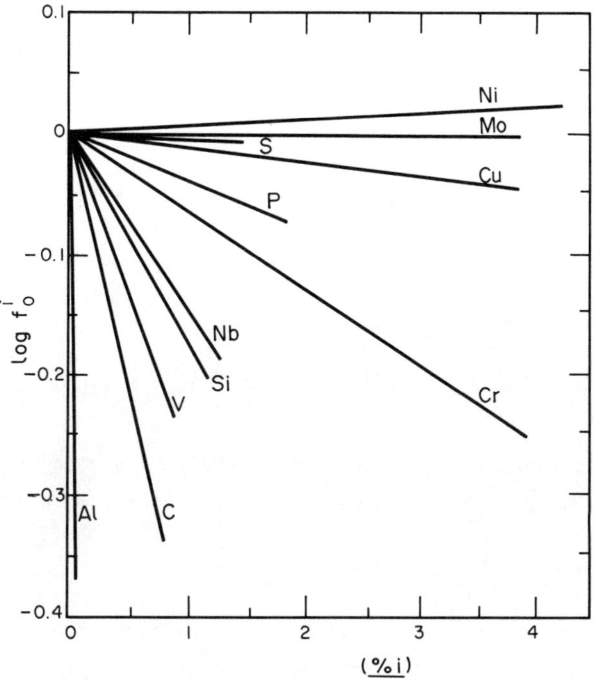

Fig. 113. Influence of a second solute on the activity coefficient of oxygen dissolved in liquid iron.(14-17)

Fig. 112 shows that the composition of iron oxide in equilibrium with oxygen-saturated liquid iron changes with temperature and we may consider that this oxide $FeO_{(1+x)}$ is a combination of stoichiometric FeO and Fe_2O_3 (3.5% at 1550 C). Additions of CaO raise the proportion of Fe_2O_3 and additions of SiO_2 decrease it. The activities of FeO in the binary system $FeO\text{-}SiO_2$, the pseudo-ternary $FeO\text{-}CaO\text{-}MgO\text{-}SiO_2$ and the pseudo-binary $FeO\text{-}nCaO \cdot SiO_2$ systems are given in Figs. 86, 90 and 92, taking the $FeO_{(1+x)}$ in equilibrium with Fe as the standard state.

Oxygen activity in steel is related to wustite activity in the slag by:

$$\frac{a_{(O)_m}}{a_{(O)_{sat.}}} = \frac{a_{(FeO)_s}}{a_{(\text{pure FeO})}} = a_{(FeO)_s} \qquad (7.18)$$

The activity of wustite in a slag can therefore be used to define the oxidizing power.

Numerical Application

According to 7.16. at 1600°C

$$a_{(O)_{sat}} = (\%O)_{sat} = 0.23\%$$

Equation (7.17a) makes it possible to deduce the pressure of O_2 in equilibrium with the metallic phase (Fe, O) and a slag of FeO. At 1600°C we find that

$$P_{O_2} = 7.7 \times 10^{-9} \text{ atm.}$$

(B) THE CARBON-OXYGEN EQUILIBRIUM

The free energy of solution of carbon in iron, according to:

$$\langle C \rangle \rightleftarrows (\underline{C})_{1\%} \tag{7.19}$$

is always negative at refining temperatures:

$$\Delta G^o = G^o_{(\underline{C})_{1\%}} - G^o_{\langle C \rangle} = 5400 - 10.1\ T \text{ (cal)} \tag{7.19a}$$

The activity coefficient of carbon in an infinitely dilute solution, i.e. γ_C^∞, in a steel can be calculated from the value of the free energy of solution:

$$\Delta G^o = RT \ln \gamma_C^\infty \cdot \frac{0.56}{12} \tag{7.19b}$$

γ_C^∞ is less than one, as is shown in Fig. 105 which gives the activity of carbon dissolved in iron as a function of its mole fraction.

The activity coefficient of carbon in steel varies considerably in the presence of a second alloying element, as shown by Fig. 114. For solutions which are dilute in carbon and other elements, (7.2) and (7.3) can be used with the interaction parameters given in Tables 20 and 21. It should be noted that the values of the interaction parameters ε_C^i for dilute solutions, change in the same direction as the interaction parameters of carbon in carbon-saturated iron; a result predicted by Shenck.[3]

For equilibrium between carbon and oxygen dissolved in the metal and carbon monoxide in the gas phase, according to:

$$(\underline{C})_m + (\underline{O})_m \rightleftarrows [CO] \tag{7.20}$$

The equilibrium constant is expressed as a function of temperature by:

$$\log K = \log \frac{p_{CO}}{(\%\ O)(\%\ C)\ .\ f_{(O)}} \cdot \frac{1}{f_{(C)}} = \frac{1056}{T} + 2.13 \tag{7.20a}$$

If, at a given temperature, the carbon and the oxygen are dilute enough for it to be assumed that they follow Henry's Law, the product (% C) . (% O) depends only on the partial pressure of CO in equilibrium with the melt:

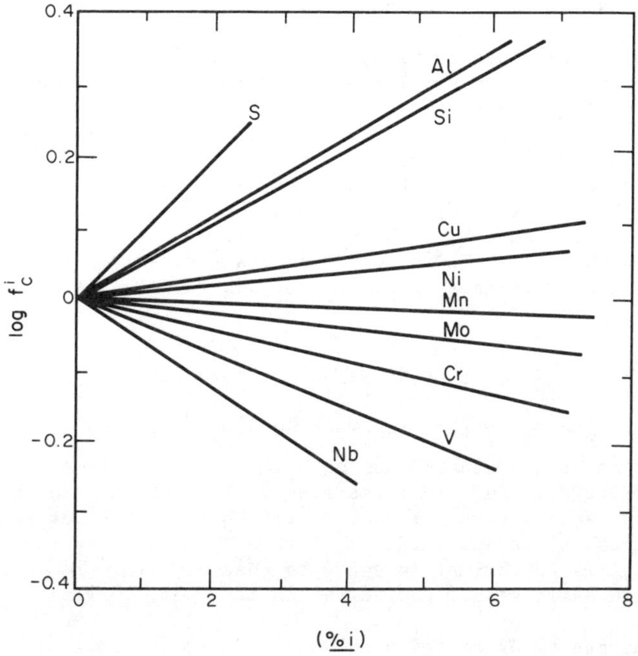

Fig. 114. Activity coefficient of carbon in solution in iron in the presence of other solutes.(18-21)

$$(\%\ C)\ .\ (\%\ O) = \frac{p_{CO}}{K} \tag{7.21}$$

This equilibrium has been evaluated experimentally by Marshall and Chipman(22) (Fig. 115) for different partial pressures of CO.

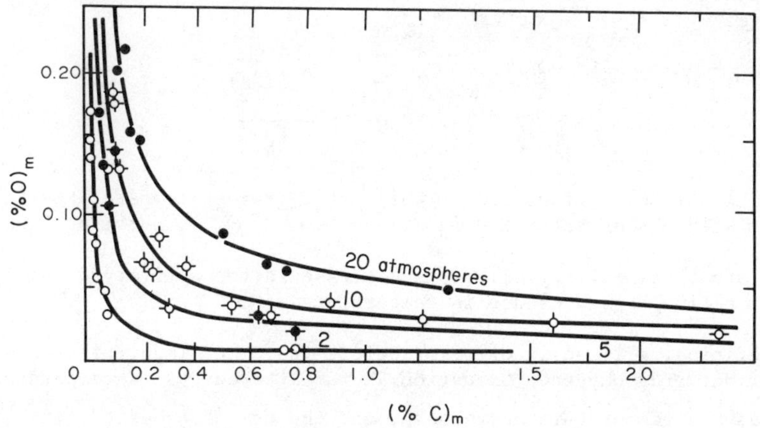

Fig. 115. Influence of CO pressure on the carbon and oxygen content in liquid iron at 1600°C.(22)

K increases as the temperature falls, with the result that the product (% C) . (% O) decreases when the steel cools. This decrease causes the generation of CO, which may escape from the ingot or remain entrapped as bubbles in the solid steel. If the amount of CO so retained is high, the ingot is "rimmed"; with less, it is "semi-killed", and with a very small amount, it is "killed". Two methods are used industrially to "kill" a steel; the first consists of eliminating the oxygen with a powerful reducing agent such as Si, Mn, or even Ca or Al. They all form oxides insoluble in iron, and the aim is to produce them in a form which can rise to the slag. This may not occur with Ca and Al. Another method consists of using a vacuum or blowing an inert gas through the steel, to decrease the partial pressure of CO and consequently, the product (% C) . (% O), as indicated by (7.21). This procedure also eliminates other gases dissolved in the steel such as hydrogen and nitrogen, which also present problems when the metal is cast and worked.

(C) SILICON-OXYGEN EQUILIBRIUM

The activity coefficient of silica in $CaO\text{-}FeO\text{-}SiO_2$ refining slags has been calculated by integration of the Gibbs-Duhem equation, starting from the experimental values of the activity of FeO in these slags. The results are shown in Fig. 91. This shows that the activity coefficient of silica is almost unaffected by changes in the mole fraction of ferrous oxide in slags with a basicity index less than 2, and that this value is approximately equal to that of SiO_2 in $CaO\text{-}SiO_2$ slags. On the contrary, in basic slags, the variation of $\gamma_{(SiO_2)}$ with $N_{(FeO)}$ is large.

The equilibrium constant for reaction:

$$(\underline{Si})_m + 2(\underline{O})_m \rightleftarrows (SiO_2)_s \tag{7.22}$$

can be deduced from the free energy of formation of SiO_2 from its elements, and from the free energy of solution of silicon and oxygen in steel. Taking the infinitely dilute solutions of silicon and oxygen expressed as weight per cent as the standard states, it is given by the following relationship:

$$\log K_{Si} = \frac{29{,}700}{T} - 11.24 \quad (\text{wt. \%}) \tag{7.22a}$$

If the percentages of $(\underline{Si})$ and $(\underline{O})$ are low enough to assume Henrian behaviour, then we have:

$$(\%\ Si)\ .\ (\%\ O)^2 = \frac{a_{(SiO_2)_s}}{K_{Si}} \tag{7.23}$$

By tracing log (% O) versus log (% Si) at different temperatures for steel in equilibrium with a slag saturated in silica ($a_{SiO_2} = 1$), straight lines of slope $-\frac{1}{2}$ are obtained (Fig. 116). A decrease in the activity of silica acts in the same direction as a decrease in temperature.

In steel refining, silicon is oxidised before the other impurities, due to its greater affinity for oxygen. Oxidation is facilitated by low temperature (K_{Si} increases as the temperature decreases), and the use of a basic slag (CaO converts SiO_2 in the slag into SiO_4^{4-} ions). Silicon is one of the most widely used elements for deoxidation, due to its large reducing power (it is usually added as ferrosilicon or silicomanganese).

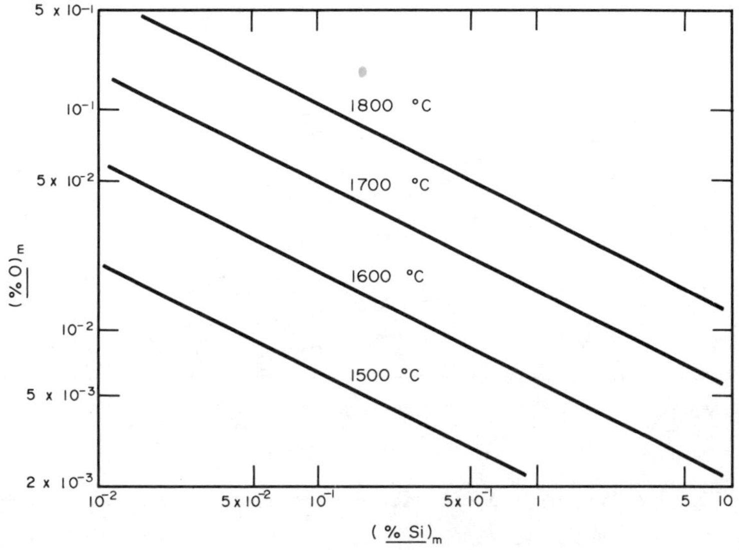

Fig. 116. Influence of oxygen on the silicon dissolved in steel in equilibrium with a slag saturated in silica.[49]

(D) MANGANESE-OXYGEN EQUILIBRIUM

The behaviour of manganese in iron has been examined earlier. In FeO-MnO binary slags, both oxides form ideal solutions in all ranges of composition. The determination of MnO activity in complex oxidised slags has been the subject of several investigations. The simplest way of expressing the results consists of taking the pseudo-ternary diagram $(SiO_2 + P_2O_5 + Al_2O_3) - (CaO + MgO) - (FeO + MnO)$ and drawing activity coefficient curves, as shown in Fig. 117.

As can be seen, the activity coefficient of the basic manganese oxide, MnO, is higher in basic slags than in acid. If Figs. 90 and 117 are compared, it can be seen that the activity of MnO decreases faster than that of FeO when the acidity is raised.

The equilibrium between manganese and oxygen dissolved in the metal, and MnO in the slag, can be analysed by considering the reaction:

$$(\underline{Mn})_m + (\underline{O})_m = (MnO)_s \qquad \log K = \frac{12,210}{T} - 5.38 \qquad (7.24)$$

or

$$(\underline{Mn})_m + (FeO)_s \rightleftarrows (\underline{Fe})_m + (MnO)_s \qquad (7.25a)$$

The equilibrium constant $(K_{(7.25a)})$ is:

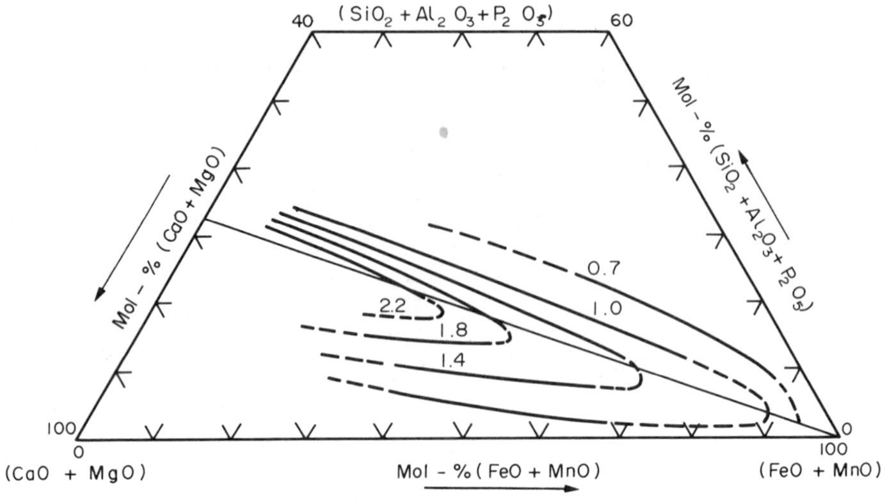

Fig. 117. Activity coefficient of manganese oxide in complex slags at 1530/1700°C.[23]

$$K = \frac{\gamma_{(MnO)_s} \cdot N_{(MnO)_s} \cdot \gamma_{(Fe)} \cdot N_{(Fe)}}{\gamma_{(FeO)_s} \cdot N_{(FeO)_s} \cdot f_{(Mn)_m} \cdot (\%\ Mn)_m} \qquad (7.25b)$$

Grouping the activity coefficients, γ_{Fe}, f_{Mn}, and the equilibrium constant K into the same term, and simplifying on the basis that N_{Fe} is close to one in steels, the "apparent constant" K' of the reaction can be obtained:

$$K' = K \frac{\gamma_{(FeO)_s}}{\gamma_{(MnO)_s}} \cdot \frac{f_{(Mn)_m}}{\gamma_{(Fe)_m}} = \frac{N_{(MnO)_s}}{N_{(FeO)_s} \cdot (\%\ Mn)_m} \qquad (7.26)$$

Apart from the change with the temperature, K' has two different values, depending on whether the slag is acid or basic, as is shown in Fig. 118.[24] This is due mainly to the change in the activity coefficient of MnO, which decreases faster than that of FeO on increasing the acidity of the slag. The metal-slag partition coefficient $(\%\ MnO)_s/(\%\ Mn)_m$ in steel refining depends on the basicity of the slag, its oxidation level, and temperature. This coefficient is represented at 1600°C as a function of the FeO content of the slag by two straight lines (Fig. 119): one for acid, and the other for basic slags. As can be deduced from this figure, deoxidation with manganese will be more effective in acid than basic slags, since for the same degree of oxidation the manganese partition coefficient is far greater in acid slags. Deoxidation by Mn is more effective when it is added as silicomanganese than when ferromanganese is the source.

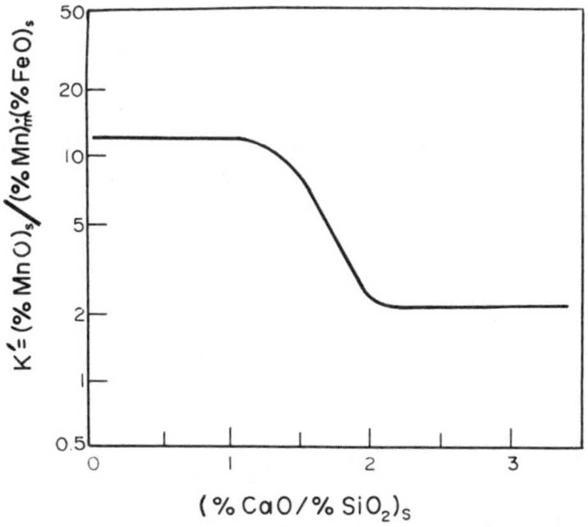

Fig. 118. Influence of the basicity of refining slags on the apparent equilibrium constant of (7.26).[24]

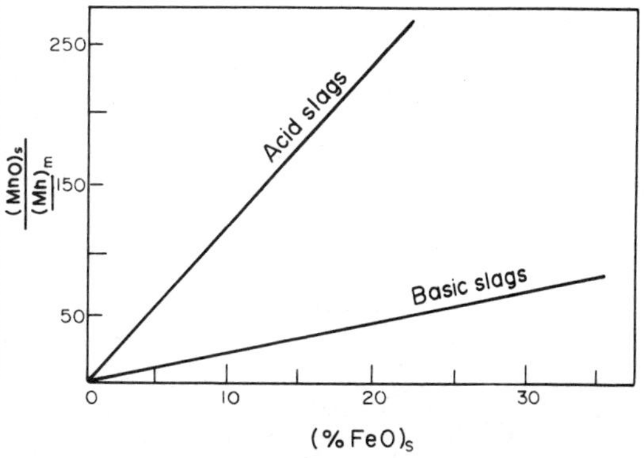

Fig. 119. Influence of the oxidation level of the slag on the partition coefficient of manganese for acid and basic slags.[24]

(E) CHROMIUM-OXYGEN EQUILIBRIUM

For practical purposes, the mole fraction of chromium in solution in iron can be taken as the activity and, like manganese, the "apparent" equilibrium constant for chromium oxide and iron oxide in slags in equilibrium with chromium in liquid steel depends on the temperature and the basicity of the slag. It has been shown that the state of oxidation of the chromium in the slag is a function of the basicity and the general state of oxidation: in an acid slag, chromium exists as chromous oxide, whereas in a basic slag, chromites^{2+} (or chromates^{3+}) are formed. Therefore, two "apparent" constants exist, corresponding to:

$$\underline{(Cr)}_m + (FeO)_s \rightleftarrows (CrO)_s + \underline{(Fe)}_m \tag{7.27}$$

and

$$2\underline{(Cr)}_m + 3(FeO)_s \rightleftarrows (Cr_2O_3)_s + 3\underline{(Fe)}_m \tag{7.28}$$

The "apparent" constants are expressed, respectively, by:

$$K_1' = \frac{(\%\ CrO)_s}{(\%\ FeO)_s \cdot (\%\ Cr)_m} \tag{7.27a}$$

and

$$K_2' = \frac{(\%\ Cr_2O_3)_s}{(\%\ Cr)_m^{\ 2} \cdot (\%\ FeO)_s^{\ 3}} \tag{7.28a}$$

Figure 120[25] indicates the values of K_1' and K_2' and their variation with basicity. The partition coefficient for chromium between metal and an acid slag is shown in Fig. 121 as a function of the percentage of FeO in the slag. This coefficient is much smaller in basic than in acid slags.

The refining of a steel containing chromium and carbon at normal temperatures results in heavy losses of chromium to the slag. In the manufacture of stainless steel from scrap, where it is necessary to have less than 0.04% C and to retain the maximum amount of alloying elements, selective oxidation of carbon relative to chromium involves operation at 1700°C with a basic slag. Other alternatives applied in the manufacture of stainless steels to oxidise carbon in preference to other elements, particularly chromium, consist of lowering the partial pressure of CO by blowing with a mixture of argon and oxygen, so as to displace the C-O equilibrium towards the formation of CO without displacing the equilibria of the other solute elements.

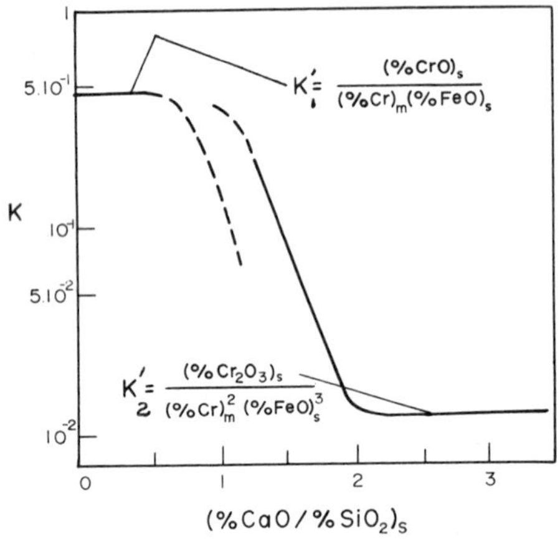

Fig. 120. Influence of slag basicity on the apparent equilibrium constant for chromium in steel in equilibrium with a CaO-FeO-SiO_2 slag at 1600°C.[25]

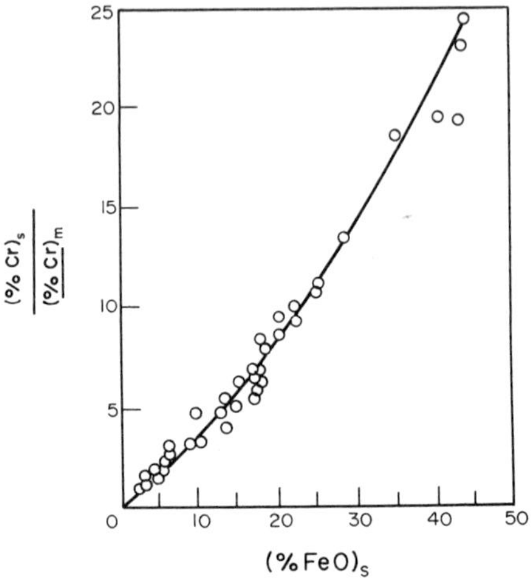

Fig. 121. Influence of the state of oxidation of the slag on the distribution of chromium between steel and acid slag at 1600°C.[26]

7.2.4. DESULPHURISATION

(A) SULPHUR IN METAL AND SLAG

The binary diagram Fe-S (Fig. 122)[27] indicates the presence of an eutectic at 988°C and 31% S. Iron and sulphur form a continuous liquid solution at all temperatures above the liquidus, in contrast to oxygen, where the saturation value and the separation of an oxide phase occur when oxygen in solution in Fe exceeds the amount given by (7.16). The solubility of sulphur in solid iron is practically nil. Therefore, when liquid iron containing sulphur cools, iron is precipitated until the temperature reaches 988°C, then the eutectic is precipitated on the grain boundaries. The solid, consisting of grains of iron embedded in the brittle low melting point eutectic, has no mechanical strength, especially at high temperatures, and it cannot be rolled. For this reason, the sulphur content of high quality steels should not exceed 0.020%. The effects of sulphur are so deleterious that further improvements have been obtained by reducing the content to below 0.010%.

The dilute solutions of sulphur in liquid iron exhibit a strong negative deviation from the ideal condition, due to the large mutual affinity of these two elements. This results in negative values for the free energy of solution of gaseous sulphur in liquid iron, according to:

$$[S_2] = 2(S)_{1\%} \qquad \Delta G^o = -63{,}040 + 10.54\ T\ \text{(cal)} \tag{7.29}$$

The activity coefficient of sulphur in an infinitely dilute solution, which can be calculated from this free energy value, $\Delta G^o = RT \ln \left(\gamma_S^\infty \cdot \frac{0.56}{32}\right)$, is then less than one.

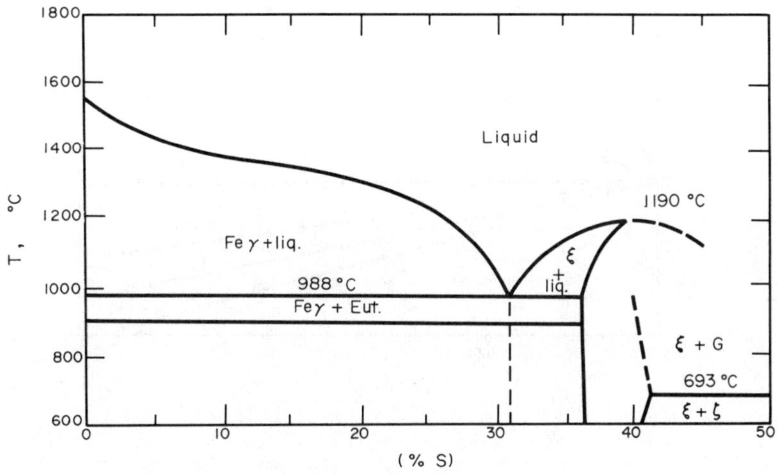

Fig. 122. The iron-sulphur phase diagram.[27]

The addition of a second solute or an increase of the sulphur content in the melt, modifies the activity coefficient, as indicated by Fig. 123. For very dilute solutions, the interaction parameters indicated in Tables 20 and 21 can be used.

Sulphur exists in slags as sulphide if the oxygen partial pressure in equilibrium with the slag is less than 10^{-5} atm, and in the form of sulphate if P_{O_2} is higher than 10^{-3}.[60] But if P_{O_2} is higher than 10^{-5}, all of the iron of the metal phase would be oxidised, as can be determined from the Fe-O equilibrium. Therefore, sulphur exists only as sulphide in slag in equilibrium with steel.

(B) SULPHUR-OXYGEN EQUILIBRIUM BETWEEN METAL AND SLAG: APPLICATION OF FLOOD'S THEORY[35]

The equilibrium between sulphur and oxygen dissolved in iron or steel and in a slag, can be expressed in an ionic form (see 6.3.1):

$$(\underline{S})_m + (O^{2-})_s \rightleftarrows (S^{2-})_s + (\underline{O})_m \tag{7.30}$$

The basic oxides participating in the equilibrium supply the O^{2-} anions. In order to displace the equilibrium in (7.30) to the right, the elements forming these oxides should have affinities for sulphur, similar to those that they have for oxygen. Taking as the standard state for sulphur and oxygen the infinitely dilute solutions expressed in weight percentage, the differences between the affinity of the elements for sulphur and for oxygen in solution (that is, $\Delta A^o = -\Delta G^o$ of formation of MS, $+\Delta G^o$ of formation of MO) are:

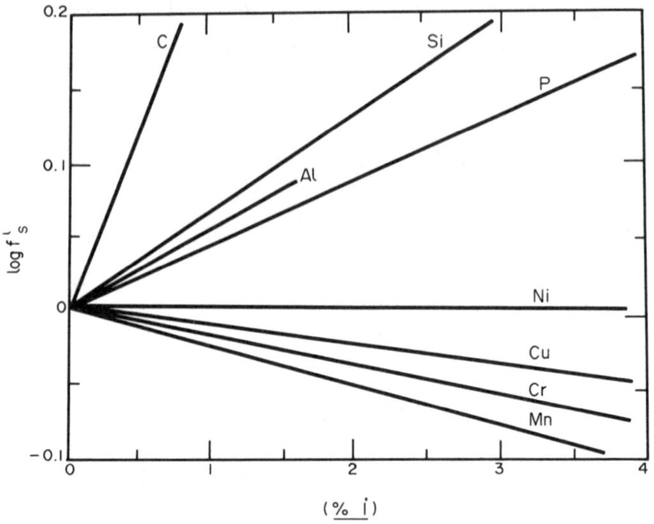

Fig. 123. Influence of a second solute on the activity coefficient of sulphur in liquid iron.[28-34]

$$(Na_2O)_s + (\underline{S})_m \rightleftarrows (Na_2S)_s + (\underline{O})_m \tag{7.31}$$

$$\Delta A^o = -5750 + 17.46\ T - 2.88\ T \log T \quad : (\Delta A^o_{1600°C} = +9300)$$

$$(CaO)_s + (\underline{S})_m \rightleftarrows (CaS)_s + (O)_m \tag{7.32}$$

$$\Delta A^o = -25{,}190 + 6.7\ T \quad : (\Delta A^o_{1600°C} = -12{,}600)$$

$$(FeO)_s + (\underline{S})_m \rightleftarrows (FeS)_s + (\underline{O})_m \tag{7.33}$$

$$\Delta A^o = -28{,}200 + 5.2\ T \quad : (\Delta A^o_{1600°C} = -18{,}450)$$

$$(MnO)_s + (\underline{S})_m \rightleftarrows (MnS)_s + (\underline{O})_m \tag{7.34}$$

$$\Delta A^o = -35{,}930 + 10.14\ T \quad : (\Delta A^o_{1600°C} = -16{,}900)$$

$$(MgO)_s + (\underline{S})_m \rightleftarrows (MgS)_s + (\underline{O})_m \tag{7.35}$$

$$\Delta A^o = -50{,}550 + 32.79\ T - 7.37\ T \log T \quad : (\Delta A^o_{1600°C} = -34{,}300)$$

At 1600°C, sodium is the only element that has a larger affinity for sulphur than for oxygen $(\Delta A^o > 0)$. Sodium oxide is used as a desulphurising agent in mixers which can contain metal from several taps of pig iron, and are used to diminish the variations in composition of the blast furnace product. This oxide is also the desulphurising agent when soda ash (Na_2CO_3) is used in the tapping ladle to remove sulphur from steel. Since Na_2O is extremely corrosive toward refractories, it cannot be used as a slag component in a furnace. In addition, Na_2O and K_2O are not desirable in the blast furnace because they are volatilised at the level of the tuyeres and condensed on the sides of the shaft, sometimes forming large accretions or "scaffolds".

The other elements in this table have higher affinities for oxygen than for sulphur, but at 1600°C it can be seen that the difference of affinities of calcium for oxygen and sulphur are of the same order of magnitude as those of Fe and Mn for these elements.

In a slag containing several basic oxides (CaO, MgO, FeO, MnO), Flood's theory (see Chapter 6, from (6.16) to (6.29)) assumes that the apparent constant K' of the complex reaction (7.30):

$$K' = \frac{N_{(S^{2-})_s}}{N_{(O^{2-})_s}} \cdot \frac{(\%\ O)_m}{(\%\ S)_m} \tag{7.36}$$

is a function of the equivalent fractions $N_{M^{2+}}$ of cations M^{2+} $(Ca^{2+}, Mn^{2+}, Fe^{2+}, Mg^{2+})$ which participate in the equilibria, and of the constants K_M of the individual reactions (7.32) to (7.35):

$$\log K' = \Sigma N'_{M^{2+}} \times \log K_M - \log g(f) \tag{7.37}$$

where $g(f)$ is the ratio of the activity coefficients $f_{(\underline{O})_m}$ and $f_{(\underline{S})_m}$.

In Fig. 124, log K' is plotted against $N_{Ca^{2+}}$. It can be seen that log K' is not a function of $N_{Ca^{2+}}$ but a constant, which can be explained in the simplest way by the fact that the individual constants $\log K_M$ are similar to each other and, as $N_{Ca^{2+}}$ increases, $N_{Mn^{2+}}$ and $N_{Fe^{2+}}$ decrease, so the term $\Sigma N_{M^{2+}} \cdot \log K_M$ changes very little.

When sulphur and oxygen in solution in iron follow Henry's Law, the term log g(f) is nil. Since the individual constants, $\log K_M$, have similar values, (7.37) may be written in the form:

$$\log K' = \log K_M \left(N_{Ca^{2+}} + N_{Fe^{2+}} + N_{Mn^{2+}}\right) \tag{7.38}$$

The sum of the equivalent cationic fractions is equal to one, and therefore $\log K' = \log K_M$. It has been found experimentally that $\log K' = -1.35$ (Fig. 124) when the concentrations of oxygen and sulphur are expressed in moles, and -1.65

when these are expressed in weight percentage. These are very close to the values of the logarithms of the individual reaction constants for (7.32) to (7.34).

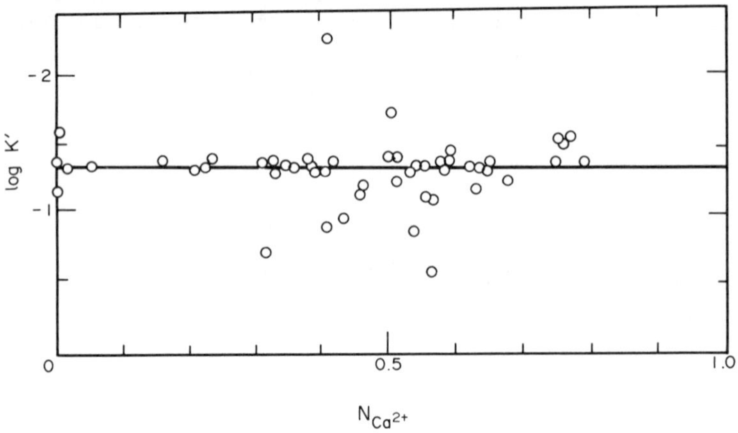

Fig. 124. Experimental relationship between log K' and $N_{Ca^{2+}}$.(35)

The partition coefficient of sulphur, $(\%\ S)_s/(\%\ S)_m$, can be calculated from the apparent equilibrium constant K'. In (7.36), it is sufficient to replace $N_{(S^{2-})}/N_{(O^{2-})}$ by $\left(n_{S^{2-}}\right)/\left(n_{O^{2-}}\right)$, and $n_{(S^{2-})}$ by $(\%\ S)_s/32$ in order to find the following relationship:

$$\frac{(\%\ S)_s}{(\%\ S)_m} = 32\ K' \cdot \frac{n_{(O^{2-})_s}}{(\%\ O)_m} = K'' \cdot \frac{n_{(O^{2-})_s}}{(\%\ O)_m} \tag{7.39}$$

$n_{(S^{2-})}$ and $n_{(O^{2-})}$ represent the number of S^{2-} and O^{2-} ions in 100 g of slag. The constant K", deduced from Fig. 124, has a value of 0.71. This result is similar to the value obtained experimentally and independently by plotting the partition coefficient of sulphur $(\%\ S)_s/(\%\ S)_m$ as a function of $n_{(O^{2-})}/(\%\ O)_m$. The slope K" of the line in Fig. 125 has a value near to that deduced from Fig. 124 (K" = 0.6, instead of 0.71). These figures indicate that all of the basic oxides are in the same category as suppliers of O^{2-} ions.

(C) CONDITIONS FOR EFFICIENT REMOVAL OF SULPHUR BY SLAG

The fact that the constants (K_{Ca}, K_{Mn}, K_{Fe}) are approximately equal does not mean that the corresponding basic oxides have the same desulphurising capacity. Iron oxide in the slag supplies O^{2-} anions in the same way as lime, but there is also an equilibrium between iron oxide in the slag and oxygen in the molten metal. An increase in FeO in the slag increases the amount of dissolved oxygen. This causes a decrease in the ratio $n_{(O^{2-})}/(\%\ O)_m$ and consequently, according to (7.39) and Fig. 125, there is a decrease of the partition coefficient. As $(\%\ O)_m$

is proportional to the FeO activity in the slag, and therefore to its mole fraction, (7.39) can be expressed in the form:

$$\log \frac{(\%\ S)_s}{(\%\ S)_m} = \text{constant} + \log n_{(O^{2-})} - \log a_{FeO} \tag{7.40}$$

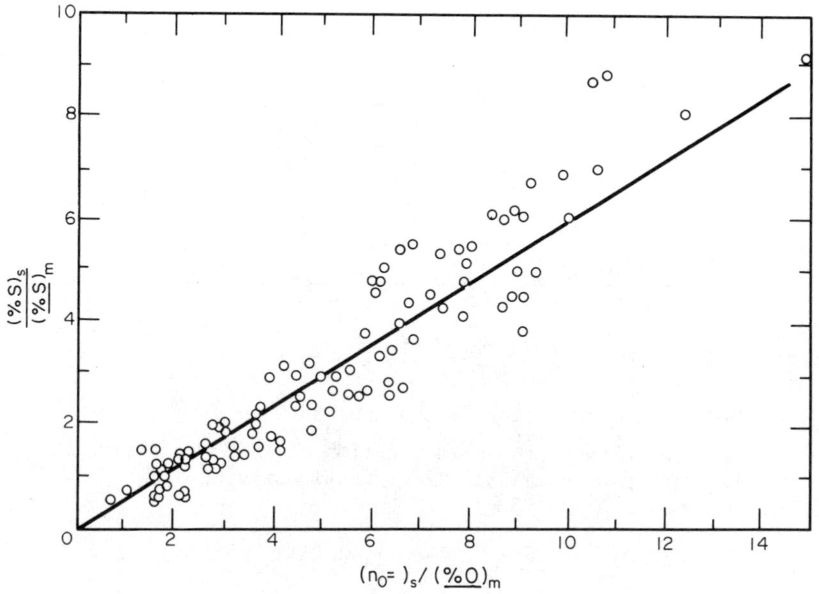

Fig. 125. Partition coefficient for sulphur between slag and metal versus $(n_{O^{2-}})/(\%\ O)$.[36]

The partition coefficient for sulphur is then a function of the basicity and the oxidising power of the slag. This function is represented in Fig. 126,[37] taking the basicity as $i = CaO/SiO_2$. In this figure it should be noted that when the mole percentage of FeO is small, $\log (\%\ S)_s/(\%\ S)_m$ decreases linearly with $\log N_{FeO}$. Starting from approximately 4 mole % FeO, the effect of this oxide on the value of $\log n_{(O^{2-})}$ is appreciable and, for a higher value of N_{FeO}, the logarithm of the partition coefficient rises again.

Desulphurisation is then most efficient with a basic slag in reducing conditions. The basic oxide could be MnO or CaO, and the reducing agent carbon or any other. As can be deduced from the affinity value of reaction (7.35), MgO is inferior to other basic oxides as a desulphurising agent.

(D) SULPHUR REMOVAL WITH CALCIUM OXIDE

In a blast furnace operating with a high lime/silica ratio, the slag in equilibrium with pig iron saturated in carbon contains a high percentage of calcium oxide. The sulphur removal is then the result of a combination of the following reactions:

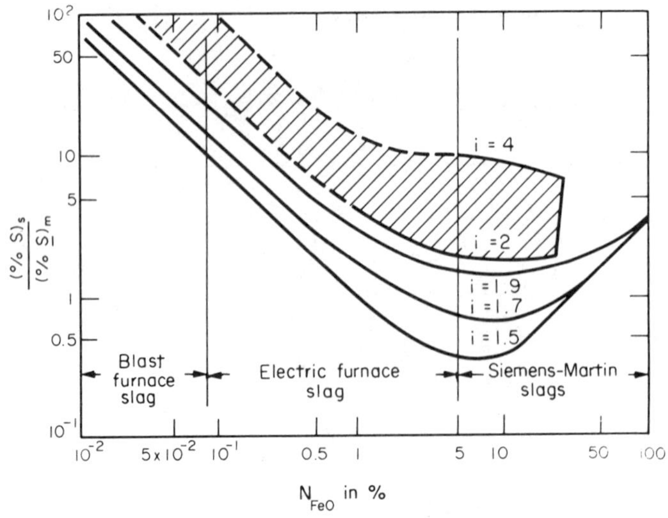

Fig. 126. Partition coefficient for sulphur between slag and metal as a function of the mole fraction of FeO in slag.[37] Hatched area: usual range for desulphurising slag

$$(CaO)_s + \underline{(FeS)}_m \rightleftarrows (FeO)_s + (CaS)_s \tag{7.41}$$

and

$$(FeO)_s + \langle C \rangle \rightleftarrows \underline{(Fe)} + [CO] \tag{7.42}$$

and the overall reaction is:

$$(CaO)_s + \underline{(FeS)}_m + \langle C \rangle \rightleftarrows (CaS)_s + \underline{(Fe)}_m + [CO] \tag{7.43}$$

the standard free energy of which is:

$$\Delta G^o = 31,140 - 34.4\ T\ (cal) \tag{7.43a}$$

This endothermic reaction requires high temperatures in order to displace the equilibrium toward the right. Figure 127 indicates the sulphur percentage in pig iron saturated in carbon in equilibrium with a $CaO\text{-}SiO_2\text{-}Al_2O_3$ slag containing 1.5% S at 1500°C. The influence of the CaO content is substantial, as can be seen from the fall in $(\% S)_m$.

(E) SULPHUR REMOVAL WITH MANGANESE

In a steel containing manganese in equilibrium with a manganiferous slag, MnO partially replaces calcium oxide and, in this case, manganese is the reducing agent:

$$(MnO)_s + (\underline{S})_m \rightleftarrows (MnS)_s + (\underline{O})_m \tag{7.44}$$

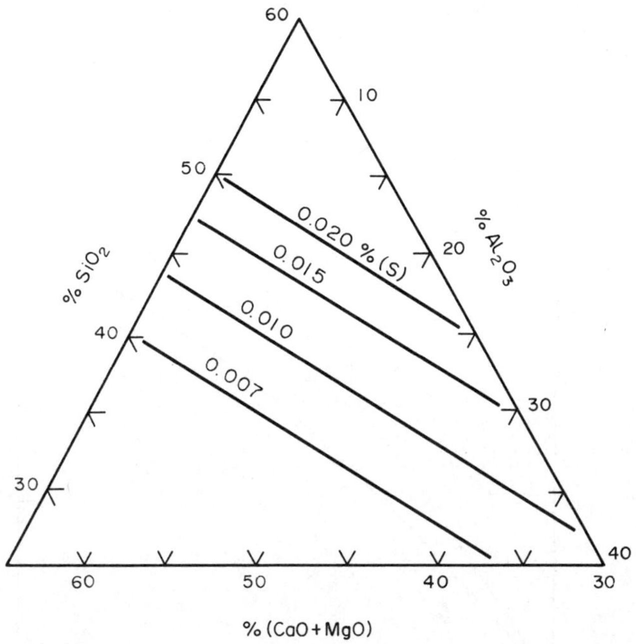

Fig. 127. Sulphur content of carbon-saturated iron in equilibrium with a $CaO-SiO_2-Al_2O_3$ slag containing 1.5% S at 1500°C.[38]

and

$$\underline{(Mn)}_m + \underline{(O)}_m \rightleftarrows (MnO)_s \tag{7.45}$$

These can be combined to give:

$$\underline{(Mn)}_m + \underline{(S)}_m \rightleftarrows (MnS)_s \tag{7.46}$$

or, applying Temkin's ionic theory:

$$\underline{(Mn)}_m + \underline{(S)}_m \rightleftarrows (Mn^{2+})_s + (S^{2-})_s \tag{7.47}$$

If the equilibrium constant of (7.47) is expressed as a function of the activity coefficients, the ionic fractions and weight percent, we have:

$$K = \frac{1}{f_{(S)_m} \cdot f_{(Mn)_m}} \cdot \frac{N_{(S^{2-})_s} \cdot N_{(Mn^{2+})_s}}{(\%\ S)_m \cdot (\%\ Mn)_m} \tag{7.48}$$

The partition coefficient for sulphur is:

$$\frac{(\%\ S)_s}{(\%\ S)_m} = K' \cdot g'(f) \cdot \frac{(\%\ Mn)_m}{(\%\ Mn)_s} \tag{7.49}$$

In Fig. 128, $(\% S)_s/(\% S)_m$ has been plotted against $(\% Mn)_m/(\% Mn)_s$ at 1500°C and 1600°C. As long as Henry's Law is followed in the metal $(g'(f) = 1)$ the relationship is represented by straight lines of slope $K' = 28$ at 1500°C, and $K' = 21$ at 1600°C

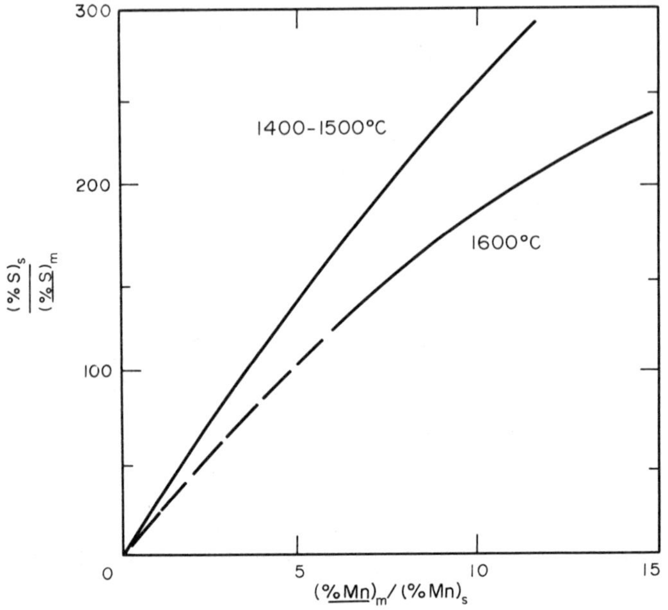

Fig. 128. Metal-slag distribution of sulphur as a function of the manganese distribution.[52]

Reducing agents such as Si and C enhance the slag-to-metal transfer of manganese, so favouring desulphurisation.

(F) SULPHUR REMOVAL WITH CaC_2 OR METALS

Complete removal of S from pig iron is possible by adding CaC_2. Metals such as Ce, La, Ca or Mg can be used for treating steels, but the steels must be deoxidised to a very low level before they are added, because they have a greater affinity for oxygen than for sulphur, as the solubility product values $K_{M_xS_y}$ and $K_{M_xO_y}$ indicate:

$$K_{M_xO_y} = (\% M)_M^x \times (\% O)_M^y \Big/ a_{(M_xO_y)_S}$$

At 1600°C $K_{MgS} = 2 \times 10^{-6}$, $K_{CaS} = 1 \times 10^{-9}$, $K_{CeS} = 2 \times 10^{-5}$, $K_{LaS} = 1 \times 10^{-6}$,

$K_{MgO} = 1 \times 10^{-6}$, $K_{CaO} = 6 \times 10^{-11}$, $K_{Ce_2O_3} = 1 \times 10^{-17}$, $K_{La_2O_3} = 1 \times 10^{-19}$.

Industrially, desulphurisation is effected by Ca or Mg (cheaper than Ca). As both are volatile, better yields can be obtained by blowing calibrated spheres 1-2 mm diameter with Argon, from the bottom of the bath, or by controlled immersion of thin walled steel tube filled with Mg or Ca in the ladle.

7.2.5. DEPHOSPHORISATION

(A) PHOSPHORUS IN SLAG AND METAL

Phosphorus has a high affinity for iron, forming Fe_2P and Fe_3P. This gives a negative deviation of the activity curve relative to Raoult's Law and a significantly negative value for the free energy of solution:

$$(P_2) = 2(P)_{1\%\ \text{in Fe}} \qquad \Delta G^o = -58{,}500 - 9.1\ T\ \text{cal} \qquad (7.50)$$

Phosphorus in slag is in the oxidised state and the equilibrium between metal and slag is represented by:

$$2(P)_m + 5(O)_m = (P_2O_5) \qquad (7.51)$$

Taking the infinitely dilute solution, expressed in weight percent, as the standard state for $(P)_m$ and $(O)_m$, and as vapour at 1 atm for P_2O_5, the standard free energy for this reaction is given by:

$$\Delta G_P^o = -179{,}200 + 131.8\ T \qquad (7.51a)$$

Since $$a_{P_2O_5} = P_{P_2O_5} = N_{P_2O_5} \times \gamma_{P_2O_5}$$

then $$\Delta G_P^o = -RT \ln K_P = -RT \ln \frac{N_{(P_2O_5)_s} \times \gamma_{(P_2O_5)_s}}{a^5_{(O)_m} \times a^2_{(P)_m}} \qquad (7.51b)$$

$P_{P_2O_5}$ is the pressure of this oxide in atmospheres in equilibrium with the slag. At working temperature, ($\sim$ 1600°C) the value of ΔG_P^o is positive, i.e. phosphorus will not form an oxide unless $\gamma_{P_2O_5}$ is reduced to a very low value. This is effected when a basic oxide is added to the slag and combines with P_2O_5 to form a stable compound. Experimental results obtained for the activity of P_2O_5 in slags can be interpreted in two ways, depending on whether the slag in equilibrium with iron is completely fluid, or contains an excess of solid lime. In the first case, the slag-metal equilibrium can be interpreted according to molecular theory, as in Problem 21, or by applying Flood's theory. When the slag is saturated with lime, it can be seen that $a_{P_2O_5}$ depends essentially on the degree of oxidation of the slag.

(B) APPLICATION OF FLOOD'S THEORY[41]

According to Flood's theory, the equilibrium ratio of reaction (7.51),

$$K = N_{(P_2O_5)_s} \Big/ (\%\ O)_m^5 \times (\%\ P)_m^2$$

can be written in the form:

$$\log K = \Sigma N_{M^{++}} \log K_M \qquad (7.52)$$

where $N_{M^{++}}$ is the ionic fraction of basic oxide in the slag and K_M is the equilibrium ratio of (7.51) when there is only the basic oxide MO in the slag. Comparing the equilibrium constant K_P in (7.51b) with the equilibrium ratio K_M in (7.52) gives:

$$\log \gamma_{P_2O_5} = \log K_P - \Sigma N_{M^{++}} \log K_M \qquad (7.53)$$

Turkdogan and Pearson[41] found the following relationship between $\log P_2O_5$, the temperature and the mole fraction of each oxide forming the slag, by plotting $\log \gamma_{P_2O_5}$ against $N_{MO} \log K_M$ from experimental results:

$$\log \gamma_{P_2O_5} = -1.12\left(22N_{CaO} + 15N_{MgO} + 13N_{MnO} + 12N_{FeO} - 2N_{SiO_2}\right) - \frac{42{,}000}{T} + 23.58 \qquad (7.54)$$

Figure 129 shows that $\log \gamma_{P_2O_5}$ increases with T and so phosphorus removal is improved at low temperature. Magnesia improves phosphorus removal, but the influence, as indicated by the value of $\log \gamma_{P_2O_5}$ is only about 2/3 of the value for CaO. This 2/3 coefficient appears in experimental expression for the basicity index i:

$$i = \frac{\%\ CaO + 2/3\ \%\ MgO}{\%\ SiO_2 + \%\ P_2O_5}$$

Iron oxide plays a double role in dephosphorisation, as oxidant and as base. A high content of FeO in the slag corresponds to highly oxidised metal which, as has been seen, favours dephosphorisation. However, (7.54) shows that CaO has a higher dephosphorising power than FeO. Therefore, an increase in the oxidation level of the metal, as a result of raising the FeO in the slag, corresponds to a dilution of CaO and therefore to a decrease in the dephosphorising power of the slag. The two effects of FeO in the slag are in opposition. Up to 16-20% FeO, the oxidising effect is dominant and the partition coefficient, tending to transfer phosphorus to the slag, increases with the FeO. Above 20% FeO, the dilution effect becomes evident and the partition coefficient decreases, as is shown in Fig. 130a for different basicity values.[42]

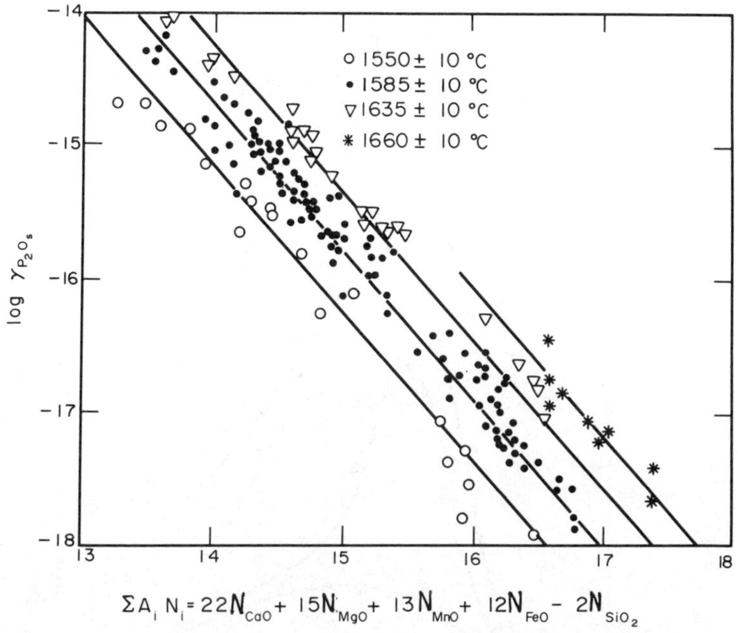

Fig. 129. Log $\gamma_{P_2O_5}$ versus $\Sigma N_{MO} \cdot \log K_M$ according to (7.54) at different temperatures.[41, 51]

(C) SLAGS SATURATED WITH CaO AND 4CaO . P_2O_5[61]

The activity of P_2O_5 in the binary CaO-P_2O_5 system has been studied by Schwertfeger and Endell[62] by an electrochemical method. The standard state selected is such that $a_{P_2O_5} = 1$ in the presence of CaO and 4CaO . P_2O_5 (or C_4P). From the standard free energy of the reaction

$$4 \langle CaO \rangle + \underset{(1\ atm)}{\{P_2O_5\}} = \langle 4CaO \cdot P_2O_5 \rangle \qquad (7.55a)$$

$$\Delta G^o = -170{,}800 + 13.2\ T \qquad (7.55b)$$

it is possible to calculate the partial pressure of P_2O_5 corresponding to this equilibrium with these two condensed phases.

The effect of composition on the activity and pressure of P_2O_5 in these slags can be calculated from the data of Schwertfeger and Endell[62] given in Table 22. The activity of P_2O_5 increases rapidly as the mole fraction increases in the slag.

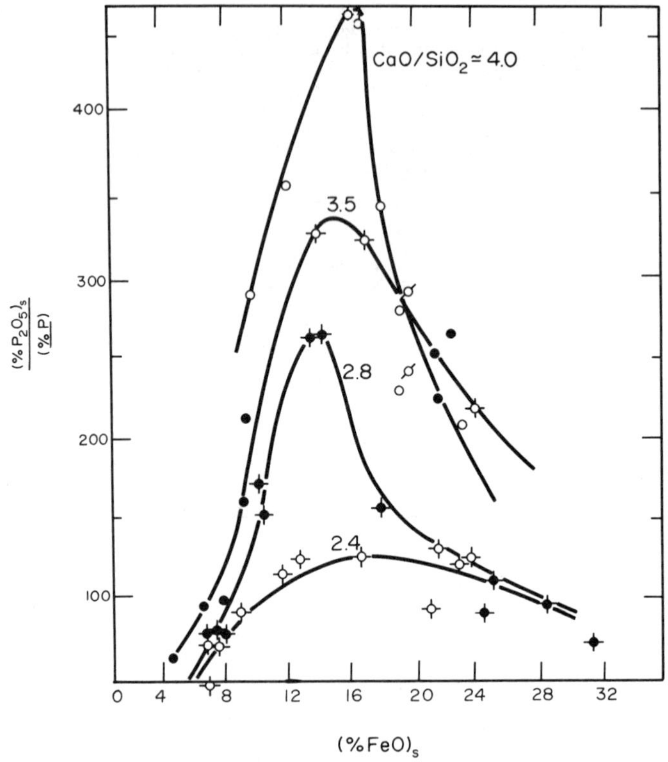

Fig. 130a. Influence of the FeO content and the basicity index of the slag on the partition coefficient of phosphorus.(42)

TABLE 22

$a_{P_2O_5}$ and $P_{P_2O_5}$ for $CaO-P_2O_5$ slags at 1650°C				
Condensed phases	CaO + C_4P	$C_4P + L_1$	$C_3P + L_1$	$C_3P + L_2$
log $a_{P_2O_5}$	0 (Std. St.)	0.63	1.7	9.28
$P_{P_2O_5}$ atm	1×10^{-17}	4.3×10^{-17}	5×10^{-16}	1.9×10^{-8}

Figure 131a gives an approximate isothermal section of the ternary $CaO-FeO-P_2O_5$ diagram at about 1600°C. In Fig. 131b, a_{FeO} and log $a_{P_2O_5}$, calculated in Ref. 61 from the results of Tromel and Fix,(63) are plotted against % FeO in a slag saturated with either $4CaO \cdot P_2O_5$ or CaO and in equilibrium with Fe at 1600°C. The activity of FeO increases rapidly up to 10% and then stabilises

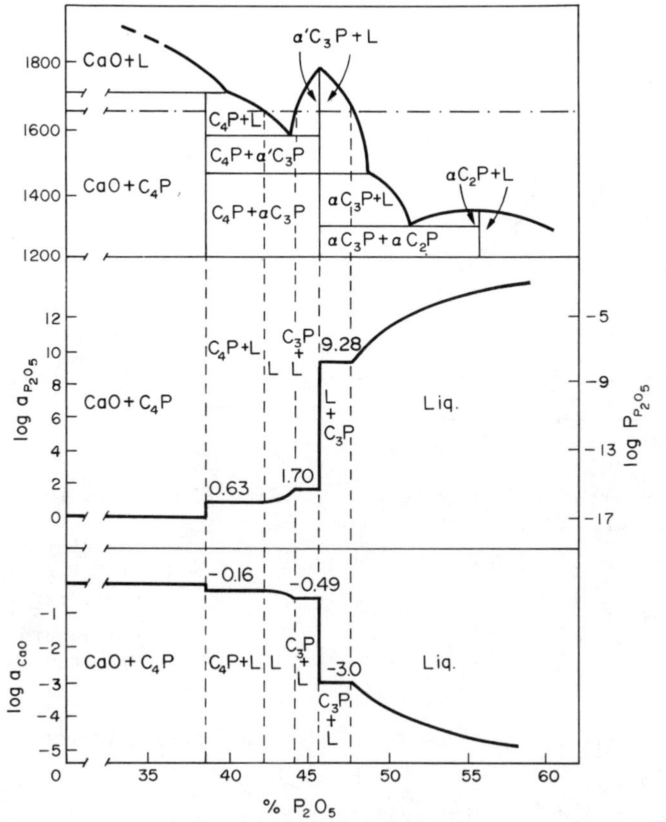

Fig. 130b. Portion of the $CaO-P_2O_5$ phase diagram and activity states at 1650 C[61, 62]

around 0.32 along the lime saturation line. Log $a_{P_2O_5}$ decreases continuously as N_{FeO} increases.

By combining the curves of Fig. 131 with previous data, the activities of $(O)_m$ and $(P)_m$ in iron in equilibrium with lime-saturated slags can be calculated. For example, when the slag contains 20% FeO, i.e. $a_{FeO} = 0.32$, $\log P_{P_2O_5} = -17.2$ (Fig. 131b). The activity of (O) can be calculated from (7.16) and (7.18):

$$(\%O) = 0.23 \times 0.32 = 0.0736.$$

Then $a_{(P)}$, from (7.51), is 0.015.

The presence of SiO_2 and MnO in such a slag enlarges the domain of the liquid in equilibrium with CaO and C_4P in Fig. 131a. These compounds reduce the activities of FeO and P_2O_5, factors acting in opposite directions relative to

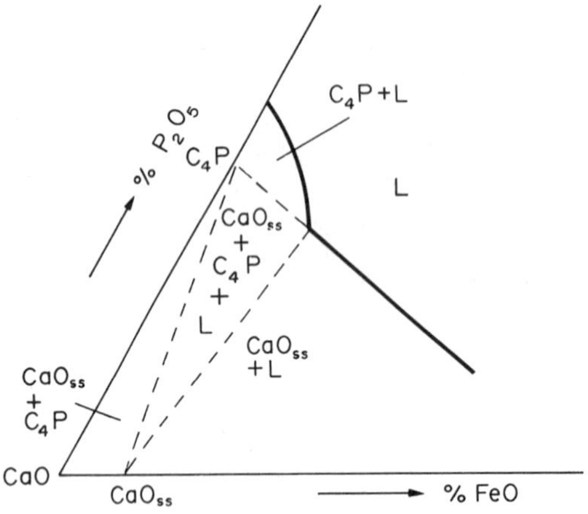

Fig. 131a. Approximate isothermal section of the $CaO-FeO-P_2O_5$ system at 1600 C[61, 63]

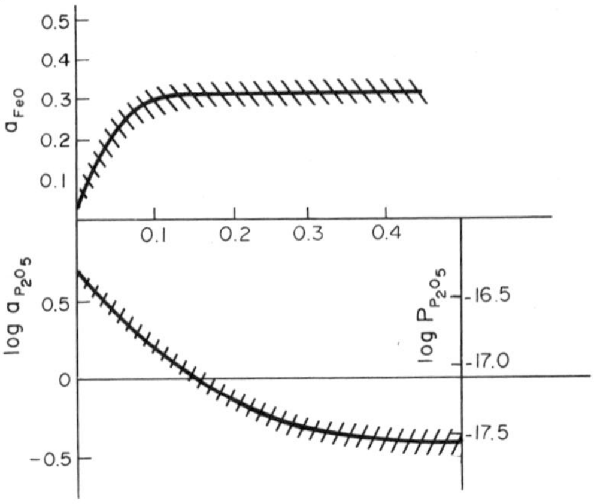

Fig. 131b. a_{FeO}, log $a_{P_2O_5}$ and log $P_{P_2O_5}$ for slag saturated with CaO or C_4P at 1600°C. Standard state for P_2O_5 is log $a_{P_2O_5} = 0$ for liquid in equilibrium with C_4P and CaO (see Table 16). (61)

removal of phosphorus from the metal, thus the overall effect of SiO_2 and MnO is unimportant.

To summarise, the best conditions for phosphorus removal are:

> A large volume of highly basic and lime-rich slag. A satisfactorily high level of oxidation in the metal. The lowest possible temperature.

These conditions are achieved in the basic open hearth process, in the basic converter with air or oxygen, and in the basic electric furnace during the oxidation period at the beginning of a cycle. It is necessary to draw attention to the fact that before deoxidation of the metal, the slag which is rich in phosphorus must be removed. If this is not done, phosphorus will be reduced back to the metal during deoxidation and carburisation. In high quality steels, phosphorus should not exceed 0.020%.

7.2.6. GASES IN STEEL

(A) GENERAL

The diatomic gases (CO, O_2, S_2, N_2, H_2, etc.) dissolve in iron and steel in the monatomic form. They are more soluble at high temperature, and in liquid iron than in solid iron. During the cooling of an ingot, the gases will be evolved, sometimes violently, projecting metal out of the mould, or they will form blowholes which will affect the subsequent working and use of the steel. Hydrogen in solution will cause defects which may lead to mechanical failure; nitrogen will limit the degree of rolling and drawing which can be applied, and oxygen may form SiO_2,

MnO, and Al_2O_3 inclusions or bubbles of CO with the elements, if they are present.

Oxygen is the only gas which reacts with liquid iron to form, at saturation, a separate liquid phase of wustite. For CO, N_2, and H_2, the only possible equilibrium is between the atmosphere and the metal, and a study of the variance of the system shows that the solubility of a gas in a metal depends on its partial pressure in the atmosphere and on the temperature (see Fig. 115 for the C-O equilibrium).

(B) HYDROGEN AND NITROGEN

Bessemer and Thomas steels contain about 0.014% N_2 as a result of absorption of this element from the air blast. This is between 3 and 4 times as much as is present in other steels. The hydrogen in steel results mainly from reaction with H_2O in the air. The equilibrium between dissolved oxygen and hydrogen and steam from the atmosphere is expressed in the form:

$$[H_2O] = 2(\underline{H}) + (\underline{O}) \qquad \Delta G^o = 47{,}500 - 35.5\ T\ (cal) \qquad (7.56)$$

For the equilibria between nitrogen and hydrogen in the atmosphere and in the metal:

$$[N_2] \rightleftarrows 2(\underline{N}) \quad \text{and} \quad [H_2] \rightleftarrows 2(\underline{H}) \qquad (7.57)$$

Sievert's law indicates that:

$$(\%\ \underline{N}) = a_{(\underline{N})} = K_N \cdot \sqrt{P_{N_2}} \quad \text{and} \quad (\%\ \underline{H}) = K_H \cdot \sqrt{P_{H_2}} \qquad (7.58)$$

In liquid iron:

$$\log K_N = -\frac{188}{T} - 1.246 \quad \text{and} \quad \log K_H = -\frac{1670}{T} - 1.68 \qquad (7.59)$$

In Figs. 132 and 133, the solubility curves for these two gases at a total pressure of 1 atm are given as functions of the temperature.

The solution of other elements in the iron alters the solubility of these gases in the manner indicated by Fig. 134 and 135 for a temperature of 1600°C and a total pressure of 1 atm of H_2 or N_2. The elements that have a large affinity for nitrogen (V, Ti, Nb) or for hydrogen (Ti, Ta) increase the solubility of these gases.

Gases dissolved in steel can be removed to low levels in industrial practice by subjecting the liquid metal to a vacuum of about 1 mm Hg or by bubbling an inert gas, such as Argon, through the liquid metal. Several types of equipment are available, some of which are shown in Fig. 136.

By applying Sievert's Law, it is possible to calculate the percentage by weight of a gas dissolved in a steel, at different pressures and temperatures. Sievert's constants for nitrogen and hydrogen at 1600°C are:

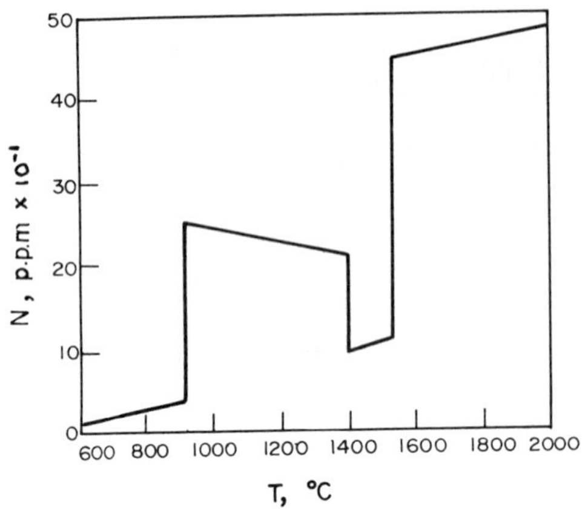

Fig. 132. Solubility of nitrogen in iron.(43)

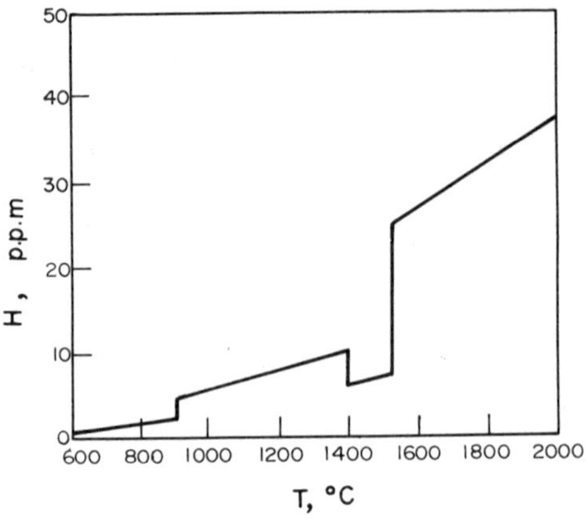

Fig. 133. Solubility of hydrogen in iron.(44)

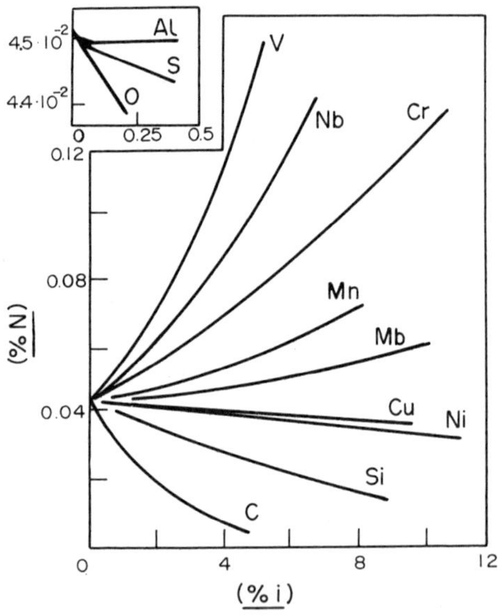

Fig. 134. Influence of a second solute on the solubility of nitrogen in iron.(45)

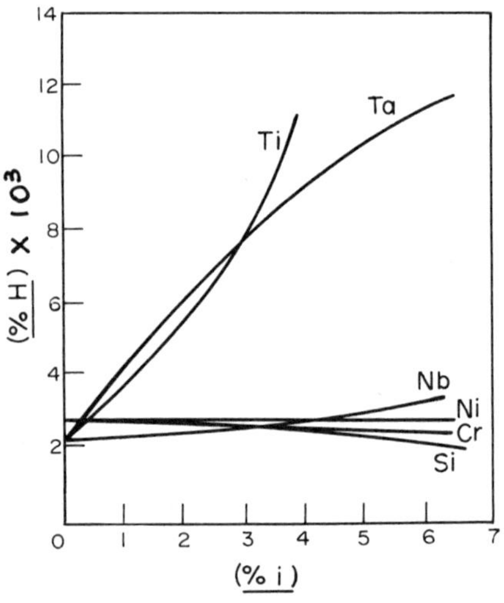

Fig. 135. Influence of a second solute on the solubility of hydrogen in steel.(46, 47, 48)

$$K_N = 4.5 \times 10^{-2} \qquad \text{and} \qquad K_H = 2.7 \times 10^{-3}$$

and the percentage of each in solution in a steel in equilibrium with N_2 and H_2 at a total pressure of 10^{-3} atm will be:

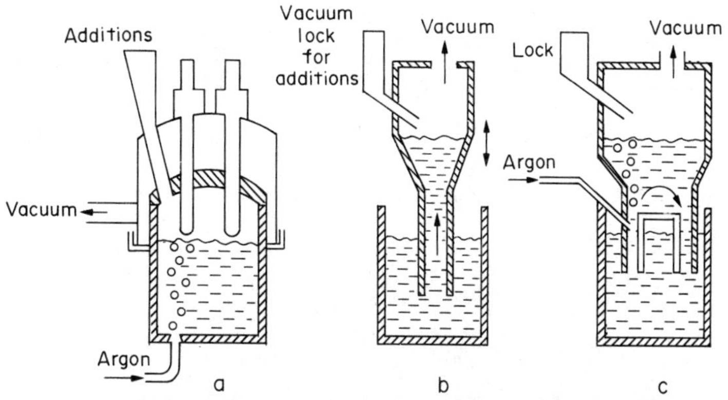

Fig. 136. Vacuum equipment for steel degasification. (a) ladle furnace; (b) D.H. vacuum lifter (intermittent); (c) R.H. argon circulator under vacuum

$$(\%\ N) = 4.5 \times 10^{-2} \times \sqrt{10^{-3}} = 1.42 \times 10^{-3} = 14.2\ \text{ppm}.$$

$$(\%\ H) = 2.7 \times 10^{-3} \quad \sqrt{10^{-3}} = 0.85 \times 10^{-4} = 0.85\ \text{ppm}.$$

These gases do not present problems in steels at these low concentrations.

REFERENCES

1. M. Hansen and K. Anderko, *Constitution of Binary Alloys*. 2nd Ed. McGraw-Hill, N.Y. (1958).

2. M. Othani and N.A. Goksen, *Trans. A.I.M.E.* 216, 533 (1960).

3. H. Schenck, *Revue de Metallurgie*, 57, 9 (1960).

4. J.F. Elliott, *Physical-Chem. of Steel Making*, p. 37, Wiley and Sons, N.Y. (1958).

5. S. Marshall and J. Chipman, *Trans. Am. Soc. Metals*, 30, 695 (1942).

6. H. Schenck, *Physical-Chem. of Steel Making*. B.I.S.R.A., London (1945).

7. F.D. Richardson and W. Dennis, *J. Iron and Steel Inst.* 175, 257-263 (1953).

8. A. Rist and J. Chipman, *Revue de Metallurgie*, 53, 796-807 (1956).

9. K.P. Abraham, M.W. Davies and F.D. Richardson, *J. Iron and Steel Inst.* 96, Part I, p. 84 (1960).

10. R.H. Rein and J. Chipman, *Trans. A.I.M.E.* 233, 417 (1965).

11. J.C. Fulton and J. Chipman, *Trans. A.I.M.E.* 200, 1136-1146 (1954).

12. R.H. Rein and J. Chipman, *Trans. A.I.M.E.* 227, 1202 (1963).

13. L.S. Darken and R.W. Gurry, *J. Am. Chem. Soc.* 68, No. 5, p. 799 (1946).

14. J. Chipman, *J. Iron and Steel Inst.* 180, 97 (1955).

15. J. Chipman and T.C.M. Pillay, *J. Iron and Steel Inst.* 180, 1277 (1955).

16. J. Pearson and E.T. Turkdogan, *J. Iron and Steel Inst.* 176, 19 (1954).

17. Z. Buzhek and A.M. Samarin, *Doklady Akad. Nauk., U.R.S.S.* 114, 97 (1957).

18. J. Chipman, *J. Iron and Steel Inst.* 180, 97 (1955).

19. T. Fuwa and J. Chipman, *Trans. A.I.M.E.* 215, 708 (1959).

20. J. Chipman and T.P. Floridis, *Acta Met.* 3, 456 (1955).

21. H. Schenck and G.H. Gerdom, *Arch. Eisenhüttenwessen*, 15, 77 (1941).

22. S. Marshall and J. Chipman, op. cit., pp. 695-41.

23. H.L. Bishop, N.J. Grant and J. Chipman, *Trans. A.I.M.E.* 212, 890 (1958).

24. E. Plöckinger, *Arch. Eisenhüttenwessen*, 22, 283-293 (1951).

25. Ibid.

26. F. Körber and W. Oelsen, *Mitt. K. Wolhelm Inst. Eisenforsch*, 17, 231-245 (1935).

27. G. Kellerud and H.S. Yoder, *Econ. Geol.* 54, No. 4, p. 561 (1959).

28. T. Fuwa and J. Chipman, *Trans. A.I.M.E.* 218, 887 (1960).

29. J. Chipman, *J. Iron and Steel Inst.* 180, 97 (1955).

30. C.W. Sherman and J. Chipman, *Trans. A.I.M.E.* 194, 597 (1952).

31. J.A. Cordier and J. Chipman, *Trans. A.I.M.E.* 203, 905 (1955).

32. T. Rosenqvist and E.M. Cox, *Trans. A.I.M.E.* 188, 1389 (1950).

33. N.R. Griffing and G.W. Healy, *Trans. A.I.M.E.* 218, 849 (1960).

34. J. Pearson and E.T. Turkdogan, *J. Iron and Steel Inst.* 176, 19 (1954).

35. H. Flood and K. Grjotheim, *J. Iron and Steel Inst.* 171, 68 (1952).

36. P.T. Carter, *Disc. Faraday Soc.* 4, 307-316 (1948).

37. R. Rocca, N.J. Grant and J. Chipman, *Trans. A.I.M.E.* 191, 319-326 (1951).

38. E.T. Turkdogan, *J. Iron and Steel Inst.* 179, 147-154 (1955).

39. A. Jackson, *Modern Steelmaking for Steelmakers*, Newnes, London, 170, 1967 Ed.

40. H. Flood and K. Grjotheim, op. cit., p. 69.

41. E.T. Turkdogan and J. Pearson, *J. Iron and Steel Inst.* 175, 398-401 (1953).

42. K. Balajaiva, A.G. Quarell and P. Vajragupta, *J. Iron and Steel Inst.* 153, 115-145 (1946).

43. L. Darken and R. Gurry, *Physical Chemistry of Metals*, p. 372, McGraw-Hill Co. (1953).

44. Von W. Geller and T. Sun, *Arch. Eissenhüttenwessen*, 21, 423-430 (1950).

45. R.D. Pehlke and J.F. Elliott, *Trans. A.I.M.E.* 218, 1088-1101 (1960).

46. H. Liang, M.B. Bever and C.F. Floe, *Trans. A.I.M.E.* 167, 395-403 (1946).

47. M.M. Karnaukhov and A.N. Morozov, *Bull. Acad. Sci., U.R.S.S., A. Sci. Tech.* pp. 1845-1855 (1948).

48. F. De Kazinczy and O. Lindberg, *Jerkontor. Ann.* 144, 288-296 (1960).

49. C. Bodsworth (see General Reading).

50. P. Herasymenko and G.E. Speight, *J. Iron and Steel Inst.* 166, 169-183 and 289-303 (1950).

51. R.G. Ward (see General Reading).

52. E.T. Turkdogan, *J. Iron and Steel Inst.* 179, 147-154 (1955).

53. J. Astier, *Rev. de Met. C.I.T.* 2, 7, 101 (1983).

54. A. Rist and N. Meysson, *Rev. de Met.* 61, 2, 121 (1964).

55. B. Wilshire, D. Homer and N.L. Cooke, *Technological and Economic Trends in the Steel Industries*, p. 125, Pineridge Press, Swansea (1982).

56. A. Berthet, *Rev. de Met. C.I.T.* 3, 213 (1981).

57. T. Veda *et al.*, *Rev. de Met. C.I.T.* 4, 361 (1981).

58. G. Denier *et al*, *Rev. de Met. C.I.T.* 3, 213 (1981).

59. J.P. Motte and R. Vasse, *Rev. de Met.* 12, 981 (1980).

60. C.J.B. Fincham and F.D. Richardson, *Proc. Roy. Soc., Series A*, 233, 40 (1954).

61. H. Gaye and P. Riboud, IRSID Report, RE 337, Jan. 1976.

62. K. Schwertfeger and H.J. Engell, *Arch. Eisen.* 34, 647 (1963).

63. G. Trömel and W. Fix, *Arch. Eisen.* 33, 745 (1962).

GENERAL READING

Bodsworth, C., *Physical Chemistry of Iron and Steel Manufacture*, Longmans, London (1963).

Darken, L.S., *Basic Open Hearth Steelmaking*, A.I.M.E., N.Y. (1951).

Darken, L.S., *The Physical Chemistry of Steelmaking*, J. Wiley and Sons, N.Y. (1956).

Darken, L. and R. Gurry, *Physical Chemistry of Metals*, McGraw-Hill Co. (1953).

Elliott, J.F., *Physical Chemistry of Steel Making*, Wiley and Sons, N.Y. (1958).

Hopkins, D.W., *Physical Chemistry and Metal Extraction*, J. Garnet-Miller, London (1954).

Kozakevitch, P. and G. Urbain, *Physical Chemistry of Steelmaking*, Proc. Conference Endicott House, Mass. (1956).

McGannon, H.E., Ed., USS *The Making, Shaping and Treating of Steels*, 8th Ed. (1964).

Mudd, S.W. Series, *Basic Open Hearth Steelmaking*, 3rd Ed., Met. Soc. A.I.M.E., N.Y. (1964).

Richardson, F.D., *Physical Chemistry of Melts in Metallurgy*, Vols. 1 and 2. Academic Press, London (1974).

Schenck, H., *Physical Chemistry of Steelmaking*, B.I.S.R.A., London (1945).

Ward, R.G., *An Introduction to the Physical Chemistry of Iron and Steel Making*, E. Arnold Ltd., London (1960).

Chapter 8

PROBLEMS

PROBLEM NO. 1

SOLID-LIQUID-VAPOUR PHASE DIAGRAMS

Construct the phase diagram for H_2O, given the sublimation, fusion, and vaporisation enthalpies:

$$\Delta H_s^{273K} = 12,150 \qquad \Delta H_f^{273K} = 1436 \qquad \Delta H_v^{373K} = 9717$$

and the specific heats, assumed constant, of the solid, liquid, and vapour:

$$C_{P_{(s)}} = 9.085 \qquad C_{P_{(l)}} = 17.996 \qquad C_{P_{(v)}} = 8.025$$

Water increases 9% by volume on changing from liquid to solid.

SOLUTION

In order to integrate the Clausius-Clapeyron equation:

$$\frac{\Delta H_t}{T\Delta V_t} = \frac{dP}{dT} \tag{1.75}$$

it is necessary to know the variations of the enthalpies of transformation with the temperature. These can be calculated, starting from relation (1.23). In this form, we have:

$$\Delta H_v^T = \Delta H_v^{373} + \int_T^{373} \Delta C_{p(l-v)}\, dT = 13,437 - 9.97\ T$$

$$\Delta H_f^T = \Delta H_f^{273} + \int_T^{273} \Delta C_{P(s-l)} \, dT = -997 + 8.911\,T$$

$$\Delta H_s^T = \Delta H_f^T + \Delta H_v^T = 12{,}440 - 1.06\,T$$

(a) Liquid-Vapour and Solid-Vapour Equilibria

For the solid-vapour and liquid-vapour equilibria:

$$\langle H_2O \rangle \rightleftarrows [H_2O] \qquad \text{or} \qquad (H_2O) \rightleftarrows [H_2O]$$

the Clausius-Clapeyron equation is expressed by:

$$d \ln P = \frac{\Delta H_t}{RT^2} \, dT \tag{1.78}$$

Since the pressure of water vapour is 1 atm at 373 K, the $P(T)$ equation for liquid-vapour equilibrium can be integrated in the form:

$$\int_P^1 d \ln P = \frac{1}{R} \int_T^{373} \frac{\Delta H_v}{T^2} \, dT$$

Using the value of ΔH_v^T, calculated above, it is possible to obtain an equation relating pressure to temperature:

$$\log P = -\frac{2940}{T} - 5.01 \log T + 20.78$$

At 0°C (273 K), the vapour pressure of water in equilibrium with water and ice is 0.606×10^{-2} atm.

For the solid-vapour equilibrium, integration of (1.78) gives:

$$\int_P^{0.606 \times 10^{-2}} d \ln P = \frac{1}{R} \int_T^{273} \frac{\Delta H_s}{T^2} \, dT$$

Substituting for ΔH_s gives:

$$\log P = -\frac{2720}{T} - 0.533 \log T + 9.05$$

(b) Solid-Liquid Equilibrium

If a solid is under a pressure P, different from 1 atm, its melting temperature will change by an amount dT that can be calculated from:

$$\frac{dP}{dT} = \frac{\Delta H_f}{T\Delta V_f}$$

where ΔH_f is the enthalpy of fusion, and ΔV_f is the volume change. The volume of 1 mole of water is 18 cm^3, and the volume of 1 mole of ice is 19.62 cm^3. Consequently:

$$\Delta V_f = V_\ell - V_s = -1.62 \text{ cm}^3 \text{ mole}^{-1}$$

In order to operate in compatible units, the volume change must be expressed in cal $mole^{-1}$ atm^{-1}. One cal atm^{-1} represents 41.3 cm^3; therefore:

$$\Delta V_f = -\frac{1.62}{41.3} = -3.92 \times 10^{-2} \text{ cal mole}^{-1} \text{ atm}^{-1}$$

and thus:

$$\frac{dP}{dT} = -\frac{1430}{273 \times 3.92 \times 10^{-2}} = -134 \text{ atm K}^{-1}$$

This means that a container with water at -1°C will have to resist a pressure of 134 atm.

Figure 137 represents the solid-liquid, solid-vapour and liquid-vapour equilibria in the vicinity of the triple point.

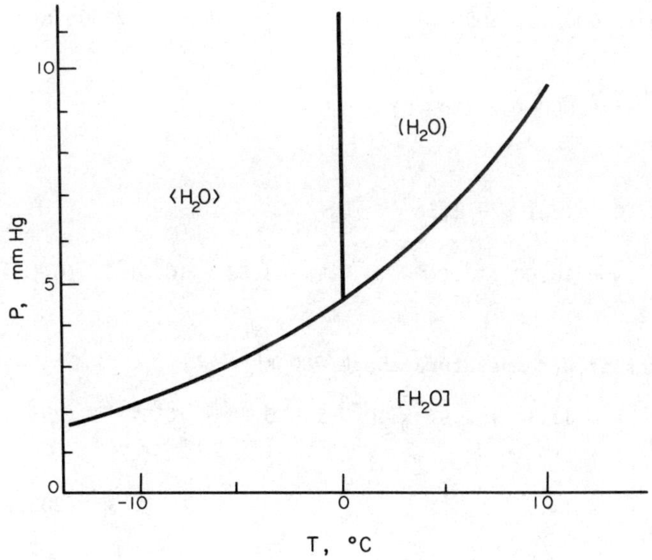

Fig. 137. Equilibrium diagram P(T) for water

PROBLEM NO. 2

ALUMINOTHERMIC REDUCTION OF MANGANESE OXIDES

Manganese metal can be produced from MnO, Mn_2O_3 and MnO_2 by aluminothermic reduction. Assuming that the input materials are at 25°C (298 K), and the products are at 1800°C (2073 K), compare the heats of reaction between Al and MnO_2 and Al and MnO. Calculate the heat loss necessary in each case in order that the maximum temperature shall not exceed 1800°C.

Data: Heats of formation at 298 K (cal $mole^{-1}$)

$\Delta H^o_{Al_2O_3} = -400,000$ $\qquad$ $\Delta H^o_{MnO_2} = -124,300$

$\Delta H^o_{MnO} = -92,000$ $\qquad$ $\Delta H^o_{Al \text{ and } Mn} = 0.0$

Latent heats of melting and transformation (cal/mole)

$\Delta H_{t_{Mn(\alpha\to\beta)}} = 540$ (at 1000 K) $\qquad$ $\Delta H_{t_{Mn(\beta\to\gamma)}} = 540$ (at 1373 K)

$\Delta H_{t_{Mn(\gamma\to\delta)}} = 430$ (at 1410 K) $\qquad$ $\Delta H_{f_{Mn}} = 3400$ (at 1517 K)

$\Delta H_{f_{MnO}} = 13,000$ (at 2058 K) $\qquad$ $\Delta H_{f_{Al}} = 2500$ (at 932 K)

$\Delta H_{f_{Al_2O_3}} = 26,000$ (at 2303 K)

Specific heats C_p (cal g - $mole^{-1}$

$C_{p_{<MnO_2>}} = 16.60 + 2.44 \times 10^{-3}\, T - 3.88 \times 10^5\, T^{-2}$ (0 to 780 K)

MnO_2 decomposes at a temperature above 780 K

$C_{p_{<MnO>}} = 11.11 + 1.94 \times 10^{-3}\, T - 0.88 \times 10^5\, T^{-2}$ (298 to 2058 K)

$C_{p_{(MnO)}} = 14.5$ (2058 to 3000 K)

$C_{p_{<Mn>_\alpha}} = 5.16 + 3.81 \times 10^{-3}\, T$ (298 to 1000 K)

$C_{p_{<Mn>_\beta}} = 8.33 + 0.66 \times 10^{-3}\, T$ (1000 to 1374 K)

$C_{P_{<Mn>_\gamma}} = 10.70$ (1374 to 1410 K)

$C_{P_{<Mn>_\delta}} = 11.30$ (1410 to 1517 K)

$C_{P_{(Mn)}} = 11.00$ (T > 1517 K)

$C_{P_{<Al_2O_3>}} = 27.38 + 3.08 \times 10^{-3}\,T - 8.20 \times 10^{5}\,T^{-2}$ (298 to 1800 K)

$C_{P_{<Al>}} = 4.94 + 2.96 \times 10^{-3}\,T$ (298 to 932 K)

$C_{P_{(Al)}} = 7.0$ (T > 932 K)

SOLUTION

The aluminothermic reduction of MnO_2 and MnO can be written as:

$$3\langle MnO_2\rangle + 4\langle Al\rangle \rightleftharpoons 3(Mn) + 2\langle Al_2O_3\rangle \quad (1)$$

and

$$\langle 3MnO\rangle + 2\langle Al\rangle \rightleftharpoons 3(Mn) + \langle Al_2O_3\rangle \quad (2)$$

The energy changes are based only on the initial and final states and not on the manner in which the changes take place. The calculation can be carried out in two ways:

(a) Assume that the reaction occurs at 298 K, and that the heat of reaction serves to heat the products to 1800°C and to compensate for the heat loss.

(b) Assume that the reactants are heated to 1800°C and that reaction takes place at this temperature. The heat of reaction is utilised in heating the reactants and compensating for the heat losses.

Method (a) will be used for reaction 1, because MnO_2 is not stable above 780 K. Method (b) will be used for reaction 2.

(i) The standard enthalpy of reaction (1) at 25°C is:

$$\Delta H^o_{(1)} = 2\Delta H^o_{Al_2O_3} - 3\Delta H^o_{MnO_2} = -\,427{,}100 \text{ cal}$$

The heat necessary to raise the temperature of one mole of $\langle Al_2O_3\rangle$ from 298 to 2073 K is:

$$\Delta H^o_{(298-2073)} = \int_{298}^{2073} \left(27.38 + 3.08 \times 10^{-3}\,T - \frac{8.20 \times 10^5}{T^2}\right) dT$$

$$= 48,423 + 6482 - 2355 = 52,550 \text{ cal}$$

$$\Delta H^o_{(298-2073)} \text{ for 2 moles of } Al_2O_3 = 105,100 \text{ cal}$$

The heat necessary to raise the temperature of one mole of Mn from 298 to 2073 K and to melt it, is:

$$\Delta H^o = \int_{298}^{1000} (5.16 + 3.81 \times 10^{-3} T)\, dT + \Delta H_t^{\alpha-\beta}$$

$$+ \int_{1000}^{1374} (8.33 + 0.66 \times 10^{-3} T)\, dT + \Delta H_t^{\beta-\gamma}$$

$$+ \int_{1374}^{1410} 10.70\, dT + H_t^{\gamma-\delta} + \int_{1410}^{1517} 11.30\, dT + \Delta H_f$$

$$+ \int_{1517}^{2073} 11.00\, dT$$

$$= 3622 + 1736 + 540 + 3115 + 293 + 540 + 385 + 430 + 1209 + 3400 + 6116$$

$$= 21,390 \text{ cal mole}^{-1}$$

$$\Delta H^o_{298-2073} \text{ for 3 moles} = 64,170 \text{ cal}$$

The heat lost will be equal to:

$$Q_P = 427,100 - 105,100 - 64,170 = 257,830 \text{ cal}$$

This loss is equal to 64,460 cal atom^{-1} of aluminium; that is, 60.3% of the total heat evolved by the reaction. It is necessary to remove a large amount of heat continuously in order to prevent the reaction from becoming explosive.

(ii) The standard enthalpy of reaction (2) at 2073 K is:

$$\Delta H^o_{2073} = \Sigma H^{o\,2073}_{\text{products}} - \Sigma H^{o\,2073}_{\text{reactants}}$$

with $$H^{o\,2073}_{\substack{\text{reactants}\\ \text{products}}} = H^o_{298} + \int_{298}^{T_f} C_{P(\text{solid})}\, dT + \Delta H_f + \int_{T_f}^{2073} C_{P(\text{liq})}\, dT$$

The enthalpies of formation of reactants and products at 2073°C per mole are then:

$$H^{o\,2073}_{(MnO)} = -\ 92{,}000 + \int_{298}^{2058} \left[11.11 + 1.94 \times 10^{-3}\ T - \frac{0.88 \times 10^5}{T^2} \right] dT$$

$$+\ 13{,}000 + \int_{2058}^{2073} 14.5\ dT$$

$$= -\ 92{,}000 + 39{,}784 = -\ 52{,}216\ \text{cal. mole}^{-1}$$

$$H^{o\,2073}_{Mn} = 21{,}390\ \text{cal mole}^{-1}$$

$$H^{o\,2073}_{Al_2O_3} = -\ 400{,}000 + 52{,}550 = -\ 347{,}450\ \text{cal mole}^{-1}$$

$$H^{o\,2073}_{Al} = \int_{298}^{932} (4.94 + 2.96 \times 10^{-3}\ T)\ dT + \Delta H_{f_{Al}} + \int_{932}^{2073} 7.0\ dT$$

$$= 14{,}773\ \text{cal atom}^{-1}$$

The enthalpy of reaction (2) at 1800°C will then be equal to:

$$\Delta H^o_{2073} = -\ 347{,}450 + 3 \times 21{,}390 - 2 \times 14{,}773 + 3 \times 52{,}216 = -\ 156{,}178\ \text{cal}$$

ΔH required in order to heat 3 moles of MnO and 2 moles of Al from 298 to 2073:

$$\Delta H = 3 \times 39784 + 2 \times 14{,}773 = 148{,}898\ \text{cal}$$

The heat loss, Q_P, is equal to the heat generated, less that required to raise the temperature of the MnO and Al to 2073:

$$Q_P = +\ 156{,}178 - 148{,}898 = 7280\ \text{cal}$$

equivalent to 3640 cal mole^{-1} of aluminium.

From method (a), we would have also found:

$$Q_P = 124{,}000 - 52{,}550 - 3 \times 21{,}390 = 7280\ \text{cal}$$

In the case of the reduction of MnO with Al, the necessary loss or surplus heat generated represents about 5% of the heat evolved, which is not enough to maintain the reaction temperature. Industrial production of manganese by aluminothermic reduction is, in fact, carried out using Mn_3O_4 as the source of Mn.

PROBLEM NO. 3

IRON-SILICON SOLUTIONS

Given the integral enthalpies and entropies of mixing H^M and S^M for the binary system Fe-Si at 1580°C, calculate:

(a) The partial and integral free energy of mixing of silicon in iron. Deduce the silicon activity in iron for different molar fractions of Si, and trace the curve $a_{Si} = f(N_{Si})$. Calculate the value of the activity coefficient, γ_{Si}^{∞} for the infinitely dilute solution of silicon in iron.

(b) Calculate the percentage of silicon dissolved in iron in a quartz crucible under an argon atmosphere which contains 10^{-10} atm of oxygen.

Data: Integral enthalpy and entropy of mixing of the system Si-Fe at 1580°C:

N_{Si}	0.01	0.1	0.2	0.3	0.4	0.5	0.6	0.7	0.8	0.9
H^M cal mol^{-1}	-373	-2840	-5560	-7420	-8660	-8980	-8410	-6950	-4950	-2540
S^M cal mol^{-1}K^{-1}	+0.0022	+0.02	-0.05	-0.35	-0.69	-0.99	-1.12	-0.94	-0.55	-0.16

The standard free energy of formation of silica is:

$$(Si) + [O_2] = \langle SiO_2 \rangle \qquad \Delta G^o = -\,218{,}600 + 47.7\ T$$

The self-interaction parameter of silicon is $\varepsilon_{Si}^{Si} = 37$.

SOLUTION

(a) The integral free energy of mixing is related to the integral enthalpy and entropy of mixing by:

$$G^M = H^M - TS^M$$

G^M at 1853 K can be calculated from the data provided. The partial free energy of mixing of silicon $\bar{G}_{Si}^M$ and the activity $\left(\bar{G}_{Si}^M = RT \log a_{Si}\right)$ can be calculated by using the graphical method (Fig. 138a) of drawing tangents to the curve of G^M against N_{Si}.

In Table 23 the values of G^M, $\bar{G}_{Si}^M$ and a_{Si} for different mole fractions of silicon in iron, are given. The way in which $a_{(Si)}$ varies as a function of $N_{(Si)}$ is given in Fig. 138b. The activity coefficient of Si in iron is very low for low concentrations of Si. This facilitates the reduction of silica to silicon by iron.

The activity coefficient γ_{Si}^{∞} for the infinitely dilute solution can be calculated by applying the relationship:

$$\ln \gamma_{Si} = \ln \gamma_{Si}^{\infty} + \varepsilon_{Si}^{Si} N_{Si}$$

TABLE 23 Values of G^M, $\bar{G}_{Si}^M$, $\log a_{Si}$, and γ_{Si} at 1580°C for Different Mole Fractions of Si

N_{Si}	0·01	0·1	0·2	0·3	0·4	0·5	0·6	0·7	0·8	0·9
G^M	-377	-2877	-5467	-6771	-7381	-7145	-6334	-5208	-3930	-2243
$\bar{G}_{Si}^M$	-35,000	-25,600	-21,800	-12,800	-8700	-4600	-3000	-1900	-1100	-550
$\log a_{Si}$	-4·128	-3·02	-2·51	-1·51	-1·03	-0·543	-0·354	-0·224	-0·13	-0·065
a_{Si}	$7{\cdot}5\times10^{-5}$	$0{\cdot}96\times10^{-3}$	$0{\cdot}27\times10^{-2}$	$0{\cdot}31\times10^{-1}$	$0{\cdot}94\times10^{-1}$	0·287	0·44	0·60	0·74	0·86
γ_{Si}	$7{\cdot}5\times10^{-3}$	$0{\cdot}96\times10^{-2}$	$1{\cdot}35\times10^{-2}$	0·103	0·23	0·57	0·73	0·86	0·93	0·96

When $N_{Si} = 10^{-2}$, $\log a_{Si} = -4.13$, and consequently, $\log \gamma_{Si} = -2.13$ or $\ln \gamma_{Si} = -4.90$. $\ln \gamma_{Si}^{\infty}$ then has a value of $-4.90 - 0.37 = -5.27$ and $\gamma_{Si}^{\infty} = 5.16 \times 10^{-3}$.

When steel containing silicon is melted in a silica crucible, there is an equilibrium between Si and SiO_2 of the form:

$$\underline{(Si)} + [O_2] = \langle SiO_2 \rangle \qquad \Delta G^o = RT \ln p_{O_2} \cdot a_{Si}$$

At 1853 K, $\Delta G^o = -130{,}700$ cal. When the oxygen partial pressure is 10^{-10} atm, the activity of silicon in iron is 4.36×10^{-6}. Taking into account that:

$$\ln a_{Si} = \ln \gamma_{Si} + \ln N_{Si} = \ln \gamma_{Si}^{\infty} + \varepsilon_{Si}^{Si} N_{Si} + \ln N_{Si}$$

the value of N_{Si} in iron in a silica crucible with $p_{O_2} = 10^{-10}$ atm can be calculated:

$$N_{Si} = 0.084 \times 10^{-2} = 0.084 \text{ at. \%}$$

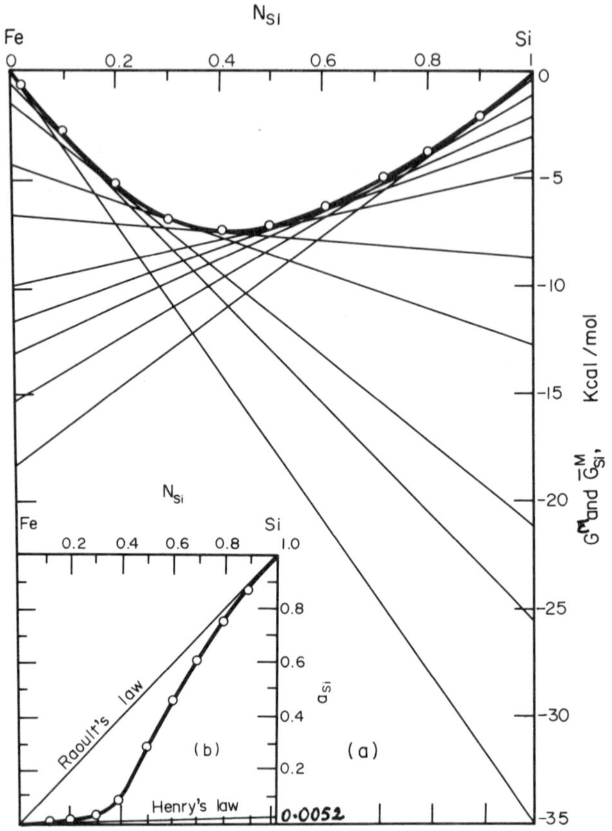

Fig. 138. (a) Partial and integral free energy of mixing for the system Fe-Si; (b) Activity of Si in the system Fe-Si

PROBLEM NO. 4

Fe-O-Mn EQUILIBRIUM

Liquid steel containing oxygen ($N_{(O)} = 3 \times 10^{-3}$) is deoxidised with Mn at 1600°C.

(a) Show that there is no slag phase before Mn is added, and then calculate the value of $N_{(Mn)}$ required to just produce a slag. FeO and MnO form ideal solid and liquid solutions in all ranges of composition.

(b) Starting from the relationship:

$$\ln \left[\frac{a_{<MO>}}{a_{(MO)}} \right] = \frac{\Delta H_{f_{MO}}}{RT \cdot T_{f_{MO}}} \left(T_{f_{MO}} - T \right)$$

calculate the mole fractions of (MnO), <MnO>, <FeO> and (FeO) in equilibrium at 1500°C and 1600°C, assuming that $\Delta H_{f_{MO}}$ does not change with T, and construct the FeO-MnO phase diagram.

(c) The slag in equilibrium with the metal (Fe, Mn, O) consists of two phases (liquid and solid), the compositions of which were determined in (b). What conclusion can be reached from the analysis of the variance of the system?

(d) Find the mole fractions of manganese and oxygen in the liquid metal in equilibrium with the slag at 1600°C. Assume that the manganese forms an ideal solution in Fe, and that the presence of oxygen does not influence its activity. The interaction parameters ε_O^O, ε_{Mn}^O and ε_O^{Mn} are assumed to be nil.

$$\frac{1}{2}\,[O_2] \rightleftarrows (\underline{O})_{Fe} \qquad \Delta G^o = -\,27{,}930 + 6.08\ T$$

The infinitely dilute solution expressed in terms of the atom fraction has been chosen as the standard state for $(\underline{O})_{Fe}$:

$$(Mn) + \frac{1}{2}\,[O_2] \rightleftarrows <MnO> \qquad \Delta G^o = -\,95{,}400 + 19.62\ T$$

$$(Fe) + \frac{1}{2}\,[O_2] \rightleftarrows (FeO) \qquad \Delta G^o = -\,59{,}830 + 11.94\ T$$

$$<MnO> \rightleftarrows (MnO) \qquad \Delta G_f = 10{,}700 - 5.2\ T$$

$$<FeO> = (FeO) \qquad \Delta G_f = 7300 - 4.56\ T$$

SOLUTION

(a) The only slag phase which can be present in the system Fe-O, when Fe is present, is "FeO" (wustite). According to the data:

$$(Fe)_m + (\underline{O})_m = (FeO)_s \qquad \Delta G^o = -\,31{,}900 + 5.86\ T\ \text{kcal mole}^{-1}$$

giving $\Delta G^o = -\,20{,}924$ cal at 1600°C, and $K = \dfrac{a_{FeO}}{N_{(O)}} \cdot a_{Fe} = 275.4$.

Since a_{FeO} and $a_{Fe} = 1$, the fraction of oxygen in equilibrium with FeO is 3.63×10^{-3}. As the actual content is less than this value, no separate FeO phase will form. Slag will form when $N_{(O)}$ exceeds 3.63×10^{-3} in the absence of Mn. When Mn is added in sufficient quantity to form MnO, but not to have a significant effect on $N_{(O)}$, then when $a_{Fe} = 1$ and $N_{(\underline{O})} = 3.0 \times 10^{-3}$,

$a_{(FeO)} = N_{(FeO)} = 0.826$ in the slag. $N_{(MnO)} = 1 - N_{(FeO)} = 0.174$. This slag is liquid. The mole fraction of Mn required to form this drop of slag can be calculated from:

$$\underline{(Mn)} + \underline{(O)} = \underline{(MnO)} \qquad \Delta G^o = -56{,}770 + 8.34\ T$$

At 1600°C, $\Delta G^o = -41{,}150$, $K = 63{,}417$ and $N_{Mn} = \dfrac{0.174}{3 \times 10^{-3} \times 63{,}417} = 9.2 \times 10^{-4}$

(b) Since the liquid and solid FeO-MnO solutions are ideal, it is enough to replace a_{MO} by N_{MO} to find:

$$\ln \frac{N_{<MO>}}{N_{(MO)}} = \frac{\Delta H_{f_{MO}}}{RT \cdot T_{f_{MO}}} \left(T_{f_{MO}} - T \right)$$

This relationship between the mole fractions of oxide in a solid and a liquid solution is valid for iron oxide as well as for manganese oxide. In order to determine $N_{(MnO)}$, $N_{<MnO>}$, $N_{(FeO)}$, and $N_{<FeO>}$, it is necessary to make use of two additional relationships. These are:

$$N_{(MnO)} + N_{(FeO)} = 1 \qquad \text{and} \qquad N_{<MnO>} + N_{<FeO>} = 1$$

The enthalpies and temperatures of melting $\Delta H_{f_{MO}}$ and $T_{f_{MO}}$ can be obtained from the data:

$$\Delta H_{f_{MnO}} = 10{,}700 \text{ cal mole}^{-1} \qquad \Delta H_{f_{FeO}} = 7300 \text{ cal mole}^{-1}$$

$$T_{f_{MnO}} = \frac{\Delta H_{f_{MnO}}}{\Delta S_{f_{MnO}}} = 2060 \text{ K}, \quad T_{f_{FeO}} = \frac{\Delta H_{f_{FeO}}}{\Delta S_{f_{FeO}}} = 1600 \text{ K}$$

Starting from these data, it is possible to calculate the mole fractions:

At 1600°C	$N_{(FeO)} = 0.515$	$N_{(MnO)} = 0.485$
	$N_{<FeO>} = 0.369$	$N_{<MnO>} = 0.631$
At 1500°C	$N_{(FeO)} = 0.724$	$N_{(MnO)} = 0.276$
	$N_{<FeO>} = 0.579$	$N_{<MnO>} = 0.421$

The MnO-FeO phase diagram (Fig. 139) can be constructed from these values.

(c) The variance of the system (FeO-MnO), <FeO-MnO> and (Fe, Mn, O) is:

$$v = 3 + 1 - 3 = 1$$

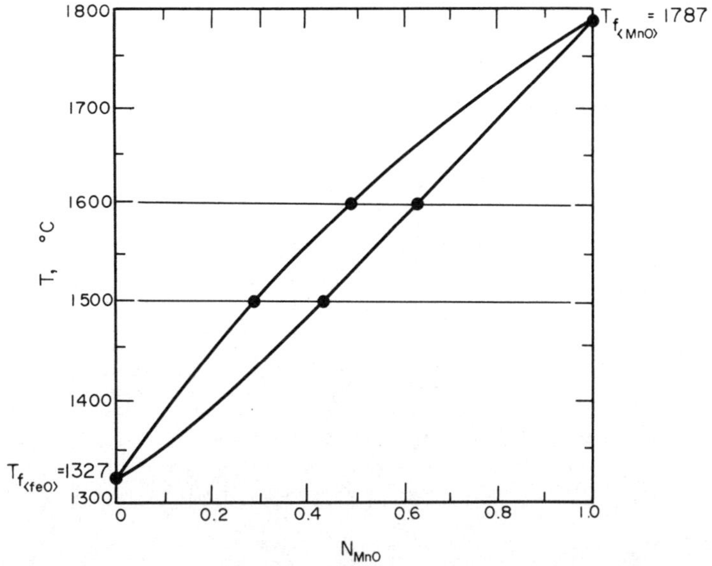

Fig. 139. MnO-FeO phase diagram

If the temperature is specified, then so are the compositions of the three phases in equilibrium.

(d) One of the following equilibria will be used to find the mole fraction of the manganese dissolved in iron:

$$(MnO)_s + (\underline{Fe})_m \rightleftarrows (FeO)_s + (\underline{Mn})_m$$

or

$$\langle MnO\rangle_s + (\underline{Fe})_m \rightleftarrows \langle FeO\rangle_s + (\underline{Mn})_m$$

or, since the free energies of formation of solid MnO and liquid FeO are known, the following equilibrium:

$$\langle MnO\rangle_s + (\underline{Fe})_m \rightleftarrows (FeO)_s + (\underline{Mn})_m$$

The standard free energy of this reaction can be calculated from available data:

$$\Delta G^o = 35{,}570 - 7.68\ T$$

At equilibrium, we have:

$$\Delta G^o = -\ RT \log \frac{N_{(Mn)_m} \cdot N_{(FeO)_s}}{N_{(Fe)_m} \cdot N_{\langle MnO\rangle_s}}$$

The values of the mole fractions $N_{(FeO)_s}$ and $N_{<MnO>_s}$ and ΔG^o at 1600 C are known. It can be assumed that $N_{(Fe)}$ is close to one. Otherwise $N_{(Fe)_m}$ + $N_{(Mn)_m}$ = 1. Inserting the numerical values, we have:

$$N_{(Mn)_m} = 0.42 \times 10^{-2}$$

The oxygen is in equilibrium with Mn as well as with FeO:

$$(\underline{Fe})_m + (\underline{O})_m \rightleftarrows (FeO)_s \quad (\text{or } <FeO>_s)$$

$$(\underline{Mn})_m + (\underline{O})_m \rightleftarrows <MnO>_s \quad (\text{or } (MnO)_s)$$

The standard free energy of the latter reaction is:

$$\Delta G^o = -67{,}470 + 13.54\ T$$

and

$$\Delta G^o = -RT \log \frac{N_{<MnO>_s}}{N_{(O)_m} \cdot N_{(Mn)_m}}$$

Inserting the values of ΔG^o, the value of $N_{(\underline{O})}$ can be found:

$$N_{(\underline{O})} = 0.19 \times 10^{-2}$$

PROBLEM NO. 5

CONSTRUCTION OF THE Bi-Cd PHASE DIAGRAM[3]

Bismuth and cadmium form a single eutectic alloy system in which there is complete miscibility in the liquid state and complete insolubility in the solid state. In addition, the liquid solution is ideal.

(a) Applying the Clausius-Clapeyron relationship, draw the liquidus line of the diagram.

(b) Calculate the constants K_{Bi} and K_{Cd} according to Raoult's Law.

Data:

$$\Delta H_{f_{Bi}} = 2600 \text{ cal g-atom}^{-1} \qquad T_{f_{Bi}} = 544.5 \text{ K}$$

$$\Delta H_{f_{Cd}} = 1450 \text{ cal g-atom}^{-1} \qquad T_{f_{Cd}} = 594 \text{ K}$$

SOLUTION

(a) When there is equilibrium between an element A in a liquid solution and the pure solid, the activity of A in solution relative to pure liquid A as the standard state is related to the temperature by the integral of the Clausius-Clapeyron equation:

$$\log a_{(A)} = -\frac{\Delta H_{f_A} \cdot (T_{f_A} - T)}{RT \cdot T_{f_A}}$$

If, furthermore, the liquid solution is ideal, the activity of A is equal to its mole fraction, i.e.:

$$\log N_{(A)} = -\frac{\Delta H_{f_A} (T_{f_A} - T)}{RT\, T_{f_A}}$$

Substituting the value of the constants, we have for the liquidus:

$$\log N_{Bi} = -\frac{568}{T} + 1.043$$

$$\log N_{Cd} = -\frac{317}{T} + 0.533$$

Starting from these equations, the values of N_{Bi} and N_{Cd} for the liquidus can be calculated (Table 24), and the diagram can be drawn (Fig. 140).

TABLE 24

Temp., °C	log N_{Bi}	N_{Bi}	log N_{Cd}	N_{Cd}
300	-	-	-0·020	0·955
250	-0·043	0·906	-0·073	0·845
200	-0·158	0·696	-0·137	0·729
150	-0·30	0·502	-0·216	0·608
100	-0·48	0·332	-0·317	0·482

The eutectic indicated by the calculated diagram at $N_{Bi} = 0.43$, and T = 137°C, differs little from the real diagram (eutectic is at $N_{Bi} = 0.45$ and T = 144°C).

(b) The constants in that version of Raoult's Law which relates the lowering of the melting point to the molar fraction of solute ($\Delta T = K_A \cdot N_B$) are given by 2-129:

$$K_A = \frac{RT_{f_A}^2}{\Delta H_{f_A}}$$

With the data given, these are:

$$K_{Bi} = 262 \qquad \text{and} \qquad K_{Cd} = 560$$

These constants represent the slope of the liquidus at points $N_{Bi} = 1$ and $N_{Cd} = 1$, respectively. It can be seen that in this diagram, the slopes are nearly constant for any value of molar fractions of N_{Bi} and N_{Cd}. Graphically, it can be found that the average slopes of the liquidus $ET_{f_{Cd}}$ and $ET_{f_{Cd}}$ are, respectively, 248 and 444. This result is surprising since Raoult's Law assumes that the molar fraction of the solute is small enough for $\ln(1 - N_B)$ to be equal to $(-N_B)$, and that the melting temperature, T, of the solution is sufficiently close to that of the pure metal, so that T can be replaced by T_f. These two errors ($\ln(1 - N_B) \neq -N_B$ and $T \neq T_f$) tend to compensate for each other, and so minimise the effect on the overall accuracy.

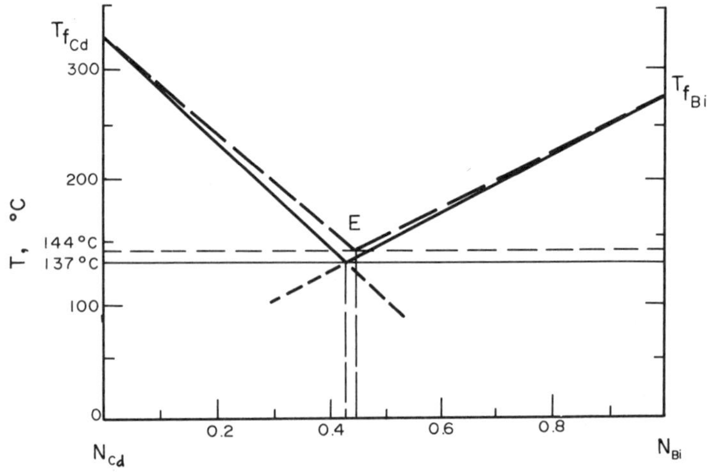

Fig. 140. Cd-Bi phase diagram. ——— calculated; — — — real

PROBLEM NO. 6

DETERMINATION OF THE ACTIVITY OF A COMPONENT IN A SOLUTION BY VAPOUR PRESSURE MEASUREMENT[(1)]

Measurements of the vapour pressure of zinc in In-Zn liquid alloys at 500°C gave the following results:

N_{Zn}	$P_{(mm\ Hg)}$
0.2	0.485
0.5	0.938
0.8	1.196

(a) Determine the vapour pressure vs temperature relationship for pure Zn.

(b) The value of P_{Zn} at 300°C for an alloy in which $N_{Zn} = 0.24$, is 1.605×10^{-3} mm Hg. Calculate the activity of Zn in the alloy.

(d) At 300°C, the liquid alloy of composition $N_{Zn} = 0.24$ is in equilibrium with pure solid, Zn. Determine the activity of Zn by applying the Clausius-Clapeyron equation, and compare it with the value calculated in (c). The standard state is pure liquid zinc.

Data: $\Delta H_v^{Zn} = 23{,}500$ cal mole^{-1} $\quad T_v^{Zn} = 907°C$

$\Delta H_f^{Zn} = 1740$ cal mole^{-1} $\quad T_f^{Zn} = 419°C$

$C_{P_{(l)}} - C_{P_{(v)}} = 5.75$ cal mole^{-1} °C^{-1}

SOLUTION

(a) By integrating the Clausius-Clapeyron equation:

$$\frac{d \ln p^o_{Zn}}{dT} = \frac{\Delta H_v^{Zn}}{RT^2}$$

the variation of the vapour pressure of pure zinc with the temperature can be found directly. It is assumed that the boiling enthalpy varies with temperature, according to the expression:

$$\Delta H_v^T = \Delta H_v^{T_v} + \left[C_{P_{(l)}} - C_{P_{(v)}} \right] (T_v - T)$$

Substituting this value in the Clausius-Clapeyron equation and integrating between the boundary conditions $T = T$ and T_v, p^o_{Zn} and 1 (at the boiling point $p^o_{Zn} = 1$ atm), one obtains:

$$\ln p^{o\ \text{Atm.}}_{Zn} = -\frac{\Delta H_v^{T_v} + \Delta C_p \cdot T_v}{RT} - \frac{\Delta C_p \ln (T/T_v)}{R} + \frac{\Delta H_v^{T_v} + \Delta C_p \cdot T_v}{RT_v}$$

Substituting the values, gives:

$$\log p^{o\,\text{Atm.}}_{Zn} = -\frac{6620}{T} - 1.255 \log T + 9.46$$

To express p^{o}_{Zn} in mm Hg, it is enough to add the term log 760;

$$\log p^{o\,\text{mm Hg}}_{Zn} = -\frac{6620}{T} - 1.255 \log T + 12.34$$

(b) At 500°C (773 K), the vapour pressure of zinc in equilibrium with pure liquid zinc is:

$$p^{o\,\text{mm Hg}}_{Zn} = 1.417 \text{ mm Hg}$$

The activity of zinc in the alloys is equal to p_{Zn}/p^{o}_{Zn} (see Table 25 and Fig. 141).

TABLE 25 Vapour Pressure and Activity of Zn in In-Zn Alloys at 500°C

N_{Zn}	$p^{\text{mm Hg}}_{Zn}$	a_{Zn}	γ_{Zn}
0·2	0·485	0·343	1·715
0·5	0·938	0·662	1·324
0·8	1·196	0·844	1·055

(c) At 300°C (573 K), the vapour pressure of zinc which would be in equilibrium with pure liquid zinc (if liquid Zn could exist at this temperature) would be equal to:

$$p^{o\,\text{mm Hg}}_{Zn} = 2.117 \times 10^{-3} \text{ mm Hg}$$

The activity referred to liquid zinc when $N_{Zn} = 0.24$ is:

$$a_{Zn} = \frac{p_{Zn}}{p^{o}_{Zn}} = 0.758$$

(d) The activity of zinc in solution in equilibrium with solid zinc, taking pure liquid zinc as the standard state, is given by:

$$\log a_{(Zn)} = -\frac{\Delta H_f \,.\, (T_f - T)}{RT \,.\, T_f}$$

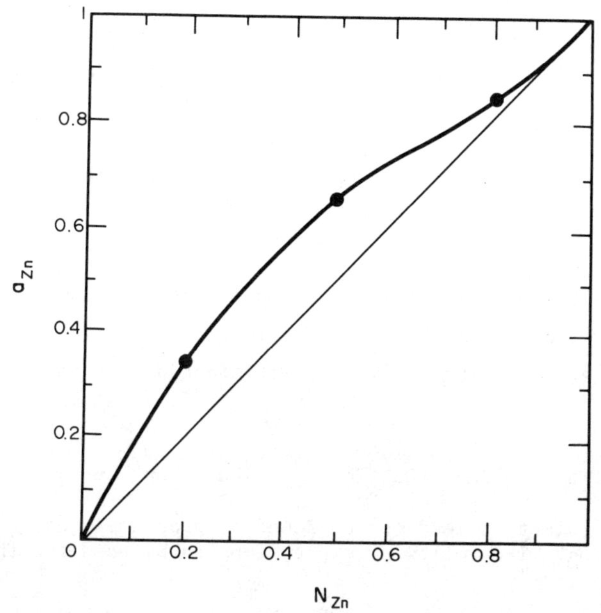

Fig. 141. Activity of zinc in In-Zn alloys at 500°C

If it is assumed that the melting enthalpy of zinc does not change with temperature at 573 K, then:

$$\log a_{(Zn)} = -0.114 \qquad\qquad a_{(Zn)} = 0.77$$

This value is close to that determined by vapour pressure measurements. The difference arises from the fact that the melting enthalpy is not a constant but varies with temperature.

PROBLEM NO. 7

EQUILIBRIUM BETWEEN TWO SLIGHTLY MISCIBLE PHASES (Pb AND Zn)(4)

Measurements of the vapour pressure at 873 K over a Pb-Zn alloy gave the following results:

N_{Zn}	P_{Zn} (mm Hg)	N_{Zn}	P_{Zn} (mm Hg)
1.000	11.62	0.70	11.47
0.995	11.57	0.50	11.47
0.99	11.55	0.40	11.47
0.985	11.50	0.20	11.25
0.98	11.47	0.154	11.20
0.97	11.47	0.10	9.05
0.90	11.47	0.05	6.40
0.80	11.47		

(a) Calculate the value of a_{Zn} and a_{Pb} in this alloy, and draw the curves a_{Zn} and a_{Pb} as functions of N_{Zn} and N_{Pb}. What can be deduced from the shapes of these curves?

(b) When there is immiscibility between two phases with two components in solution, the integral free energy of mixing G^M, is represented by a straight line for any composition between the two solutions. Determine the equation of this line, and plot the curve of G^M for all ranges of zinc concentration in the alloy.

SOLUTION

(a) Starting from the vapour pressure of Zn in equilibrium with pure metal, (p^o_{Zn} = 11.62 mm Hg), and with the Pb-Zn alloy, both at 873 K, it is possible to determine the activity and the activity coefficient of zinc as a function of the mole fraction:

$$a_{Zn} = p_{Zn}/p^o_{Zn} \quad \text{and} \quad \gamma_{Zn} = a_{Zn}/N_{Zn}$$

The activity coefficient of lead can be found by graphical integration of the Gibbs-Duhem equation. The following expression can be used for this purpose:

$$\ln \gamma_{Pb} = - \int_{N_{Zn}=0}^{N_{Zn}} \frac{N_{Zn}}{N_{Pb}} \, d \ln \gamma_{Zn}$$

The calculation of the areas under the integral (Fig. 142(a)) must be carried out carefully, since the errors are cumulative, and a small error in the value of $\log \gamma_{Pb}$ can produce a very large error in the value of γ_{Pb}. Table No. 26 and Figs. 142(a), (b), and (c), summarise the results.

From the activity curves of Fig. 142(b), the following can be deduced: between N_{Zn} = 0.15 and N_{Zn} = 0.98, the activities of zinc and lead are constant, which indicates the presence of two phases in equilibrium; one high in lead, and one high in zinc. Zinc is more soluble in lead than lead in zinc. The variable value of a_{Pb} between N_{Zn} = 0.98 and N_{Zn} = 0.3 results from the lack of accuracy of the graphical method.

Henry's Law is followed up to 5% of Zn in lead, and up to 2% of Pb in zinc. Raoult's Law is followed in the same ranges of concentration by the other constituent in each case.

(b) The integral free energy of mixing is calculated from:

$$G^M = N_{Zn} \cdot \bar{G}^M_{Zn} + N_{Pb} \cdot \bar{G}^M_{Pb} = RT \, (N_{Zn} \cdot \log a_{Zn} + N_{Pb} \cdot \log a_{Pb})$$

When the activities of zinc and lead are constant, the integral free energy of mixing is a linear function of N_{Zn}:

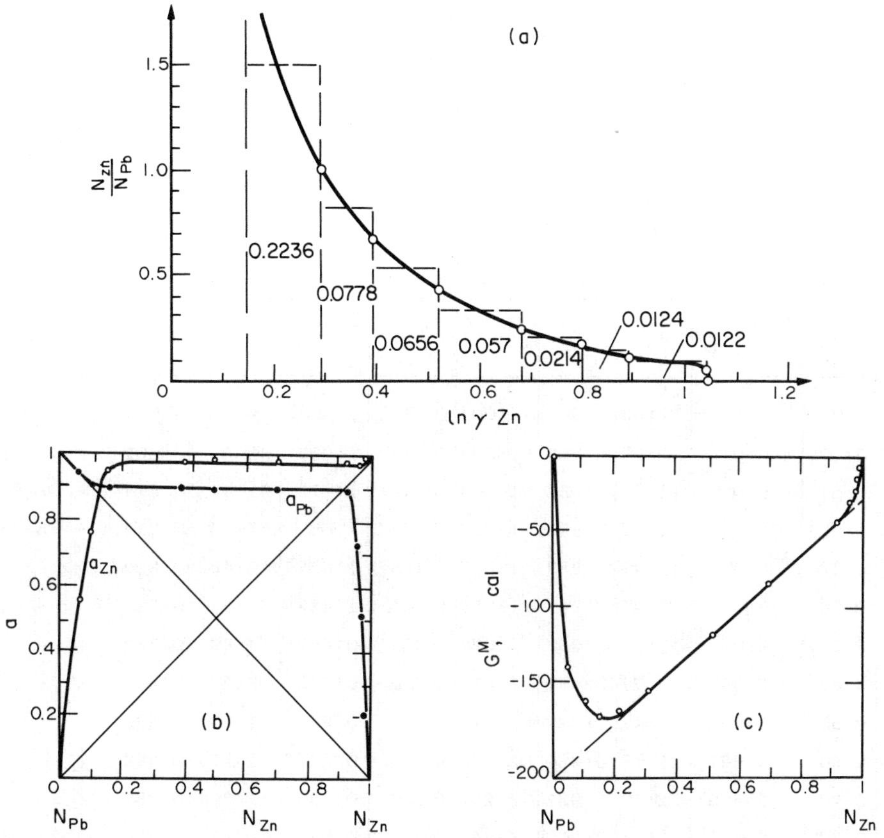

Fig. 142. (a) N_{Zn}/N_{Pb} versus log $_{Zn}$

(b) a_{Zn} and a_{Pb} versus N_{Zn}

(c) G^M versus N_{Zn} for the binary Pb-Zn

$$G^M = RT \left[N_{Zn} \log a_{Zn} + (1 - N_{Zn}) \log a_{Pb} \right]$$

$$= RT \left[\log a_{Pb} + N_{Zn} (\log a_{Zn} - \log a_{Pb} \right]$$

When a_{Zn} and a_{Pb} are constant, this is represented by a straight line. Replacing RT, a_{Zn}, and a_{Pb} by the values given in Table 26, between $N_{Zn} = 0.15$ and $N_{Zn} = 0.97$, one obtains:

$$G^M = -200 + 173.5\ N_{Zn}$$

Table 26 gives the values of the integral free energy of mixing corresponding to the values of N_{Zn} lower than 0.20 or higher than 0.98. The curve

$G^M = f(N_{Zn})$ is plotted in Fig. 142(c). The values of the partial free energies of mixing of Zn and Pb when there are two phases in equilibrium are equal, according to the equation relating G^M and N_{Zn}, to:

$$\bar{G}^M_{Pb} = -200 \text{ cal mole}^{-1} \qquad \bar{G}^M_{Zn} = -26.5 \text{ cal mole}^{-1}$$

TABLE 26

N_{Zn}	N_{Pb}	N_{Zn}/N_{Pb}	P_{Zn}	a_{Zn}	γ_{Zn}	$\log\gamma_{Zn}$	$\log\gamma_{Pb}$	γ_{Pb}	a_{Pb}	G^M_{cal}
1	0	∞	11·62	1·00	1·00	0·00	-	-	0	0
0·995	0·005	199·0	11·57	0·996	1·001	0·0004	1·907	80·3	0·402	- 14·8
0·99	0·01	99·0	11·55	0·994	1·004	0·0017	1·713	51·4	0·514	- 21·8
0·985	0·015	65·7	11·50	0·990	1·005	0·0022	1·68	47·65	0·715	- 25·8
0·98	0·020	49·0	11·47	0·987	1·007	0·0030	1·65	44·48	0·890	-
0·97	0·003	32·3	11·47	0·987	1·017	0·0073	1·487	30·57	0·917	-
0·90	0·10	9·0	11·47	0·987	1·097	0·040	0·897	7·87	0·787	-
0·70	0·30	2·33	11·47	0·987	1·410	0·149	0·470	2·95	0·885	-
0·50	0·50	1·0	11·47	0·987	1·974	0·295	0·2464	1·764	0·882	-
0·40	0·60	0·667	11·47	0·987	2·467	0·392	0·1686	1·474	0·884	-
0·30	0·70	0·428	11·47	0·987	3·290	0·517	0·103	1·268	0·887	-
0·20	0·80	0·25	11·25	0·968	4·84	0·685	0·046	1·112	0·890	-173·0
0·154	0·846	0·182	11·20	0·964	6·26	0·797	0·0246	1·058	0·895	-173·0
0·10	0·90	0·111	9·05	0·779	7·79	0·892	0·0122	1·0284	0·926	-163·0
0·05	0·95	0·0525	6·40	0·551	11·02	1·042	0·00	1·0	0·95	-136·0
0·00	1·00	0·00	0·00	0·00	11·02	1·042	0·00	1·0	1·00	0

PROBLEM NO. 8

DETERMINATION OF THE ACTIVITIES OF COMPONENTS OF A BINARY SOLUTION (ELECTROCHEMICAL METHOD)[5]

Measurement of the e.m.f. of the cell:

$$Sn/Sn^{2+}/(\underline{Sn})_{Hg}$$

at 30°C gave the following results:

$(\underline{Sn})$ (mol. %)	E (mV)
0.0295	46.50
0.110	29.60
0.333	16.50
0.430	13.45
0.962	5.15
1.195	2.75
1.360	1.70
1.487	1.56
1.615	1.66
1.784	1.63
2.010	1.63

(a) Calculate the activity of Sn in Hg-Sn alloys at 30°C, taking as the standard states:

(i) the pure solid Sn:
(ii) the pure liquid Sn.

$\Delta H_f^{Sn} = 1690 \text{ cal mole}^{-1}$ $\qquad$ $T_f^{Sn} = 232°C$

(b) With the aid of these experimental data, determine the maximum solubility of Sn in Hg at 30°C.

(c) For a given Hg-Sn alloy, the e.m.f. of the $Sn/Sn^{2+}/(\underline{Sn})_{Hg}$ cell at different temperatures was as follows:

E (mV)	T (°C)
24.3	1.3
31.5	20
35.4	30
40.0	40
43.4	50
48.4	60
51.9	70
54.3	80
58.1	90

Determine the value of $\bar{H}_{Sn}^{M}$ and $\bar{S}_{Sn}^{M}$ for this alloy, and calculate the manner in which $\bar{G}_{Sn}^{M}$ varies with the temperature, taking as the standard states:

(a) Pure solid Sn
(b) Pure liquid Sn

SOLUTION

(a) Solid tin dissolves at the anode and accumulates in the Hg-Sn cathode solution. The reaction in the cell corresponds to the solution of Sn in mercury, according to:

$$\langle Sn \rangle \rightleftarrows (\underline{Sn})_{Hg}$$

The partial free energy of mixing of tin in mercury is related to the e.m.f. of the cell, and to the activity referred to the solid standard state, by:

$$\bar{G}^M_{<Sn>} = \bar{G}_{(\underline{Sn})} - G^o_{<Sn>} = -n\mathcal{F}E$$

and

$$\bar{G}^M_{<Sn>} = RT \ln a_{<Sn>}$$

E is expressed in volts, G in calories, $\mathcal{F}$ = 23,060 cal volt^{-1}, and n = 2, since two electrons are exchanged in the reactions of solution and deposition. Equating the above expressions results in:

$$\log a_{<Sn>} = -\frac{n\mathcal{F}E}{RT} = -33.3\ E$$

The activity values for Sn for different mole fractions of Sn in Hg, referred to the solid state, are given in Table 27.

The partial free energy of mixing referred to the liquid state is:

$$\bar{G}^M_{(Sn)} = \bar{G}_{(\underline{Sn})} - G^o_{(Sn)}$$

This differs from $\bar{G}^M_{<Sn>}$ by the free energy of melting:

$$\bar{G}^M_{(Sn)} = \bar{G}_{(\underline{Sn})} - G^o_{<Sn>} + G^o_{<Sn>} - G^o_{(Sn)} = \bar{G}^M_{<Sn>} - \Delta G_{f_{Sn}}$$

with:

$$\Delta G_f = \frac{\Delta H_f(T_f - T)}{T_f}$$

Finally, the activity of Sn referred to the liquid standard state, is obtained from:

$$\log a_{(Sn)} = \log a_{<Sn>} - \frac{\Delta H_f(T_f - T)}{RT\ T_f} = \log a_{<Sn>} - 0.488$$

The values for the activity of Sn corresponding to different mole fractions of Sn in Hg, referred to the liquid standard state, are also given in Table 27.

(b) The maximum solubility of tin in mercury corresponds to the mole fraction above which the activity of tin does not change, i.e. when the e.m.f. of the cell attains a constant value. According to Fig. 143, the maximum solubility of Sn in Hg is 1.35 mol. %.

(c) The partial entropy of mixing $\bar{S}^M_{<Sn>}$ of Sn in Hg referred to the solid state, is equal to the change of $\bar{G}^M_{<Sn>}$ with the temperature:

$$-\bar{S}^M_{<Sn>} = \left[\frac{d\,\bar{G}^M_{<Sn>}}{dT}\right] \quad \text{or} \quad \bar{S}^M_{<Sn>} = n\mathcal{F}\left[\frac{dE}{dT}\right]_{N_{Sn}}$$

TABLE 27

(Sn) (mol-%)	E (mV)	$\log a_{\langle Sn \rangle}$	$a_{\langle Sn \rangle}$	$\log a_{(Sn)}$	$a_{(Sn)}$
0·0295	46·50	- 1·548	0·0284	- 2·036	0·0093
0·110	29·60	- 0·985	0·103	- 1·473	0·034
0·333	16·50	- 0·550	0·282	- 1·038	0·092
0·430	13·45	- 0·448	0·357	- 0·936	0·116
0·962	5·15	- 0·171	0·674	- 0·659	0·220
1·195	2·75	- 0·0915	0·810	- 0·579	0·264
1·360	1·70	- 0·0566	0·878	- 0·544	0·286
1·487	1·56	- 0·0520	0·887	- 0·540	0·289
1·615	1·66	- 0·0552	0·881	- 0·542	0·287
1·784	1·63	- 0·0543	0·883	- 0·542	0·288
2·010	1·63	- 0·0543	0·883	- 0·541	0·288

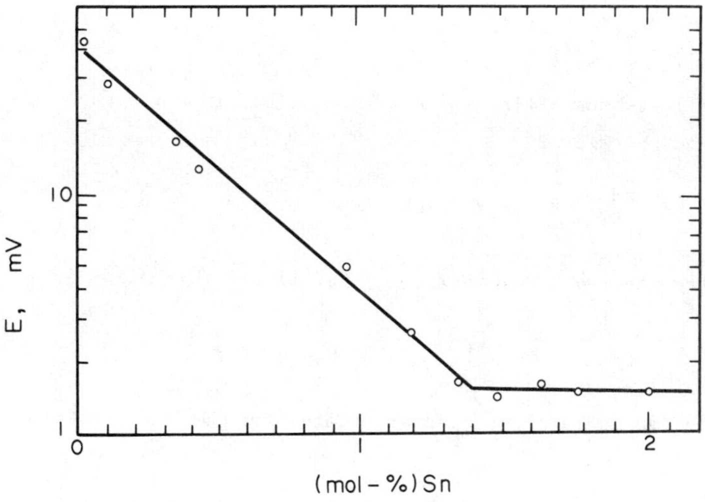

Fig. 143. Determination of the maximum solubility of Sn in Hg

Figure 144(a) is a plot of E versus T. The slope of the line is equal to 0.382×10^{-3}; therefore:

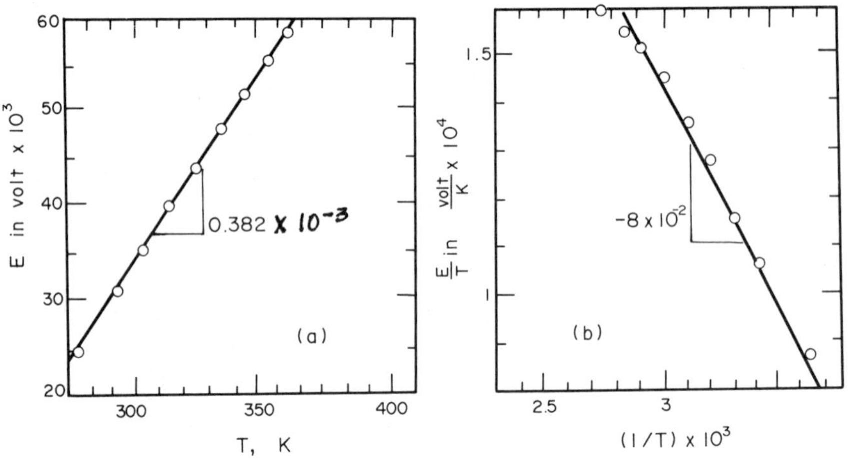

Fig. 144. Graphical determination of: (a) $\bar{S}^M_{\langle Sn \rangle}$ and (b) $\bar{H}^M_{\langle Sn \rangle}$

$$\bar{S}^M_{\langle Sn \rangle} = 2 \times 23{,}060 \times 0.382 \times 10^{-3} = 17.6 \text{ EU}$$

The partial enthalpy of mixing $\bar{H}^M_{\langle Sn \rangle}$ of Sn in Hg referred to the solid state can be determined from the Clausius-Clapeyron equation:

$$\bar{H}^M_{\langle Sn \rangle} = \frac{d\left(\bar{G}^M_{\langle Sn \rangle}/T\right)}{d(1/T)} \qquad \text{or} \qquad \bar{H}^M_{\langle Sn \rangle} = -\, n\mathcal{F}\, \frac{d(E/T)}{d(1/T)}$$

Plotting (E/T) versus (1/T) gives Fig. 144(b), in which the slope is equal to -8×10^{-2}, and therefore:

$$\bar{H}^M_{\langle Sn \rangle} = -\, 2 \times 23{,}060\; (-8 \times 10^{-2}) = 3690 \text{ cal mole}^{-1}$$

The partial free energy of mixing $\bar{G}^M_{\langle Sn \rangle}$ of tin in this alloy is:

$$\bar{G}^M_{\langle Sn \rangle} = 3690 - 17.6\, T$$

If pure liquid tin is taken as standard state, the partial free energy of mixing is expressed by:

$$\bar{G}^M_{(Sn)} = \bar{G}^M_{\langle Sn \rangle} - \frac{\Delta H_f (T_f - T)}{T_f}$$

$$= 3690 - 17.6\, T - 1690 + \frac{1690}{505}\, T = 2000 - 14.25\, T$$

Consequently: $\bar{H}^M_{(Sn)} = 2000$ cal mole^{-1} and $\bar{S}^M_{(Sn)} = 14.25$ EU.

PROBLEM NO. 9

CALCULATION OF THE ACTIVITY OF A COMPONENT IN AN ALLOY (GIBBS-DUHEM INTEGRATION)[2]

The vapour pressure of zinc in equilibrium with a Cu-Zn alloy at 1060°C has been measured with the following results:

N_{Zn}	1	0.8	0.6	0.45	0.30	0.20	0.15	0.10	0.05
p_{Zn} (mm Hg)	3040	2400	1640	970	456	180	90	45	22

(a) In what range of concentration is Henry's Law followed, and what is the value of the activity coefficient of zinc in an infinitely dilute solution?

(b) By integration of the Gibbs-Duhem equation, determine the activity of copper in the alloy.

(c) Trace the curve of the excess integral free energy for this alloy.

SOLUTION

(a) Taking pure liquid zinc as standard state, the activity and the activity coefficient of this metal can be determined by means of:

$$a_{(Zn)} = \frac{p_{Zn}}{p^o_{Zn}} \qquad \text{and} \qquad \gamma_{(Zn)} = \frac{a_{(Zn)}}{N_{(Zn)}}$$

The pressure of zinc in equilibrium with pure liquid zinc, p^o_{Zn}, is equal to 3040 mm Hg at 1060°C. The results are summarised in Table 28. Henry's Law is followed up to a mole fraction of Zn of approximately 0.1. The activity coefficient for the infinitely dilute solution is equal to $\gamma^\infty_{Zn} = 0.146$.

(b) Integration of the Gibbs-Duhem equation can be carried out starting from the expression:

$$\log \gamma_{Cu} = - \int_{N_{Zn}=0}^{N_{Zn}} \frac{N_{Zn}}{N_{Cu}} \, d \log \gamma_{Zn}$$

The values of N_{Zn}/N_{Cu} and $\log \gamma_{Zn}$ (Table 28) were calculated for different values of N_{Zn}, and the curve N_{Zn}/N_{Cu} has been plotted as a function of $\log \gamma_{Zn}$ (Fig. 145(b)). The area between the curve, the axis of the abscissa and the vertical at the abscissa = $-\log \gamma^\infty_{Zn}$, corresponds to $\log \gamma_{Cu}$. γ_{Cu} and a_{Cu} were calculated from these values.

TABLE 28

$N_{(Zn)}$	1	0·8	0·6	0·45	0·30	0·20	0·15	0·10	0·05
$a_{(Zn)}$	1	0·789	0·539	0·319	0·15	0·0592	0·0296	0·0148	0·00724
$\gamma_{(Zn)}$	1	0·987	0·899	0·709	0·50	0.296	0·197	0·148	0·146
N_{Zn}/N_{Cu}	∞	4·0	1·5	0·818	0·429	0·25	0·1765	0·111	0·0526
$\log \gamma_{Zn}$	0	-0·0057	-0·0046	-0·149	-0·301	0·529	-0·706	-0·830	-0·839
$\log \gamma_{Cu}$		-0·44	-0·3319	-0·2199	-0·1314	-0·0564	-0·0188	-10^{-3}	0
γ_{Cu}		0·363	0·466	0·603	0·739	0·878	0·958	0·998	
a_{Cu}	0	0·0726	0·186	0·332	0·517	0·702	0·814	0·898	0·95
$\bar{G}^{ex}_{Zn}$	0	-32	-280	-908	-1835	-3226	-4306	-5062	-5098
$\bar{G}^{ex}_{Cu}$		-2683	-2024	-1341	- 801	- 344	- 115	-6·1	0
G^{ex}	0	- 562	- 978	-1146	-1111	- 920	- 744	- 512	- 255

(c) The excess integral free energy is expressed as a function of the activity coefficient by:

$$G^{ex.} = N_{Zn}\bar{G}^{ex.}_{Zn} + N_{Cu}\bar{G}^{ex.}_{Cu} = RT(N_{Zn} \cdot \log \gamma_{Zn} + N_{Cu} \cdot \log \gamma_{Cu})$$

The values of $G^{ex.}$ versus N_{Zn} (Fig. 145(c)) are given in Table 28. $G^{ex.}$ is always negative (Fig. 145(c)), since γ_{Cu} and γ_{Zn} are both less than one.

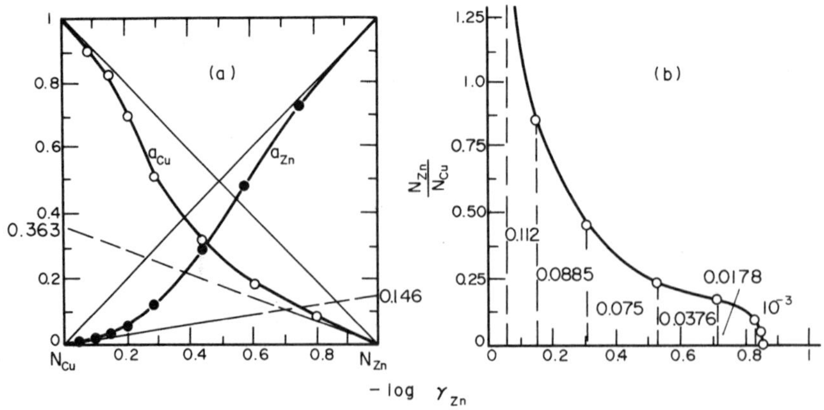

Fig. 145. (a) a_{Cu} and a_{Zn} versus N_{Cu} and N_{Zn}; (b) N_{Zn}/N_{Cu} versus $\log \gamma_{Zn}$

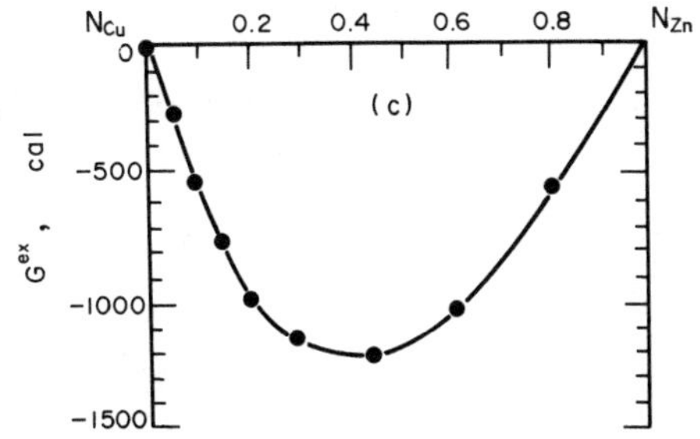

Fig. 145. (c) $G^{ex.}$ versus N_{Zn} for the binary alloy Cu-Zn

PROBLEM NO. 10

EXTRACTION OF SILVER FROM LEAD-SILVER ALLOYS

Silver can be extracted from lead by a solvent such as zinc. A certain amount of zinc is added to the Pb-Ag alloy, forming two phases: one of lead, containing a small amount of Zn, and another of Zn containing nearly all the silver. It is assumed that both solutions are liquid and that the amount of Ag is too small to form a definite compound with Zn.*

(a) Assuming that Zn and Pb are immiscible at 427°C, and using the data given below, calculate the relationship between the partition coefficient of Ag between Zn and Pb and the temperature.

Data: The dilute solution of Ag in Zn can be considered regular and the partial energy of mixing is:

$$\bar{H}^{M}_{\underline{(Ag)}_{Zn}} = -3400 \text{ cal}$$

The excess partial free energy of solution of Ag in Pb is:

$$\bar{G}^{ex.}_{(Ag)_{Pb}} = 7500 - 7.0\ T \text{ (cal)}$$

Henry's Law is followed up to a mole fraction of silver of 0.05 in both solutions.

NUMERICAL APPLICATION

The mole fraction of Ag is 10^{-3} in the initial Pb solution, and the fraction of Zn added is 0.1. Calculate the mole fractions of Ag in both solutions at 427°C when equilibrium is reached.

*For strict accuracy, it must be noted that richer Ag/Zn solutions will give rise to compounds of the form Ag_nZn_m and the balance of Ag between Pb and Zn will be different.

(b) Purification of Pb. Zinc is slightly soluble in lead, as is indicated by the binary diagram Pb-Zn. In order to extract Zn from Pb, chlorine is bubbled through the alloy at 427°C, forming two phases: a metallic phase almost free from Zn, and a slag phase, assumed ideal, which contains $PbCl_2$ and $ZnCl_2$. Find the percentage of $PbCl_2$ in $ZnCl_2$ at 427°C when the mole fraction of Zn in Pb is 10^{-4} (100 ppm).

Data: $(Pb) + Cl_2 \rightleftarrows (PbCl_2)$ $\Delta G^o = -83{,}600 - 14.12\,T \log T + 73.3\,T$

$(Zn) + Cl_2 \rightleftarrows (ZnCl_2)$ $\Delta G^o = -93{,}950 + 27.35\,T$

For dilute solutions of Zn in Pb, the excess partial enthalpy and entropy are:

$$\bar{H}^{ex.}_{(Zn)_{Pb}} = 5000 \text{ cal mole}^{-1} \qquad \bar{S}^{ex.}_{(Zn)_{Pb}} = 2 \text{ cal mole}^{-1}\ {}^\circ C^{-1}$$

SOLUTION

(a) When the two solutions Pb-Ag and Zn-Ag are in equilibrium, the activity of silver is the same in both, then:

$$\gamma_{(Ag)_{Zn}} \cdot N_{(Ag)_{Zn}} = \gamma_{(Ag)_{Pb}} \cdot N_{(Ag)_{Pb}}$$

The partition coefficient K_d is, by definition:

$$K_d = \frac{N_{(Ag)_{Zn}}}{N_{(Ag)_{Pb}}} \quad \text{therefore} \quad K_d = \frac{\gamma_{(Ag)_{Pb}}}{\gamma_{(Ag)_{Zn}}}$$

In order to calculate K_d, it is necessary to know the value of the activity coefficients. As the Ag-Zn solution is regular, the excess partial enthalpy $\bar{H}^{ex.}_{(Ag)_{Zn}}$ is equal to the excess partial free energy, and consequently:

$$\bar{H}^{ex.}_{(Ag)_{Zn}} = \bar{G}^{ex.}_{(Ag)_{Zn}} = RT \ln \gamma_{(Ag)_{Zn}}$$

or

$$\log \gamma_{(Ag)_{Zn}} = \frac{-3400}{RT} = \frac{-744}{T}$$

In the same way, we have:

$$\bar{G}^{ex.}_{(Ag)_{Pb}} = RT \ln \gamma_{(Ag)_{Pb}}$$

or

$$\log \gamma_{(Ag)_{Pb}} = \frac{7500 - 7\,T}{RT} = \frac{1640}{T} - 1.53$$

Thus the partition coefficient can be expressed as a function of the temperature by:

$$\log K_d = \log \gamma_{(Ag)_{Pb}} - \log \gamma_{(Ag)_{Zn}} = \frac{2384}{T} - 1.53$$

APPLICATION

At 700 K (427°C), the partition coefficient has a value of 74.5. As the mole fraction of Ag is small, we have:

$$N_{(Ag)_{Zn}} \simeq \frac{n_{(Ag)_{Zn}}}{n_{(Zn)}} \qquad \text{and} \qquad N_{(Ag)_{Pb}} \simeq \frac{n_{(Ag)_{Pb}}}{n_{(Pb)}}$$

and

$$K_d = \frac{N_{(Ag)_{Zn}}}{N_{(Ag)_{Pb}}} = \frac{n_{(Ag)_{Zn}}}{n_{(Zn)}} \cdot \frac{n_{(Pb)}}{n_{(Ag)_{Pb}}}$$

In this expression, n represents the number of atoms of the element considered. 10^{-1} moles of Zn is added to 1 mole of Pb containing 10^{-3} moles of Ag. Substituting these values and using the distribution expression $n_{(Ag)_{Pb}} + n_{(Ag)_{Zn}} = 10^{-3}$, gives:

$$n_{(Ag)_{Zn}} = 0.882 \times 10^{-3} \qquad \text{and} \qquad n_{(Ag)_{Pb}} = 0.118 \times 10^{-3}$$

or else, expressing the quantities in molar fractions:

$$N_{(Ag)_{Zn}} = \frac{n_{(Ag)_{Zn}}}{n_{(Zn)}} = 0.882 \text{ (at. \%)} \qquad \text{and} \qquad N_{(Ag)_{Pb}} = \frac{n_{(Ag)_{Pb}}}{n_{(Pb)}} = 0.0118 \text{ (at. \%)}$$

Thus, by using Zn equal to 10% (molar) with respect to lead, it is possible to extract 88.2% of Ag. If the lead is given a second treatment with the same proportion of zinc, 88.2% of the remaining Ag in the alloy, i.e. 10.3% more, or a total of 98.7%, can be extracted. Low temperatures favour the extraction, and the process is carried out industrially at temperatures close to the melting point of zinc.

(b) The equilibrium $(\underline{PbCl_2}) + (\underline{Zn}) \rightleftarrows (\underline{ZnCl_2}) + (\underline{Pb})$ can be used to determine the composition of the phases present (Pb, Zn, and chlorides) after bubbling chlorine through the residual zincy lead. The standard free energy of this reaction can be calculated from:

$$\Delta G^o = -10{,}350 + 14.12\, T \log T - 45.95\, T$$

At 700 K, $\Delta G^o = -14{,}400$ cal, and when there is equilibrium between the metallic phase (in which the activity of Pb is approximately equal to 1), and the solution of chlorides assumed ideal, this is equal to:

$$\Delta G^{o} = - RT \ln \frac{N_{ZnCl_2}}{N_{PbCl_2}} \cdot \frac{1}{\gamma_{(Zn)_{Pb}} \cdot N_{(Zn)_{Pb}}} = - RT \ln K_e$$

$K_e = 2.92 \times 10^4$ from data and results obtained earlier. The activity coefficient of zinc in a dilute solution $\gamma^{\infty}_{(Zn)_{Pb}}$ can be calculated from:

$$\bar{G}^{ex.}_{(Zn)_{Pb}} = \bar{H}^{ex.}_{(Zn)_{Pb}} - T\bar{S}^{ex.}_{(Zn)_{Pb}} = 5000 - 2\ T$$

and $$\bar{G}^{ex.}_{(Zn)_{Pb}} = RT \log \gamma^{\infty}_{(Zn)_{Pb}}$$

At 700 K, $\gamma^{\infty}_{(Zn)} = 13.1$. To determine the mole fraction of chlorides in equilibrium with the metallic phase which contains a mole fraction of zinc equal to 10^{-4}, two equations are available:

$$\frac{N_{ZnCl_2}}{N_{PbCl_2}} = K_e \gamma^{\infty}_{(Zn)_{Pb}} \cdot N_{(Zn)_{Pb}} = 2.92 \times 10^4 \times 13.1 \times 10^{-4} = 38.3$$

and $$N_{PbCl_2} + N_{ZnCl_2} = 1$$

The solution of these equations gives:

$$N_{PbCl_2} = 0.0254 \qquad \text{and} \qquad N_{ZnCl_2} = 0.9746$$

This means that the separation of zinc from a Zn-Pb solution by oxidation with chlorine is thermodynamically possible and that the lead losses are small. This is due to two factors acting in the same direction: zinc has a higher affinity for chlorine than lead, and the activity coefficient of zinc in lead is large (if the Pb-Zn solution is ideal, i.e. $\gamma^{\infty}_{(Zn)} = 1$, the chloride phase will contain 25.5 mol. % $PbCl_2$). If the temperature is allowed to rise, the lead losses in the chloride phase will be greater, as K_e and $\gamma^{\infty}_{(Zn)_{Pb}}$ will decrease. For this reason, chlorination must be carried out at the lowest possible temperature.

PROBLEM NO. 11

CRACKING OF NATURAL GAS

Natural gas, CH_4, mixed with an excess of steam, is injected into a reactor heated to 850°C and filled with porcelain spheres containing 2% Ni, that serves as a catalyst for the cracking reaction. Methane reacts with steam according to:

$$[CH_4] + [H_2O] \rightleftarrows [CO] + 3[H_2]$$

Other reactions occur, and the gas at the exit of the reactor contains, in addition to CO, H_2 and excess H_2O, small amounts of CO_2 and CH_4. Depending on the excess of steam, solid carbon may be deposited and poison the catalyst. To avoid this troublesome deposit, an excess of 40% of steam relative to the stoichiometric amount is added, i.e. 1.4 moles of H_2O per mole of CH_4.

(a) Starting from the following data, calculate the partial pressures of all of the gases:

(i) at the exit of the reactor;
(ii) after condensing the excess of steam at 20°C.

The pressure in the reactor is maintained at 1 atm.

(b) The mixture obtained after condensing the steam is used to reduce hematite to iron at 850°C, and when all of the oxide is reduced, the gas carburises the iron formed. What is the value of the carbon activity in the iron when equilibrium is reached?

Data: Standard free energies of formation, starting from the elements:

$$CO_2 : \Delta G^o = -94{,}200 - 0.2\ T$$

$$CH_4 : \Delta G^o = -16{,}520 + 12.25\ T \log T - 15.62\ T$$

$$CO : \Delta G^o = -26{,}700 - 20.95\ T$$

$$H_2O : \Delta G^o = -58{,}800 + 13.1\ T$$

Saturated vapour pressure of steam:

$$\log p_{H_2O}^{(Atm.)} = -\frac{2900}{T} - 4.65 \log T + 19.733$$

SOLUTION

(a) (i) If it is assumed that no carbon is deposited, the unknowns are five: p_{CO}, p_{H_2}, p_{CH_4}, p_{H_2O}, p_{CO_2}. Of the five equations needed for the determination of these five unknowns, three are equations of balance and two are thermodynamic relationships.

The Mass Balances

The numbers n of moles of C, O, and H that leave the reactor are equal to those that enter, n*. It can then be written:

For carbon: $$n_{CO} + n_{CO_2} + n_{CH_4} = n^*_{CH_4} = 1$$

For oxygen: $$n_{CO} + 2n_{CO_2} + n_{H_2O} = n^*_{H_2O} = 1.4$$

For hydrogen: $n_{H_2} + n_{H_2O} + 2n_{CH_4} = n^*_{H_2O} + 2n^*_{CH_4} = 1.4$

In these expressions, additional unknowns have been introduced, which are the numbers of moles of product: n_{CO}, n_{CO_2}, etc. Nevertheless, there exists a simple relationship between these numbers and the partial pressures:

$$\frac{n_i}{\Sigma n_i} = \frac{p_i}{p}$$

Thermodynamic Relationships

Two thermodynamic relationships can be developed from the standard free energies of two independent reactions, which can be:

$$CH_4 + H_2O \rightleftarrows CO + 3H_2 \qquad (1)$$

and

$$CO + H_2O \rightleftarrows CO_2 + H_2 \qquad (2)$$

At equilibrium, $\Delta G^o = -RT \ln K$. The equilibrium constants K_1 and K_2 are calculated from the data:

$$\log K_1 = \log \frac{p^3_{H_2} \times p_{CO}}{p_{H_2O} \times p_{CH_4}} = -\frac{\Delta G^o_1}{RT} = \frac{14040}{4.575 \times 1123} = 2.73$$

$$\log K_2 = \log \frac{p_{H_2} \times p_{CO_2}}{p_{CO} \times p_{H_2O}} = -\frac{\Delta G^o_2}{RT} = \frac{109}{4.575 \times 1123} = 0.021$$

Therefore $K_1 = 537$ and $K_2 = 1.05$. The equations of mass balance are linear with respect to the number of moles of each species, but in the thermodynamic equations, the partial pressures are multiplied in pairs, and it is necessary to solve the equations by successive iterations.

The value of K_1 is large, and since there is an excess of water, it can be assumed as a first approximation that p_{CH_4} is negligible compared with the partial pressures of the other gaseous species at the exit of the reactor. From the mass balance equations, the approximate value of Σn can be calculated as well as the partial pressures of CO, H_2, H_2O, as functions of p_{CO_2}.

$$\Sigma n = n_{CO} + n_{CO_2} + n_{H_2} + n_{H_2O} + n_{CH_4} = 4.4 - 2n_{CH_4} \simeq 4.4$$

$$n_{CO} = 1 - n_{CO_2} - n_{CH_4} \simeq 1 - n_{CO_2}$$

or $$p_{CO} = \frac{1 - n_{CO_2}}{\Sigma n} \simeq 0.227 - p_{CO_2}$$

$$n_{H_2O} = 1.4 - n_{CO} - 2n_{CO_2} = 0.4 - n_{CO_2} + n_{CH_4} \simeq 0.4 - n_{CO_2}$$

or $$p_{H_2O} = \frac{0.4 - n_{CO_2}}{\Sigma n} = 0.0910 - p_{CO_2}$$

$$n_{H_2} = 3.4 - n_{H_2O} - 2n_{CH_4} = 3 + n_{CO_2} - 3n_{CH_4} \simeq 3 + n_{CO_2}$$

or $$p_{H_2} = \frac{3 + n_{CO_2}}{\Sigma n} = 0.682 + p_{CO_2}$$

By inserting these values in the expressions for K_2 and K_1, the approximate values of p_{CH_4} and p_{CO_2} can be found:

$$K_2 = \frac{p_{CO_2} \cdot p_{H_2}}{p_{H_2O} \cdot p_{CO}} = \frac{p_{CO_2}(0.682 + p_{CO_2})}{(0.0910 - p_{CO_2})(0.227 - p_{CO_2})} = 1.05$$

The only possible value is: $p_{CO_2} = 0.0218$.

The constant K_1 is then expressed in the form:

$$K_1 = \frac{p_{CO} \cdot p_{H_2}^3}{p_{CH_4} \cdot p_{H_2O}} = \frac{(0.227 - p_{CO_2})(0.682 + p_{CO_2})^3}{p_{CH_4} \cdot (0.0910 - p_{CO_2})} = 537$$

The approximate value of p_{CH_4} is then:

$$p_{CH_4} = \frac{0.2052 \times 0.3486}{537 \times 0.0692} = 1.92 \times 10^{-3}$$

Inserting this value of p_{CH_4} in the expressions for the mass balance, gives:

$n_{CH_4} = p_{CH_4}(4.4 - 2n_{CH_4})$ and $n_{CH_4} = 0.84 \times 10^{-2}$

$\Sigma n = 4.4 - 2n_{CH_4} = 4.383$

$n_{CO_2} = p_{CO_2}\ \Sigma n = 9.555 \times 10^{-2}$ $p_{CO_2} = 0.0218$

$n_{CO} = 1 - 0.0955 - 0.0084 = 0.0896$ $\qquad p_{CO} = 0.204$

$n_{H_2O} = 0.4 - 0.0955 + 0.0084 = 0.313$ $\qquad p_{H_2O} = 0.0714$

$n_{H_2} = 3 + 0.00955 - 0.0252 = 3.07$ $\qquad p_{H_2} = 0.70$

Inserting these values in K_1 and K_2, we find:

$$K_1 = 510 \qquad K_2 = 1.047$$

More accurate values of the partial pressures can be found by a second iteration. These will be close to the correct ones, which are:

$p_{CH_4} = 1.81 \times 10^{-3}$ $\qquad p_{CO_2} = 0.0217$

$p_{CO} = 0.202$ $\qquad p_{H_2O} = 0.0714$

$p_{H_2} = 0.70$

Corrections for larger divergence can be made by increasing the number of iterations.

(a) (ii) When the gas is cooled to 20°C, part of the water condenses. The pressure of the saturated steam can then be calculated from $\log p_{H_2O} = f(T)$. This gives:

$$\log p_{H_2O} = -1.635 \text{ atm} \quad \text{or} \quad p_{H_2O} = 0.0232 \text{ atm}$$

From this, (0.0714 - 0.0232) = 0.0482 atm of water are condensed. If the partial pressures of the gases, after condensing some of the water, are designated by p', we obtain:

$$p'_i = \frac{p_i}{1 - 0.0482} = 1.0506\, p_i$$

and therefore:

$p'_{H_2O} = 0.0232$ atm; $p'_{CO_2} = 0.0229$ atm; $p'_{CH_4} = 0.00191$ atm;

$p'_{CO} = 0.214$ atm; $p'_{H_2} = 0.735$ atm.

(b) In the above calculations, the risk of a carbon deposit on the catalyst has not been taken into account. This risk will exist if the carbon activity in the gas is near to one. The activity of carbon in the gas after condensing the water can be calculated from the equilibrium between C in solution in solid iron and a $CO-CO_2$ atmosphere:

$$(C) + [CO_2] \rightleftarrows 2[CO]$$

The free energy of this reaction can be calculated from $\Delta G^o = 40{,}800 - 41.7\ T$. At 850°C (1123 K), $\Delta G^o = -6020$ cal. At equilibrium, $\Delta G^o = -RT \log p^2_{CO}/(p_{CO_2} \times a_C)$. Substituting the values of p_{CO} and p_{CO_2} found in (a)(ii), we find that $a_{<C>} = 0.136$. The activity of C in the gas after condensing the excess water is very much less than one, and there is no danger of carbon deposition on the iron, except at below 740°C when carbon is produced by the reaction $2CO \rightleftarrows CO_2 + C$. The probability of this is small, because the kinetics are unfavourable below 900°C. The risk of deposition of C on the catalyst is yet less because there is excess water.

PROBLEM NO. 12

PRODUCTION OF CALCIUM BY ALUMINOTHERMIC REDUCTION OF LIME

The reduction of CaO by aluminium, at atmospheric pressure, must be carried out at a temperature higher than about 2300°C, since below this temperature the affinity of aluminium for oxygen is less than that of calcium (Fig. 146). Calcium is produced as a vapour, and as the total pressure of the system is decreased, the inversion temperature of the reduction also decreases.

(a) Find the inversion temperature of the reaction:

$$\frac{4}{3}(Al) + 2\langle CaO\rangle \rightleftarrows 2\ Ca + \frac{2}{3}\langle Al_2O_3\rangle$$

for a total pressure of 0.76×10^{-3} mm Hg, assuming that:

$$\frac{a^{2/3}_{Al_2O_3}}{a^2_{CaO}} = 1$$

(b) Aluminium boils at 2440°C (2713 K), and its enthalpy of fusion is 77,400 cal mole^{-1}. Assuming that ΔH_v does not change with the temperature, calculate the relationship between vapour pressure and temperature in the form:

$$\log P_{Al} = f\left(\frac{1}{T}\right)$$

Determine the partial pressure of Al in the gas phase and the contamination of Ca by Al, in mol. % of Al in Ca after condensation from the temperature found in (a).

(d) Using Fig. 146, comment on the effect of the temperature and pressure on the contamination of Ca by Al.

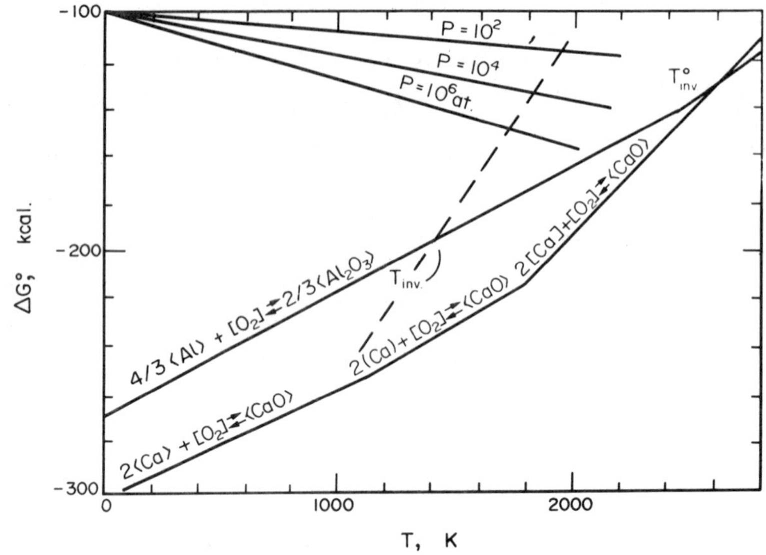

Fig. 146. Free energies of formation of CaO and Al_2O_3
The dotted line represents the free energy of formation of CaO when $p_{Ca} = 10^{-6}$ atm.

SOLUTION

(a) The ΔG_T^o curves are plotted in the Ellingham diagram for $p_{O_2} = 1$ atm and $p_{Ca} = 1$ atm. In this standard state, ($p_{Ca} = 1$), the inversion temperature for the reduction of CaO by Al is of the order of 2600 K. When the total pressure is 10^{-6} atm (0.76×10^{-3} mm Hg), it can be assumed as a first approximation that this pressure is that of the calcium, and the variation in the free energy of oxidation of gaseous Ca to lime ($\Delta G - \Delta G^o$) will be equal to:

$$\Delta G - \Delta G^o = RT \log \frac{1}{p_{Ca}^2} = RT \log 10^{12} = 12\ RT$$

The plot of the free energy of oxidation of calcium will shift upwards by an amount equal to 12 RT (Fig. 146, segmented line). The inversion temperature of the aluminothermic reduction process decreases considerably, and is then equal to 1200 K.

(b) Integration of the Clausius-Clapeyron equation:

$$\frac{\partial \ln P}{\partial\ (1/T)} = -\frac{\Delta H_v}{R}$$

between the temperature T (for which the value of P is p_{Al}) and T_v where P is equal to 1 atm, gives:

$$\log p_{Al} = -\frac{\Delta H_v}{RT \cdot T_v}(T_v - T)$$

Replacing ΔH_v and T_v by their values, results in:

$$\log p_{Al} = -\frac{16{,}980}{T} + 6.24$$

(c) At 1200 K, $\log p_{Al}$ is -7.86, which corresponds to $p_{Al} = 1.4 \times 10^{-8}$ atm. Since the calcium pressure is near to 10^{-6} atm, the contamination of calcium with aluminium is large and, after condensing, reaches 1.4%.

(d) From the values of the standard free energies of formation of Al_2O_3 and CaO (Fig. 146):

$$\frac{4}{3}(Al) + O_2 = \frac{2}{3}\langle Al_2O_3\rangle \qquad \Delta G^o = -270{,}000 + 52\ T\ \text{cal mole}^{-1}$$

$$2[Ca] + [O_2] = 2\langle CaO\rangle \qquad \Delta G^o = -405{,}000 + 109\ T\ \text{cal mole}^{-1}$$

ΔG for the aluminothermic reduction of CaO is:

$$\Delta G = 135{,}000 - 57\ T + 2RT \ln p_{Ca}\ \text{cal}$$

and, at equilibrium:

$$\log p_{Ca} = -\frac{135{,}000}{2RT} + \frac{57}{2R} = -\frac{14{,}750}{T} + 6.23$$

The ratio p_{Al}/p_{Ca} is obtained by subtracting $\log p_{Ca}$ from $\log p_{Al}$:

$$\log \frac{p_{Al}}{p_{Ca}} = -\frac{2170}{T} + 0.01$$

i.e. the contamination of Ca by Al increases with the temperature.

PROBLEM NO. 13

REDUCTION OF A BED OF IRON OXIDE PARTICLES[6]

A fixed bed of particles of hematite is reduced with a mixture of $CO\text{-}CO_2$ at atmospheric pressure and constant temperature. The fraction of CO in the gas that enters the reactor is N_o at a flow rate n_g (in moles of gas/unit area). The bed contains 9 g-atoms of oxygen per unit volume. The reactions are assumed to be instantaneous and wustite is assumed to be stoichiometric:

$$6FeO_{1.5} + CO \rightleftarrows 6\ FeO_{1.33} + CO_2 \qquad \text{(fraction of CO at equilibrium} = N_3^*)$$

$3FeO_{1.33} + CO \rightleftarrows 3FeO + CO_2$ (fraction of CO at equilibrium = N_2^*)

$FeO + CO \rightleftarrows Fe + CO_2$ (fraction of CO at equilibrium = N_1^*)

In general:

$$\frac{1}{x_i - x_{i-1}} FeO_{x_i} + CO \rightleftarrows \frac{1}{x_i - x_{i-1}} FeO_{x_{i-1}} + CO_2 \quad \text{(fraction of CO at equilibrium} = N_i^*)$$

Each oxide is reduced to the one following at its own velocity, which depends on the amount of CO available for the reduction and the number of atoms of oxygen eliminated in each reaction. Each reaction interface will then move at a certain velocity, as indicated by Fig. 147. This figure also shows the fractions of CO in equilibrium with two condensed phases at 800° and 1000°C (deduced from Chaudron's diagram).

(a) (i) Construct an expression for the velocity of reduction of an oxide to the next in sequence in the form of g-atom of oxygen carried by the gas per unit time ($r_i = f(N_i^*, N_{i-1}^*)$).

(ii) If H is the height of the bed, calculate the total reduction time for each oxide to the next, as well as the linear velocities of the transformation fronts (r_i' in cm sec^{-1}) for the interfaces Fe-FeO, FeO-$FeO_{1.33}$, and $FeO_{1.33}$-$FeO_{1.5}$ ($r_i' = g(N_i^*, x_i)$).

(b) Given the following data, taken from an actual reduction unit:

$$H = 100 \text{ cm}, \; qV = 150 \text{ (g-atom O)}$$

$$n_g S = 5 \times 10^{-2} \text{ moles sec}^{-1}; \text{ and } N_o = 1$$

Calculate the time necessary for complete reduction of each oxide to that next in order, at 800° and 1000°C, using the values for $p_{CO}/p_{CO} + p_{CO_2}$ from Fig. 147.

Compare the velocities of advance of the interfaces at each temperature. What comments can be made on the velocities of advance of the interfaces at 1000°C? Make corrections where necessary.

(c) (i) Starting from the values found in (b), trace the curve N_s (fraction of CO at the exit of reactor) versus time, at 800° and 1000°C.

(ii) F being the atom fraction of the oxygen eliminated from the ore in the reduction of hematite to iron, and F_i the fraction of oxygen eliminated in the reduction of one oxide to the next one (or to iron), trace the curve F versus time.

SOLUTION

Since reaction is assumed to be instantaneous, the zones of transformation are reduced to an interface of zero thickness. At the wustite-iron interface, the percentage of CO in the gas phase changes instantaneously from N_o to N_1^*; at the magnetite-wustite interface from N_1^* to N_2^*, etc. The values of N_i^* deduced from Chaudron's diagram at 800° to 1000° are indicated in Fig. 147.

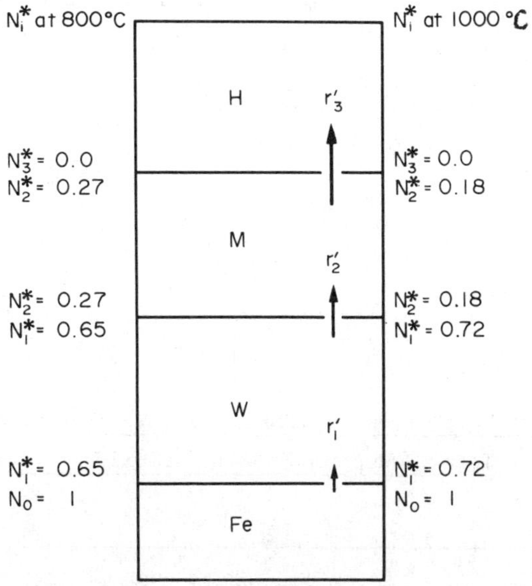

Fig. 147. Diagrammatic representation of interface movement in reactor

(a) (i) The amount of CO oxidised to CO_2 (or the number of oxygen atoms eliminated from an oxide) per unit of time is equal to the product of the gas flow $n_g S$ moles sec^{-1} and the fraction of gas transformed in the reduction of oxide i:

$$r_i = n_g S(N_{i-1}^* - N_i^*) \text{ moles sec}^{-1}$$

(ii) The total number of oxygen atoms n_i to be eliminated from an oxide in order to reduce it to the next in reaction i, is equal to the total number of oxygen atoms in the bed qV, multiplied by the fraction of oxygen $(x_i - x_{i-1})/1.5$ eliminated in the reaction under consideration, i.e.:

$$n_i = \frac{qV}{1.5}(x_i - x_{i-1}) \text{(oxygen atoms)}$$

The total reduction time for one oxide to the next (or to iron) is then equal to:

$$t_i = \frac{n_i}{r_i} = \frac{qV(x_i - x_{i-1})}{1.5\, n_g S(N_{i-1}^* - N_i^*)} \text{ (sec)}$$

and the velocity of advance r_i' of each interface:

$$r_i' = \frac{H}{t_i} = \frac{1.5\, n_g SH(N_{i-1}^* - N_i^*)}{qV(x_i\ x_{i-1})} \text{ (cm sec}^{-1}\text{)}$$

(b) Application: From these equations, the values of r_i' and t_i at 800° and 1000°C can be calculated for each reaction (Table 29).

From Table 29 it can be seen that at 1000°C the displacement velocity of the wustite-magnetite interface is greater than that for the hematite-magnetite interface, which is impossible, since reduction must be effected in the order: hematite→magnetite→wustite (i.e. the interface magnetite-wustite cannot move more rapidly than the interface hematite-magnetite). The two interfaces must then be displaced at the same velocity, r'_{3-2}:

$$r'_{3-2} = \frac{1.5\ n_g SH(N_1^* - N_3^*)}{qV(x_3 - x_i)} \quad (\text{cm sec}^{-1})$$

and the reduction time for hematite to wustite t_{3-2} is equal to H/r'_{3-2}. The values of r'_{3-2} and t_{3-2} are given in Table 29.

TABLE 29 Velocities of Advance of Each Interface r'_i and Total Reduction Time of an Oxide to the Next, t_i

Reaction	Temperature °C			
	800		1000	
	r'_i (cm/sec)	t_i (sec)	r'_i (cm/sec)	t_i (sec)
Haemat.-Magnet.	0·0795	1260	0·05 } 0·072	2000 } 1390
Magnet.-Wust.	0·0575	1740	0·0834	1200
Wust.-Iron	0·0175	5710	0·014	7140

(c) (i) During time t_3 (or t_{3-2} at 1000°C), while there is still hematite in the reactor, the gas leaves with the composition $N_s = N_3^*$, i.e. 0% of CO. Between t_3 and t_2 (at 800°C), N_s is equal to N_2^*. Starting from t_2 (or t_{3-2} at 1000°C), the percentage of CO at the exit is N_1^*. The curves of Fig. 148 show how N_s varies as a function of time at 800° and 1000°C.

(ii) The overall fraction of oxygen eliminated is the sum of the fractions F_i of oxygen eliminated in each reaction, i.e.:

$$F = \Sigma F_i$$

F can also be expressed as a function of N_s (fraction of CO in the exit gas):

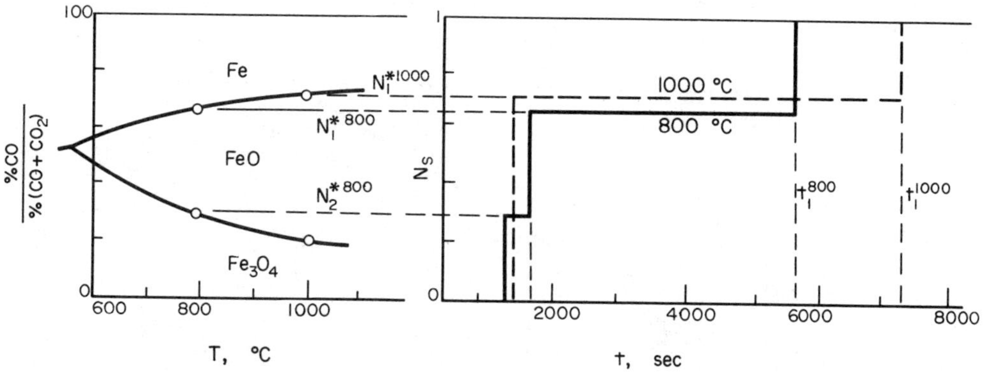

Fig. 148. CO content of exit gas versus time

$$F = \frac{n_g S}{qV} \int_0^t (N_o - N_s) dt$$

The slope of the plot of F against time changes suddenly when an oxide is eliminated by reduction. Integrating this relation between t_{i-1} and $t (t_{i-1} < t < t_i)$ gives:

$$F(t) = F(t_{i-1}) + \frac{(N_o - N_s) n_g S}{qV} (t - t_{i-1})$$

Application: Taking the data from Table 29 and Fig. 147:

At 800°C: From t = 0 to 1260 sec $\quad N_s = 0$

$F(t) = 3.33 \times 10^{-4}\ t$: $\underline{F(1260) = 0.42}$

From t = 1260 to 1740 sec $\quad N_s = 0.27$

$F(t) = 0.42 + 2.43 \times 10^{-4}\ (t-1260)$: $\underline{F(1740) = 0.537}$

From t = 1740 to 5710 sec $\quad N_s = 0.65$

$f(t) = 0.537 + 1.165 \times 10^{-4}\ (t-1740)$: $\underline{F(5710) = 1}$

At 1000°C: From t = 0 to 1390 sec $\quad N_s = 0$

$f(t) = 3.33 \times 10^{-4}\ T$: $\underline{F(1390) = 0.463}$

From t = 1390 to 7140 sec $\quad N_s = 0.72$

$f(t) = 0.463 + 0.933 \times 10^{-4}\ (t-1390)$: $\underline{F(7140) = 1}$

The values of $F(t)$ at 800° (OABC) and 1000°C (OA'C') are plotted in Fig. 149.

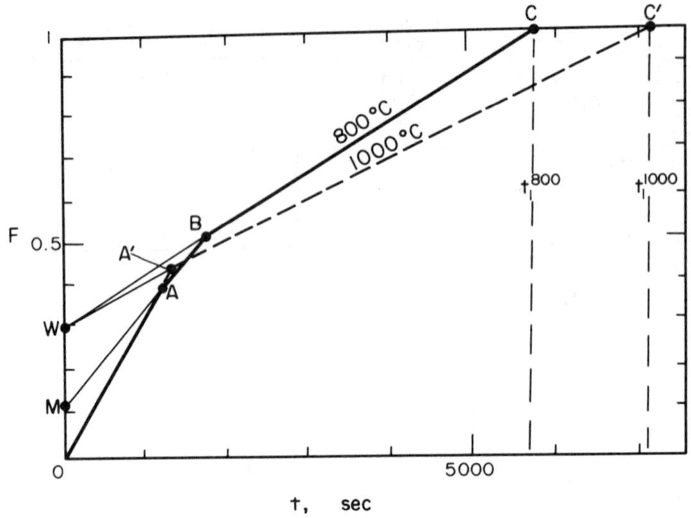

Fig. 149. Fraction of oxide reduced versus time

It can be seen that, by extrapolation of each curve back to the time ordinate, two points, M and W, are obtained, with ordinates (1.5 - 1.33)/1.5 = 0.11 and (1.5 - 1)/1.5 = 0.33, respectively. The reduction of an oxide to the next starts from the origin of the time: from O for hematite; from M for magnetite; from W for wustite. The line WC (or WC') represents the reduction of FeO when it is the only oxide.

The curves obtained in actual conditions (i.e. when kinetics are not instantaneous) are always below the ideal curves of Fig. 149. In a bed of iron ore, with reducing gas flowing counter-current, the rate of feed must be calculated from the velocity of advance of the wustite-iron interface, in order to obtain completely reduced iron at the bottom. Then the composition of the exit gas is given by the operating line of Fig. 68c (curve a).

PROBLEM NO. 14

MOLYBDENITE ROASTING IN A MULTIPLE HEARTH FURNACE[7]

Molybdenite concentrate containing 94% MoS_2 is roasted in a multiple hearth furnace to obtain molybdenum trioxide MoO_3 at 625°C.

(a) Given the following data, construct a predominance area diagram for molybdenum compounds in the presence of oxygen and sulphur dioxide at 625°C. Using the diagram, state which is the most stable molybdenum compound in the presence of air, and calculate the partial pressure of oxygen in equilibrium with this compound and molybdenite.

Data:

(1) $\langle Mo\rangle + [S_2] \rightleftarrows \langle MoS_2\rangle$ $\Delta G_1^o = -85{,}870 + 37.33\ T$

(2) $\langle Mo\rangle + [O_2] \rightleftarrows \langle MoO_2\rangle$ $\Delta G_2^o = -140{,}100 - 4.6\ T \log T + 55.8\ T$

(3) $\langle MoO_2\rangle + \frac{1}{2}[O_2] \rightleftarrows \langle MoO_3\rangle$ $\Delta G_3^o = -38{,}700 + 19.5\ T$

(4) $\langle S_2\rangle + 2[O_2] \rightleftarrows 2\langle SO_2\rangle$ $\Delta G_4^o = -173{,}240 + 34.62\ T$

(b) Molybdenite enters a roaster at a rate Q, (moles hr^{-1}) onto the upper hearth. It is then caused to travel downward by the action of the scrapers, falling from hearth to hearth. For simplification, it is assumed that the surface area, S, of each hearth is completely covered by the concentrate to a thickness z_o. At the entry point to hearth i, the fraction of MoS_2 untransformed is F_{i-1}, and F_i at its exit. The same amount of air is blown over each hearth, and the gases leave the roaster by the flue.

(i) It has been found experimentally[7] that when the untransformed fraction of MoS_2 is higher than about 14%, the rate of transformation is controlled by diffusion within the bed, and the amount of MoS_2 oxidised in each hearth $Q(F_{i-1} - F_i)$, is constant. This amount is proportional to the surface area, S, of each hearth and to the oxygen partial pressure, p_{O_2}, over the solid, which is assumed constant throughout the furnace and equal to that of the exit gases. The proportionality constant, k_d, has a value of 160 (moles $hr^{-1}\ m^{-2}\ atm^{-1}$). The quantity of air blown into each hearth is equal to n_g moles hr^{-1} and it contains 20% of oxygen.

Assuming that the total number of moles of gas at the exit is the same as at the entrance to the furnace, that is, there is no change in the total number of gas moles during the roasting, find a relation between the amount of molybdenite roasted in each hearth per unit time and unit surface area of hearth, and the gas flow rate.

(ii) Given that the oxidation of one mole of MoS_2 produces 270 kcal; that the heat losses through the walls, W_{CR}/S are 1000 kcal m^{-2} of hearth; that the air is blown in at 25°C; that the gases leave the furnace at 625°C; and that the heat capacity of the gas is 8 cal °C^{-1} $mole^{-1}$, construct a thermal balance for one hearth and find another relation between $Q(F_{i-1} - F_i)$ and the gas flow rate.

(iii) Starting from relationships developed in (i) and (ii), determine graphically the optimum working conditions (solid feed and gas flow rate) necessary to keep a constant temperature without heating or cooling the furnace, assuming that the first part of the transformation (from 94% to 14% MoS_2) uses six hearths. Explain what happens when these conditions are not satisfied.

SOLUTION

(a) The thermodynamically possible equilibria between compounds of Mo are: Mo-MoO_2, MoO_2-MoO_3, Mo-MoS_2, MoS_2-MoO_2, MoS_2-MoO_3. The value of the free energies at 900 K for the different equilibria, the logarithms of the equilibrium constants, and $\log p_{O_2}$ as a function of $\log p_{SO_2}$ are summarised in Table 30, which makes it possible to draw Fig. 150. If p_{O_2} is higher than 10^{-10} atm, the only stable compound at 900 K is MoO_3.

TABLE 30 Mo-S-O Equilibria

Equilibrium	ΔG^o at 900 K (cal)	log K	$\log p_{O_2}$
$Mo + O_2 \rightleftarrows MoO_2$	- 102,100	24·8	-24·8
$MoO_2 + \frac{1}{2}O_2 \rightleftarrows MoO_3$	- 21,150	5·1	-10·3
$Mo + 2\,SO_2 \rightleftarrows MoS_2 + 2\,O_2$	89,830	-21·8	$-10{\cdot}9 + \log p_{SO_2}$
$MoS_2 + 3\,O_2 \rightleftarrows MoO_2 + 2\,SO_2$	- 191,920	46·6	$-15{\cdot}5 + \frac{2}{3}\log p_{SO_2}$
$MoS_2 + 3{\cdot}5\,O_2 \rightleftarrows MoO_3 + 2\,SO_2$	- 213,070	51·7	$-14{\cdot}8 + 0{\cdot}571 \log p_{SO_2}$

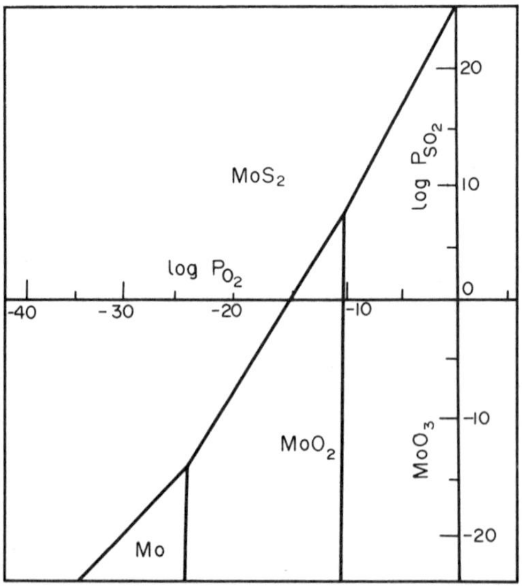

Fig. 150. Predominance area diagram for the system Mo-S-O at 900 K

(b) (i) The amount of MoS_2 roasted in one of the upper hearths, i, is equal (according to the data) to:

$$Q(F_{i-1} - F_i) = k_d \cdot S \cdot p_{O_2} \qquad \text{(a)}$$

p_{O_2} over the solid can be expressed as a function of the number of moles of oxygen n_{O_2}, the total number of moles of gas Σn leaving the hearth, and the total pressure P by:

$$p_{O_2} = \frac{n_{O_2} \cdot P}{\Sigma n} \qquad \text{(b)}$$

n_{O_2} is equal to the number of oxygen moles blown, minus the oxygen reacted, according to:

$$\langle MoS_2 \rangle + 3.5\ [O_2] \rightleftarrows \langle MoO_3 \rangle + 2[SO_2] \qquad \text{(c)}$$

that is:

$$n_{O_2} = 0.2\ n_g - 3.5\ Q(F_{i-1} - F_i) \qquad \text{(d)}$$

At the furnace exit, the total number of moles of gas (neglecting the change due to the reaction) is the same as that at the entrance, that is:

$$\Sigma n = n_g \qquad \text{(e)}$$

Inserting the values of n_{O_2}, Σn and p_{O_2} in equation (a), one obtains:

$$Q(F_{i-1} - F_i) = \frac{k_d SP}{n_g}\ [0.2\ n_g - 3.5\ Q(F_{i-1} - F_i] \qquad \text{(f)}$$

or, taking $Q(F_{i-1} - F_i)$ as a factor:

$$\frac{Q}{S}(F_{i-1} - F_i) = \frac{0.2\ k_d p n_g/S}{n_g/S + 3.5\ k_d P} \qquad \text{(g)}$$

This is the amount of MoS_2 roasted per unit area of hearth in unit time. It is proportional to the heat evolved by the reaction taking place, and is represented by the curve $(Q/S)(F_{i-1} - F_i) = f(n_g/S)$ of Fig. 151 for $k_d = 160$.

(ii) In a permanent regime, the heat generated in a hearth is enough to compensate for the losses, i.e.:

$$\Delta HQ(F_{i-1} - F_i) = W_{CR} + n_g C_p \Delta T \qquad (q)$$

W_{CR} represents the heat losses by convection and radiation, and $n_g C_p \Delta T$ the latent heat of gases. The thermal balance per unit area of hearth is:

$$\frac{Q}{S}(F_{i-1} - F_i) = \frac{W_{CR}}{S\Delta H} + bn_g/S \qquad (r)$$

where $b = \dfrac{C_p \Delta T}{270 \times 10^3} = 1.78 \times 10^{-2}$ and $\dfrac{W_{CR}}{S\Delta H} = 3.7$

in this particular case. $(Q/S)(F_{i-1} - F_i)$ is proportional to the heat loss per unit area of hearth in unit time, and is represented by the straight line:

$$\frac{Q}{S}(F_{i-1} - F_i) = g\left(\frac{n_g}{S}\right)$$

(iii) The straight line and the curve of Fig. 151, which represent the heat loss and the heat of reaction, respectively, intercept at two points, B and B', which represent the optimum operating conditions of the furnace on the upper hearths, i.e conditions where it is not necessary to heat or cool the upper hearths in a permanent regime. The coordinates of point B', for example, are: $(n_g/S) = 945$ and $Q(F_{i-1} - F_i) = 20$. The amount of air to be blown onto each hearth per unit of time and area, is then equal to about 945 moles m^{-2} hr^{-1}, or else 21.2 m^3 of air m^{-2} hr^{-1}.

From 94% to 14% MoS_2, the roasting velocity does not depend on the untransformed fraction of MoS_2, and therefore the same amount of molybdenite is roasted on each floor. In order to decrease from 94% to 14% MoS_2 in six hearths, 13.3% must be transformed on each. As $(Q/S)(F_{i-1} - F_i)$ is equal to 20, the feed rate Q/S should be equal to 20/0.133 = 150 moles of MoS_2 m^{-2} hr^{-1}.

If the air flow rate is increased above the value indicated by point B', the heat losses will be higher than the heat of reaction, and the furnace will become cold. Decreasing the air flow rate will cause the furnace to become hot. On increasing or decreasing the MoS_2 feed rate, the amount of molybdenite transformed on each hearth, which depends on the air flow rate, will not change. As the amount $(Q/S)(F_{i-1} - F_i)$ is constant, a change in Q/S will change $(F_{i-1} - F_i)$ in the opposite direction. If the feed rate is increased, more hearths will be necessary to achieve the same degree of transformation, and vice versa.

From (iii), since $(Q/S)(F_{i-1} - F_i)$ and n_g/S are known, it is possible to calculate the amount of oxygen which reacts at each hearth (20 × 3.5 = 70 moles m^{-2} hr^{-1}), of SO_2 in the exit gas (40 moles m^{-2} hr^{-1}), the flow of exit gas (945 - 70 + 40 = 915 moles m^{-2} hr^{-1}), and the composition of this gas (O_2 = 13%, SO_2 : 4.4%, N_2 : 82.6%). The error due to the assumption that the volume of gas at the exit is the same as at the entrance to the furnace is about 3% $\left(\dfrac{70 - 40}{945}\right) \simeq 3\%$.

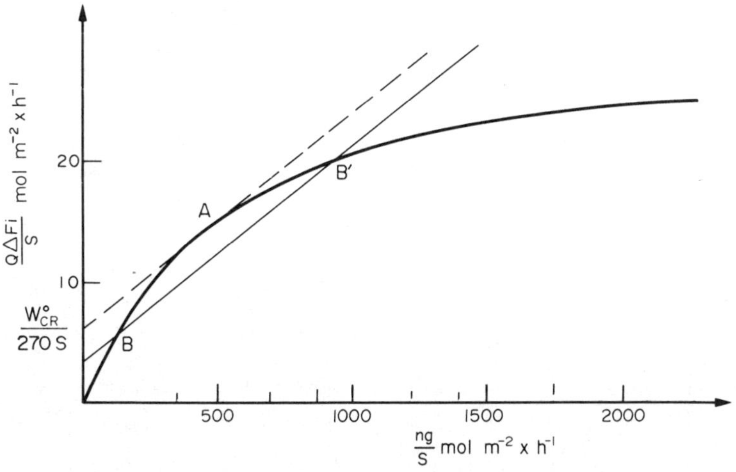

Fig. 151. Graphical determination of feed rate as a function of air flow rate

Figure 151 shows that if the losses by convection and radiation W_{CR} exceed a certain value W^o_{CR} (here $W^o_{CR}/270S$ = 6.5 moles hr^{-1} m^{-2}), the heat supplied by the roasting of MoS_2 is not sufficient to compensate the total losses. In this condition it is possible either to burn hydrocarbon fuel at each hearth (but in this case the fraction of O_2 in the gas decreases as well as the reaction velocity), or to enrich the entering gas with oxygen (which increases the heat supplied to the hearth and diminishes the heat losses in the exit gases). The above calculations are not appropriate for the first modification.

PROBLEM NO. 15

REDUCTION OF ZINC OXIDE WITH CARBON MONOXIDE (Data from Ref(20))

Solid ZnO can be reduced to Zn vapour by reaction with CO, and the free energy change is given by:

$$\langle ZnO\rangle + [CO] \rightleftarrows [Zn] + [CO_2] \quad \Delta G^o = 47{,}920 + 10.35\,T\log T - 61.63\,T \quad (a)$$

(a) What is the variance of the system, and what can be deduced from it?

(b) Calculate the partial pressures of Zn, CO and CO_2 in equilibrium with solid ZnO at 700°, 900°, 1100°, and 1300°C for a total pressure of one atmosphere, and determine if liquid zinc will be produced at any stage.

(c) If the reactor is operated at a maximum temperature of 1100°C and 1 atm pressure, what could be the effects of lowering the temperature in the upper part of the shaft?

(d) What is the maximum theoretical recovery of zinc in a condenser operating at 500°C?

Data: $$\log p^{o\ \text{Atm.}}_{Zn} = -\frac{6850}{T} - 1.255 \log T + 9.45 \qquad \text{(b)}$$

SOLUTION

(a) The system has three independent components: Zn, C, and O, and two phases: solid ZnO, and the zinc vapour/gas phase. The variance should be:

$$v = 3 + 2 - 2 = 3$$

Nevertheless, a relation exists between the partial pressures of Zn and CO_2 in that $p_{CO_2} = p_{Zn}$ from stoichiometry. Therefore the variance decreases by one unit to 2. It is then possible to operate by specifying the temperature and total pressure.

(b) When equilibrium is reached, $\Delta G^o = -RT \ln K$, and as $p_{CO_2} = p_{Zn}$ and $(p_{CO} + p_{CO_2} + p_{Zn}) = P$, then:

$$-\log K = \log \frac{P - 2p_{Zn}}{p^2_{Zn}} = \frac{10,474}{T} + 2.26\ T \log T - 13.47 \qquad \text{(c)}$$

Table 31 summarises the values of p_{Zn}, p_{CO}, and p_{CO_2} calculated for various temperatures, when the total pressure is 1 atm. In the same table, the values of the saturated vapour pressure, p^o_{Zn}, of zinc at these temperatures are indicated. Since p^o_{Zn} is higher than p_{Zn}, at all of these temperatures there is no risk of zinc condensation within the reactor if the temperature is uniform.

TABLE 31

Temperature, K	973	1173	1373	1573
$p_{Zn} = p_{CO_2}$	0.0095	0.06	0.187	0.345
p_{CO}	0.981	0.88	0.626	0.31
p^o_{Zn}	0.047	0.59	3.41	12.361

(c) A lowering of the temperature in any part of the furnace can have two effects. One is the oxidation of Zn vapour by CO_2, because of the reversal of reaction (a), and the other is the condensation of zinc. (Table 31 shows that P_{Zn} and P_{CO_2} decrease rapidly as the temperature is lowered.)

At 1100°C, P_{Zn} due to reduction of ZnO is 0.187 atm and condensation will only take place if the saturated vapour pressure of Zn is less than 0.187 atm. From equation (b) this must be below 1073 K (800°C) and the risk of condensation is much less than the risk of oxidation by carbon dioxide.

(d) At 500°C the saturated vapour pressure of Zn is 9.2×10^{-4} atm and the maximum recovery possible is $(0.187 - 9.2 \times 10^{-4})/0.187 = 99.5\%$. The remainder can be scrubbed out from the exit gases as "blue powder".

PROBLEM NO. 16

REDUCTION OF ZINC OXIDE WITH CARBON

Liquid zinc can be produced by heating together zinc oxide and carbon in a sealed vessel, capable of operating at high temperature and pressure. By adjustment of the temperature and pressure, it is possible to produce liquid zinc in the presence of a gas phase containing CO, CO_2, and zinc vapour, and the solid phases ZnO and C.

(a) Calculate the variance of the system:

(i) when zinc liquid and vapour are present;

(ii) when all of the zinc is vapour.

What can be deduced from the study of the variance?

(b) Starting from the data given below:

(i) Plot the saturated vapour pressure curve for zinc from 1200 to 1400 K.

(ii) Assuming that zinc does not condense, find the pressure of CO, CO_2 and Zn in equilibrium with ZnO and C at 1200, 1300, and 1400 K, and plot the curve $p_{Zn} = f(T)$ on the same diagram as the previous curve. At what temperature will zinc start to condense?

(c) Taking into account the condensation of zinc, recalculate the pressures of CO, CO_2 and the total pressure that exists in the vessel in the presence of ZnO, C, and liquid Zn and trace the curve of the total pressure as a function of the temperature.

Data: Standard free energies of reaction: (cals $mole^{-1}$)

$\langle C\rangle + [CO_2] \rightleftarrows 2[CO]$ (1) $\Delta G_1^o = 40{,}800 - 41.7\ T$

$\langle ZnO\rangle + [CO] \rightleftarrows [Zn] + [CO_2]$ (2) $\Delta G_2^o = 47{,}920 - 61.63\ T + 10.35\ T \log T$

$(Zn) \rightleftarrows [Zn]$ (3) $\Delta G_3^o = 31{,}340 - 43.30\ T + 5.74\ T \log T$

SOLUTION

(a) The components of the system are three (Zn, C, O):

(i) When liquid zinc exists, the number of phases is four (gas, solid ZnO, solid C, liquid Zn). The variance is then one.

(ii) If the liquid Zn phase disappears, the variance should be increased by one unit. Nevertheless, in the same way as in the previous problem, there exists a relationship between p_{Zn}, p_{CO} and p_{CO_2}, since the system is closed. The reduction of ZnO by carbon produces CO or CO_2, depending on the condition:

$$\langle ZnO\rangle + \langle C\rangle \rightleftarrows [Zn] + [CO]$$

and:

$$2\langle ZnO\rangle + \langle C\rangle \rightleftarrows 2[Zn] + [CO_2]$$

from which:

$$p_{Zn} = p_{CO} + 2p_{CO_2}$$

Under these conditions the variance decreases by one unit to become equal to 1, as in the previous case. In both cases, (i) and (ii), only the temperature needs to be fixed.

(b) (i) The saturated vapour pressure of zinc can be calculated from the free energy of vaporisation ΔG^o_3, which is equal, at equilibrium, to $-RT \log p^{o\,Atm.}_{Zn}$, and consequently:

$$\log p^{o\,Atm.}_{Zn} = -\frac{6.850}{T} - 1.255 \log T + 9.46$$

The values of p^o_{Zn} at 1200, 1300, and 1400 K, are summarised in Table 32. The curve of $p^{o\,Atm.}_{Zn} = f(T)$ is given in Fig. 152.

(ii) Three equations are necessary for calculation of the pressures of CO, CO_2 and Zn, in equilibrium with ZnO and C, assuming that no zinc condenses. One is given by the mass balance, and the other two are thermodynamic relationships calculated from the data: (ΔG^o_1 and ΔG^o_2 are equal, at equilibrium, to $-RT \log K_1$ and $-RT \log K_2$, respectively). The equations are then:

$$p_{Zn} = p_{CO} + 2p_{CO_2}$$

$$\log K_1 = \log \frac{p^2_{CO}}{p_{CO_2}} = -\frac{8.920}{T} + 9.12$$

$$\log K_2 = \log \frac{p_{CO_2} \cdot p_{Zn}}{p_{CO}} = -\frac{10,474}{T} - 2.26 \log T + 13.47$$

An approximate calculation can be made, assuming that p_{CO_2} is negligible compared with p_{CO} and p_{Zn}. In this case, p_{CO} is equal to p_{Zn}, $K_2 = p_{CO_2}$ and

$p_{CO} = \sqrt{K_1K_2}$. For a more rigorous calculation, p_{CO_2} and p_{Zn} are replaced by:

$$p_{CO_2} = \frac{p_{CO}^2}{K_1} \qquad \text{and} \qquad p_{Zn} = p_{CO} + \frac{2p_{CO}^2}{K_1}$$

Inserting these in K_2, gives a cubic equation for p_{CO}:

$$2p_{CO}^3 + K_1 p_{CO}^2 - K_1^2 K_2 = 0$$

The solutions for the values of K_1 and K_2 at 1200, 1300, and 1400 K, are given in Table 32, which makes it possible to plot the vapour pressure curve of Zn without taking into consideration the condensation which may occur above a certain temperature. In Fig. 152 it can be seen that, starting from 1293 K, the calculated vapour pressure of Zn in equilibrium with ZnO and C is higher than p^o_{Zn}, which is physically impossible. Zinc vapour will condense to liquid until its pressure is that in equilibrium with liquid Zn, i.e. p^o_{Zn}.

(c) The equilibrium pressures found in (b)(ii) at 1300 and 1400 K will change because of zinc condensation. Since the zinc pressure is equal to p^o_{Zn}, the new CO pressures can be calculated by eliminating p_{CO_2} and p_{Zn} from the constants K_1 and K_2. It can then be written:

$$\log \frac{p_{CO_2}}{p_{CO}} = \log K_2 - \log p^o_{Zn}$$

and $$\log p_{CO} = \log K_1 + \log \frac{p_{CO_2}}{p_{CO}} = \log K_1 + \log K_2 - \log p^o_{Zn}$$

Substituting for $\log K_1$, $\log K_2$, and $\log p^o_{Zn}$ by their values as functions of T, the relationship for temperatures higher than 1293 K is:

$$\log p_{CO} = -\frac{12{,}544}{T} - 1.055 \log T + 13.13$$

The values of p_{CO_2} are deduced from those of p_{CO} and K_1. The results obtained are summarised in Table 32, and the total pressure has been plotted as a function of the temperature in Fig. 152.

The minimum temperature necessary to obtain liquid Zn is then about 1020°C, and the total pressure must be above 3.8 atm. The principal source of error arises from the fact that, at high pressures, the fugacity must be used instead of the partial pressure, especially for components such as zinc, which have boiling temperatures near to that used in the process (T_V^{Zn} = 1180 K).

In industrial practice, the sintered zinc oxide is reduced at 1100-1200°C and at atmospheric pressure. In addition to the three equations in (b)(ii), we have:

$$p_{Zn} + p_{CO} + p_{CO_2} = 1$$

(Surplus to our needs for most calculations). ΔG_4 and ΔG_4^o for the reaction $ZnO + C \rightarrow Zn + CO$ calculated from ΔG_1^o and ΔG_2^o ($\Delta G_4 = \Delta G_4^o + RT \log p_{Zn} \cdot p_{CO}$) are always negative and there is no equilibrium. Since $p_{CO_2} \simeq 0$ in the presence of C at 1100-1200°C, we have $p_{Zn} \simeq p_{CO} \simeq 0.5$ atm as the theoretical position.

TABLE 32

Question	Temperature, K	1200	1300	1400
	Atm.			
(b) (i)	p^o_{Zn}	0.772	1.92	4.15
(b) (ii)	K_1	48.4	180	556
(b) (ii)	K_2	6.0×10^{-3}	2.3×10^{-2}	7.4×10^{-2}
(b) (ii)	p_{CO}	0.53	2.01	6.34
(b) (ii)	p_{CO_2}	5.9×10^{-3}	0.022	0.072
(b) (ii)	p_{Zn}	0.55	2.05	6.48
(c)	p_{CO} corrected	-	2.24	10.2
(c)	p_{CO_2} corrected	-	0.027	0.182
(c)	p total	1.09	4.19	14.53

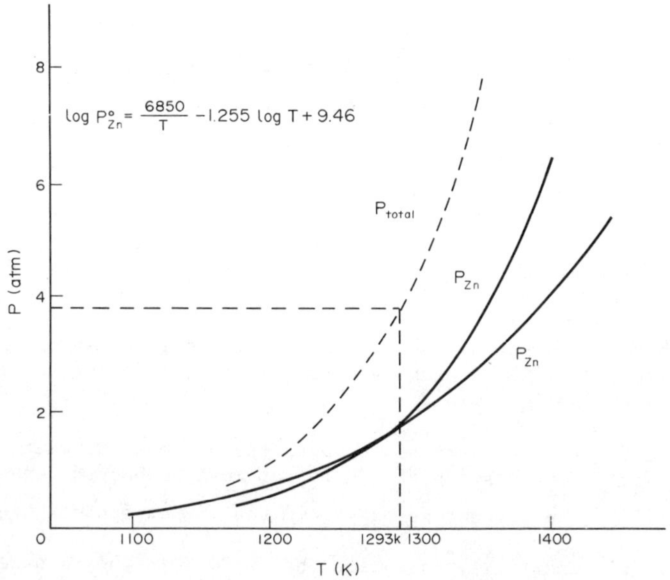

Fig. 152. Temperature and pressure necessary to obtain liquid Zn by reducing ZnO with C

PROBLEM NO. 17

ROASTING AND SMELTING OF CHALCOPYRITE (Cu Fe S_2)

A dry flotation concentrate, containing copper as chalcopyrite, has the following composition:

Cu = 26.1%	Fe = 23.1%	S = 26.4%
SiO_2 = 16.0%	CaO = 7.8%	Al_2O_3 = 0.6%

I Thermodynamics.

The diagram provided indicates the areas of predominance of the liquid constituents in the Cu-S-O (Heavy full lines) and Fe-S-O (Heavy dotted lines) systems at 1500 K and it is drawn for conditions in which a_{Cu_2S}, a_{FeS}, a_{FeO} and $a_{Cu_2O} = 1$. (The chalcopyrite decomposes to Cu_2S, FeS and S_2 at below 1500 K.) Construct the diagram for $a_{Cu_2S} = 0.7$, $a_{FeS} = 0.3$, $a_{FeO} = 0.5$ and $a_{Cu_2O} = 0.1$.

The concentrate is charged continuously into a smelting furnace and air is blown through the melt in such quantities that 65% of the sulphur is oxidised to SO_2. The other products are Cu-Fe-S matte and slag containing the gangue and the FeO formed by oxidation of FeS.

From the diagram, which constituents are present in each phase, assuming ideal behaviour? What is the role of SiO_2? Calculate the % S removed in the domains (i) Cu_2S - FeS, (ii) Cu_2S - FeO, (iii) Cu - Fe_3O_4.

II Mass Balances

(a) What quantity, expressed in moles, and what percentage of each constituent is there in each phase? Is it possible to produce copper directly? If so, what is the expected loss of copper in the slag?

(b) What quantity of air (21% O_2) must be blown in and what is the composition of the exit gas, assuming that oxygen is in equilibrium with both of the condensed phases?

III Heat Balance

(a) Construct a heat balance for the furnace, assuming that convection/radiation losses are equal to 10% of the total. Gas, matte and slag leave the furnace at 1500 K. Is the heat generated by the combustion of Fe and S enough for the process?

(b) If there is not enough heat from this source, it may be provided by (i) heating the air, (ii) burning natural gas, or (iii) enriching the blast with oxygen. Calculate the following:

(i) the temperature to which the air must be heated,
(ii) the quantity of CH_4 that must be burned, and
(iii) the percentage oxygen in the air/oxygen mixture.

In the case of (ii) and (iii), calculate the % SO_2 in the furnace exit gases. Air is assumed to contain 21% oxygen and 79% nitrogen.

TABLE 33 Thermodynamic Data (Kubaschewski and Alcock)[20]

Reaction	ΔG^o_{1500}(cal)	$\log_{10}K$	$\log_{10}P_{O_2}$
$Cu_2S + O_2 = 2Cu + SO_2$	-40,460	5.9	$-5.9 + \log P_{SO_2} - \log a_{Cu_2S}$
$2Cu + \frac{1}{2}O_2 = Cu_2O$	-13,745	2.0	$-4.0 + 2\log a_{Cu_2O}$
$Cu_2S + \frac{3}{2}O_2 = Cu_2O + SO_2$	-54,210	7.9	$-5.27 + \frac{2}{3}\log P_{SO_2} + \frac{2}{3}\log \frac{a_{Cu_2O}}{a_{Cu_2S}}$
$FeS + O_2 = Fe + SO_2$	-43,530	6.34	$-6.34 + \log P_{SO_2} - \log a_{FeS}$
$Fe + \frac{1}{2}O_2 = FeO$	-39,200	5.71	$-11.42 + 2\log a_{FeO}$
$3FeO + \frac{1}{2}O_2 = Fe_3O_4$	-31,810	4.64	$-9.28 - 6\log a_{FeO}$
$FeS + \frac{3}{2}O_2 = FeO + SO_2$	-82,730	12.05	$-8.03 + \frac{2}{3}\log P_{SO_2} + \frac{2}{3}\log \frac{a_{FeO}}{a_{FeS}}$
$3FeS + 5O_2 = Fe_3O_4 + 3SO_2$	-279,970	40.8	$-8.16 + \frac{3}{5}\log P_{SO_2} - \frac{3}{5}\log a_{FeS}$

Heats of Reaction and Decomposition at 298 K

$2CuFeS_2 = Cu_2S + FeS + \frac{1}{2} S_2$ $\Delta H_{298} = 39,975$ cal mole^{-1}

$S_2 + 2 O_2 = 2SO_2$ $\Delta H_{298} = -172,630$ cal mole^{-1}

$FeS + \frac{3}{2} O_2 = FeO + SO_2$ $\Delta H_{298} = -110,140$ cal mole^{-1}

$CH_4 + 2 O_2 = CO_2 + 2H_2O$ $\Delta H_{298} = -191,750$ cal mole^{-1}

TABLE 34 Heats of Transformation, Fusion and Temperature Rise[20]

	$L_{Trans.}$ (cal mole^{-1})	L_f (cal mole^{-1})	$\int_{298\,K}^{1500\,K} C_p\,dT$ (cal mole^{-1})
Cu_2S	920 : 200	2,600	25,070
FeS	570 : 120	7,730	17,200
FeO	-	7,400	15,910
SiO_2	175 : 600	2,600	16,555
CaO	-	19,000	14,840
Al_2O_3	-	25,700	33,360
SO_2	-		14,840
N_2			9,110
CO_2			14,470
H_2O			11,405

The mean specific heat of air up to 1000°C, $C_p^{Air} = 7.5$ cal °C mole^{-1}. The heat of formation of silicates and sulphides will be neglected.

SOLUTION

I. A decrease in the activity of a constituent enlarges its domain (see Fig. 153) (light lines). The FeO/Fe_3O_4 boundary moves to the right as a_{FeO} decreases below 1 and the Cu/Cu_2O boundary moves to the left as a_{Cu_2O} decreases to less than 1. The other boundaries change very little. SiO_2 lowers the activity of FeO and stabilises it as $2FeO \cdot SiO_2$ instead of allowing it to form Fe_3O_4, which is refractory and dense and tends to accumulate on the furnace bottom. As Fig. 153 shows, for a pressure of SO_2 between 0.1 and 0.3 atm ($-0.5 > \log P_{SO_2} > -1$).

there are successively (i) a domain with FeS and Cu_2S, proceeding from the oxidation of 25% of the S of the $CuFeS_2$; (ii) a domain with FeO and Cu_2S (with very little FeS and Cu_2O): the oxidation of all of the FeS to FeO corresponds to removal of 75% of the S; (iii) a domain with FeO or Fe_3O_4 and Cu or Cu_2O, corresponding to the oxidation of 100% of the S.

With 65% of the S oxidised, we are on the equilibrium line between FeS and FeO ($a_{FeO} = 0.5$, $a_{FeS} = 0.3$). With more than 75% of the S oxidised, Cu metal can be produced, but there are high (10% or more) losses of Cu in the slag.

II. (a) If 65% of the S is oxidised, then for 100 g of concentrate, 17.16 g forms SO_2, leaving 9.24 g. Of this, 6.576 g is combined with the 26.1 g of Cu to make 32.676 g of Cu_2S. The balance of 2.664 g is combined with 4.662 g Fe to make 7.326 g of FeS. The remainder of the Fe is oxidised to FeO (23.706 g FeO).

TABLE 35 Mass Balance for 100 g of Concentrate

Phase	Compound	Mass, g	Weight %	No. of moles	Mol. %
Matte	Cu_2S	32.676	81.7	0.206	71.3
	FeS	7.326	18.3	0.083	28.7
		40.002			
Slag	FeO	23.706	49.3	0.329	44.4
	SiO_2	16.0	33.3	0.267	36.0
	CaO	7.8	16.2	0.139	18.8
	Al_2O_3	0.6	1.2	0.006	0.8
		48.106			

II.(b) 0.701 moles of O_2 are needed per 100g of concentrate, 0.536m for SO_2 and 0.165 for FeO, N_2 = 2.637m. The exit gas contains 17% SO_2.

III(a) Heat Balance

Heat Generated

Chalcopyrite decomposes in the following manner:

$$2CuFeS_2 = Cu_2S + 2FeS + \frac{1}{2} S_2$$

100 g of the concentrate specified contains 0.412 mole of $CuFeS_2$ and this will produce 0.412/4 mole of S_2 when heated. This will burn to SO_2 and from the data, this will produce:

	cal
0.103 × 172,630 cal	= 17,780

The balance of the S oxidised will come from the FeS:

$$FeS + \frac{3}{2} O_2 = FeO + SO_2$$

		cal
This will produce $0.329 \times 110,140$ cal		= 36,236
Total heat produced by combustion of Fe and S		= 54,016

Heat Absorbed

Decomposition of $CuFeS_2$	$0.412 \times 39,375/2$	= 8111
Heating, transformation and fusion of matte, slag and gas		
Cu_2S : 0.206 (920 + 200 + 2600 + 25,070)		= 5931
FeS : 0.083 (570 + 120 + 7730 + 17,200)		= 2127
FeO : 0.329 (7,400 + 15,190)		= 7169
SiO_2 : 0.267 (175 + 600 + 2600 + 16,555)		= 5321
CaO : 0.139 (19,000 + 14,480)		= 4664
Al_2O_3: 0.006 (25,700 + 33,360)		= 354
SO_2 : 0.536 (14,840)		= 7954
N_2 : 2.637 (9110)		= 24,023
		65,654
	Heat loss at 10%	6565
	Total	72,219

Deficit = 72,219 - 54,016 = 18,203 cal/100 g.

III. (b) If this is to be compensated for by heating the air, the rise in temperature must be 18,203/(0.701 + 2.637) 7.5 = 727°C. The actual air temperature will be 727° + 25° = 752°C. If the heat is to be supplied by burning natural gas, each mole of CH_4 produces 191,750 cal, but there is increased loss in the additional N_2 leaving the system, as well as the 1 mole of CO_2 and 2 moles of H_2O produced.

$$CH_4 + 2\ O_2 + 7.523\ N_2 = CO_2 + 2\ H_2O + 7.523\ N_2$$

Heat loss in effluent gases = $1 \times 14,470 + 2 \times 11,405 + 7.523 \times 9110$ = 105,815 cal. Net heat gain = 85,935 cal $mole^{-1}$ of CH_4. CH_4 required = 18,203/85,935 = 0.212 mole, or 47.5 m^3 $tonne^{-1}$. The exit gas will be increased by 0.212 mole CO_2, 0.424 mole H_2O and 1.593 moles N_2, to a total of 5.402 moles, containing 9.9% SO_2.

The deficit can be eliminated by reducing the heat lost in nitrogen by 18,203 cal, i.e. from 24,023 to 5820 cal. This amount of heat is contained in 5820/9110 = 0.639 mole N_2. The enriched air will contain 0.701/(0.701 + 0.639) = 52.3% O_2.

O_2 from air = $0.639 \times 21/79$ = 0.17 mole, O_2 added = 0.701 - 0.17 = 0.531 mole/100 g of concentrate, = 118.9 m^3 $tonne^{-1}$ of concentrate.

SO_2 in exit gas = 0.536/(0.536 + 0.639) = 45.6%.

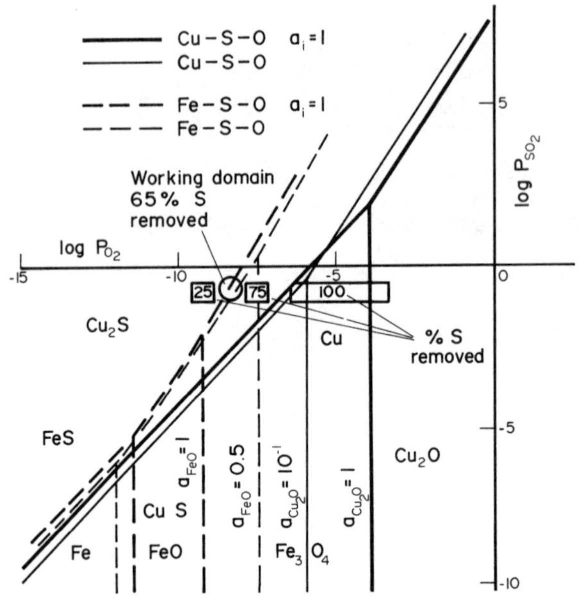

Fig. 153. Predominance area diagram at 1500 K. ——— Cu-O-S; – – – Fe-O-S. X% S removed

PROBLEM NO. 18a

MASS BALANCE IN AN IRON BLAST FURNACE

The coke rate in an iron blast furnace is 500 kg per tonne of pig iron. The sintered burden contains two hematite ores: one (A) has a high silica content and the other (B) is high in limestone. Given the analyses of the ores after sintering of the coke and of the pig iron and the information that the slag basicity (CaO/SiO_2) must be 1.4 in order to achieve the specified sulphur content of the pig iron and that the CO/CO_2 ratio in the exit gases is 1, calculate the following, all per tonnes of pig iron:

(a) The weight of each ore required.
(b) The weight and sulphur content of the slag.
(c) The volume of blast.
(d) The volume of exit gases.

Assume that there is no loss of sulphur or fine ore in the exit gases.

SOLUTION

(a) Assuming that X kg of A and Y kg of B are used per tonne of pig iron, two equations are required, based on the iron balance and the slag basicity:

Iron balance: $1000 \times 0.94 = 500 \times 0.007 + 0.571X + 0.547Y$

TABLE 36

Analyses of Charge Materials and Pig Iron (wt.%)				
	Ore A	Ore B	Coke	Pig iron
Fe	57.1	54.7	0.7	94
C	-	-	86	4.0
Mn	0.10	0.07	-	0.15
Si	-	-	-	1.50
P	0.04	0.08	-	0.10
S	0.15	0.05	0.5	0.05
SiO_2	13.4	4.8	3.0	-
CaO	3.9	16.8	5.6	-

Slag basicity:
$$1.4 = \frac{500 \times 0.056 + 0.039X + 0.168Y}{500 \times 0.03 + 0.134X + 0.048Y - 60/28 \times (1000 \times 0.015)}$$

$$(SiO_2/Si = 60/28)$$

From the iron balance: $936.5 = 0.571X + 0.547Y$

From the basicity: $52 = 0.149X - 0.101Y$

Giving $X = 884$ kg and $Y = 789.3$ kg.

(b) The slag contains all of the CaO in the charge, as well as the SiO_2 and S which do not pass into the pig iron:

CaO to slag $= 500 \times 0.056 + 884 \times 0.039 + 789.3 \times 0.168 = 28 + 34.5 + 132.6 = 195.0$

SiO_2 to slag $= 500 \times 0.03 + 884 \times 0.134 + 789.3 \times 0.048 - 60/28 \times (1000 \times 0.015)$

$= 15 + 118.5 + 38 - 32 = 139.5$

S to slag $= 500 \times 0.005 + 884 \times 0.0015 + 789.3 \times 0.005 - 1000 \times 0.0005$

$= 2.5 + 1.33 + 0.4 - 0.5 = 3.73$ kg.

Weight of slag = 338.2 kg, basicity = 195/139.5 = 1.4. Sulphur = 3.73/338 = 1.1%.

(c) and (d) The volumes of blast and exit gas must be calculated by way of the carbon and oxygen balances:

Carbon balance: $C_{gas} = C_{coke} - C_{pig\ iron}$

$= 500 \times 0.86 - 1000 \times 0.04 = 390$ kg or 32,500 moles.

Since $n_{CO} = n_{CO_2}$, the exit gas contains:

16,250 moles of CO and 16,250 moles of CO_2, or 364 m^3 of each.

Oxygen balance: O_2 from air blast + O_2 from ores = O_2 in exit gas

n_{O_2} in exit gas = n_{CO_2} + 0.5 n_{CO} = 24,375 moles

n_{O_2} from ores = n_{O_2} from Fe_2O_3 + n_{O_2} from SiO_2 reduced

= 0.75 n_{Fe} in metal + n_{Si} in metal

$$= \frac{0.75 \times 940 \times 10^3}{56} + \frac{15 \times 10^3}{28} = 13{,}125 \text{ moles}$$

n_{O_2} from air blast = 24,375 - 13,125 = 11,250 moles or 252 m^3

Volume of air = 1200 m^3, volume of N_2 = 948 m^3 (= 42,321 moles)

Volume of exit gas = 1676 m^3

CO = 364 m^3 (21.7%), CO_2 = 364 m^3 (21.7%), N_2 = 948 m^3 (56.6%)

Ratio of exit gas to blast = 1.4.

PROBLEM NO. 18b

HEAT BALANCE IN THE IRON BLAST FURNACE

Using the same data and the results of Problem 18a, construct a heat balance and determine, by difference, the heat losses of the furnace under steady conditions. Ore and coke are charged at 25°C, air is blown in at 1025°C and the exit gases leave at 125°C. Pig iron and slag are tapped at 1425°C.

TABLE 37

Enthalpies of Reaction, Solution and Change of State in kcal	
$C + O_2 = CO_2$	$\Delta H^o = -94$
$C + \frac{1}{2} O_2 = CO$	$\Delta H^o = -26.4$
$Si + O_2 = SiO_2$	$\Delta H^o = -205$
$2\,Fe + \frac{3}{2} O_2 = Fe_2O_3$	$\Delta H^o = -196$
$2\,Fe + O_2 = 2FeO$	$\Delta H^o = -126.4$
$2(CaO) + SiO_2 = 2CaO \cdot SiO_2$	$\Delta H^o = -24.4$
$\langle Fe \rangle = (Fe)$	$\Delta H_f = 3.6$

$\langle CaO\rangle = (CaO)$	$\Delta H_f = 12$	
$\langle SiO_2\rangle = (SiO_2)$	$\Delta H_f = 2$	
$\langle C\rangle = (C)_{1\%}$	$\Delta H_f \quad 5.4$	
$\langle Si\rangle = (Si)_{1\%}$	$\Delta H_f = -28$	

Assume that any CaO is present as such or as $2CaO \cdot SiO_2$

Mean specific heats, in cal $mole^{-1}$ °C.

$$\overline{Cp}_{Fe} = 6.1; \quad \overline{Cp}_{Si} = 5.3; \quad \overline{Cp}_{SiO_2} = 10.6; \quad \overline{Cp}_{C} = 1.45; \quad \overline{Cp}_{CaO} = 10.2$$

These specific heats are all assumed to be independent of T, but those below depend on T in the manner indicated:

$$\overline{Cp}_{CO,O_2,N_2} = 6.8 + 0.6 \times 10^{-3}\ T\ (°C). \qquad \overline{Cp}_{CO_2} = 8.3 + 0.5 \times 10^{-3}\ T\ (°C)$$

SOLUTION

In the steady state, the heat input from the blast, plus that generated by the exothermic reactions exactly balances the heat absorbed by the endothermic reactions plus that in the products and the losses by condustion and convection. Accurate values of calculable input and output can be obtained in three ways:

(a) The CO and CO_2 in the exit gas are assumed to be the amounts generated at the tuyeres, which have passed unchanged through the furnace, no intermediate reactions being taken into account. Fe and Si are produced by thermal decomposition of the oxides. The balance is obtained by relating quantities which are much larger than those actually produced or consumed.

(b) The complete sequence of reactions in the furnace, from tuyeres to offtake and tapholes, is followed, taking account of (1) reduction of Fe_2O_3 to FeO by CO; (2) reduction of part of the FeO to Fe by CO; (3) reduction of the remaining FeO by C; (4) reaction between CO_2 and C to produce CO; (5) reduction of SiO_2 by C; (6) oxidation of C to CO at the tuyeres; (7) fusion of metal and slag. (Even this is a simplified form of calculation.) The calculation in which there is distinction of the FeO reduced by C from that by CO requires knowledge of the operating line and a mass balance.

(c) In the simplest method of treatment, it is assumed that all of the FeO is reduced by CO and that the excess CO_2 reacts with C to form CO (Boudouard).

The standard temperature is taken as 25°C and there is no contribution to the balance from materials entering at this temperature.

Heat Input

This arises mainly from the combustion of carbon and the heat in the blast, but small amounts will be contributed by silicate formation, solution of Si in Fe and reduction of FeO by CO.

Combustion of coke: At the tuyeres, 11,250 moles of O_2 produce 22,500 moles of CO

Heat generated = 22,500 × 26.4 = 594,000 kcal

Sensible heat of air blast: Quantity of air per tonne of pig iron = 53,571 moles

Specific heat = $6.8 + 0.6 \times 10^{-3} \times 1025$

= 7.415 cal $mole^{-1}\ {}^{o}C^{-1}$

Sensible heat of 53,571 moles = $53{,}571 \times 7.415 \times 10^{-3}$ (1025 - 25) = 397,229 kcal

Formation of silicates: Assuming that all of the CaO (195 kg or 3482 moles) is present in the slag as $2CaO \cdot SiO_2$, we have:

Heat generated = 3482/2 × 24.4 = 42,480 kcal

Heat of solution of Si: Heat generated = $15/28 \times 10^3 \times 28$ = 15,000 kcal

Reduction of FeO by CO: $FeO + CO = Fe + CO_2$ $\Delta H = -4.4$ kcal $mole^{-1}$

If 940 kg or 16,786 moles of iron are produced:

Heat generated = 16,786 × 4.4 = 73,858 kcal

Heat Output

Heat is absorbed by the reduction of silica and of Fe_2O_3 to FeO, by the Boudouard reaction, as sensible heat in the metal, slag and gas and by losses to cooling water and the surroundings.

Reduction of Fe_2O_3 by CO according to: $Fe_2O_3 + CO = 2FeO + CO_2$ $\Delta H = 2$ kcal

For 16,786 moles of FeO, the heat absorbed is $\frac{16{,}786}{2} \times 2 = 16{,}786$ kcal

Reduction of SiO_2 by C: $2C + SiO_2 = 2CO$ $\Delta H = 152.2$ kcal

For 15 kg of Si, the heat absorbed is $\frac{15{,}000}{28} \times 152.2 = 81{,}536$ kcal

The Boudouard reaction: $C + CO_2 = 2CO$ $\Delta H = 41.2$ kcal

The number of moles of CO_2 reduced is equal to the CO_2 produced by the reduction of Fe_2O_3 and FeO by CO, less the CO_2 in the exit gas. This is equal to:

$$16{,}786/2 + 16{,}786 - 16{,}250 = 8929 \text{ moles}$$

The heat absorbed by the Boudouard reaction is 8929 × 41.2 = 367,875 kcal

Solution of C in Fe: Heat absorbed = $40/12 \times 10^3 \times 5.4$ = 18,000 kcal

Sensible heat of pig iron: Assuming that the specific heats of Fe, Si, and C are constant:

$$\Delta H_M = \overline{Cp}_M \Delta T + \Delta H_{f_M} \qquad (\Delta T = 1400°C)$$

$$\text{Sensible heat of 940 kg Fe} = \frac{940}{56} \times 10^3(6.1 \times 1400 + 3600) \times 10^{-3} = 203{,}778 \text{ kcal}$$

$$\text{Sensible heat of 40 kg C} = \frac{40}{12} \times 10^3(1.45 \times 1400) \times 10^{-3} = 6767 \text{ kcal}$$

$$\text{Sensible heat of 15 kg Si} = \frac{15}{28} \times 10^3(5.3 \times 1400 + 11{,}000) \times 10^{-3} = 9868 \text{ kcal}$$

$$\text{Total} = 220{,}413 \text{ kcal}$$

Sensible heat of slag:

$$\text{Sensible heat of CaO} = \frac{195}{56} \times 10^3(10.2 \times 1400 + 12{,}000) \times 10^{-3} = 91{,}511 \text{ kcal}$$

$$\text{Sensible heat of } SiO_2 = \frac{139.3}{60} \times 10^3(10.6 \times 1400 + 2000) \times 10^{-3} = 39{,}097 \text{ kcal}$$

$$\text{Total} = 130{,}608 \text{ kcal}$$

Sensible heat of gases: The specific heat of CO and N_2 at 125°C is 6.88 cal mole^{-1} °C^{-1} and of CO_2 is 8.36 cal mole^{-1} °C^{-1}. The sensible heat in the exit gases will then be:

$$[(42{,}321 + 16{,}250) \times (6.88 \times 10^{-3} + 16{,}250 \times 8.36 \times 10^{-3})](125 - 25) = 53{,}882 \text{ kcal.}$$

TABLE 38 Heat Balance in the Iron Blast Furnace

Reaction	Heat generated kcal × 10^{-3}	Heat absorbed kcal × 10^{-3}
Reduction of Fe_2O_3 to FeO by CO		16.80
Reduction of FeO to Fe by CO	73.9	
The Boudouard reaction		367.9
Reduction of SiO_2 by C		81.5
Combustion of C to CO	594.0	
Sensible heat of air blast at 1025°C	397.2	
Heat of formation of silicates	42.5	
Heat of solution of Si	15.0	
Heat of solution of C		18.0
Sensible heat of pig iron		220.4
Sensible heat of slag		130.6
Sensible heat of gases		53.9
Losses (by difference)		233.5
Totals	1122.6	1122.6

PROBLEM NO. 19

MANGANESE DISTRIBUTION BETWEEN STEEL AND SLAG

The slag of a Siemens-Martin refining furnace has the following composition (mol.%):

$CaO + MgO$	: 45
$SiO_2 + Al_2O_3 + P_2O_5$	: 25
FeO	: 25
MnO	: 5

Calculate the percent of manganese and oxygen in the metal in equilibrium with this slag at 1600°C.

Data:

Standard free energies of reaction:

(a) $(Mn) + \frac{1}{2}[O_2] \rightleftarrows \langle MnO \rangle$ $\qquad \Delta G^o_{(a)} = -95{,}400 + 19.7\ T$

(b) $(Fe) + \frac{1}{2}[O_2] \rightleftarrows (FeO)$ $\qquad \Delta G^o_{(b)} = -56{,}900 + 11.82\ T$

(c) $\frac{1}{2}[O_2] \rightleftarrows (\underline{O})_{1\%}$ $\qquad \Delta G^o_{(c)} = -27{,}790 - 0.79\ T$

(d) $(Mn) \rightleftarrows (\underline{Mn})_{1\%}$ $\qquad$ The Fe-Mn solution is ideal.

$\Delta H_{f_{MnO}} = 10{,}700$ cal $\qquad T_{f_{MnO}} = 2056$ K

The activity coefficient of MnO and the activity of FeO in a pseudoternary slag, are indicated in Figs. 117 and 90.

Solubility of oxygen in iron at saturation:

$$\log\ (\%\ \underline{O})_{sat} = -\frac{6320}{T} + 2.734$$

Interaction parameters:

$$\varepsilon_O^O = -0.20 \qquad \varepsilon_O^{Mn} = 0 \qquad \varepsilon_{Mn}^O = 0$$

SOLUTION

The equilibrium involving FeO and MnO in the slag, and Mn in the metallic bath, can be written:

$$(\underline{Mn})_m + (FeO)_s \rightleftarrows (MnO)_s + (\underline{Fe})_m \qquad (g)$$

The standard free energy of this reaction, $\Delta G^o_{(g)}$ (standard state for Mn is infinite dilution expressed in weight %), is obtained by adding or subtracting the standard free energies of reactions (a), (b), (d), and ΔG_f^{MnO}, the free energy of fusion of MnO:

$$\Delta G^o_{(g)} = \Delta G_f^{MnO} + \Delta G^o_{(a)} - \Delta G^o_{(b)} - \Delta G^o_{(d)}$$

$$\Delta G_f^{MnO} = \Delta H_f^{MnO} \cdot \frac{(T_f - T)}{T_f} = 10{,}700 - 5.21\ T$$

Since the Mn-Fe solution is ideal, the change of free energy corresponding to the change of the standard state: (pure Mn $\rightarrow$ $(Mn)_{1\%}$) is equal to (from (2.96)):

$$\Delta G^o_{(d)} = RT \ln \gamma^{\infty}_{Mn} \frac{M_{Fe}}{100\ M_{Mn}} = RT \ln \frac{56}{5500} = -\ 9.12\ T$$

The standard free energy of reaction (g) is then equal to:

$$\Delta G^o_{(g)} = -\ 27{,}800 + 11.8\ T$$

At 1600°C (1873 K), $\Delta G^o_{(g)}$ is equal to -5720 cal. $\Delta G^o_{(g)}$ can be expressed as a function of the activities, the activity coefficients, and the weight percent, by the relationship:

$$\Delta G^o_{(g)} = -\ RT \ln \frac{a_{Fe} \cdot \gamma_{MnO} \cdot N_{MnO}}{(\%\ \underline{Mn}) \cdot a_{FeO}}$$

$$\ln (\%\ Mn) = \frac{\Delta G^o_{(g)}}{RT} + \ln \frac{a_{Fe} \cdot \gamma_{MnO} \cdot N_{MnO}}{a_{FeO}}$$

The activity of FeO in this slag is 0.60 (Fig. 90), and the activity coefficient of MnO is 1.60 (Fig. 117). The iron activity can be taken as equal to 1.

The Mn content of the steel can be found by inserting these values in this equation:

$$\log (\%\ Mn) = \frac{-\ 5720}{4.575 \times 1873} + \log \frac{1.6 \times 0.05}{0.60} = -\ 1.542$$

$$(\%\ \underline{Mn}) = 2.9 \times 10^{-2}$$

The oxygen in the steel can be calculated from the equilibrium with the FeO in the slag:

$$(\underline{Fe})_m + (\underline{O})_m \rightleftarrows (FeO)_s$$

(7.23) relates $a_{(\underline{O})}$ to FeO by:

$$a_{(\underline{O})} = a_{(\underline{O})_{sat.}} \cdot a_{(FeO)}$$

$$\log f_O + \log (\%\ \underline{O}) = \log f_{(\underline{O})_{sat.}} + \log (\%\ \underline{O})_{sat.} + \log a_{FeO}$$

or $$-0.20(\%\ \underline{O}) + \log (\%\ \underline{O}) = -0.20(\%\ \underline{O})_{sat.} + \log (\%\ \underline{O})_{sat.} + \log a_{FeO}$$

At 1873 K:

$$a_{FeO} = 0.60 \quad \text{and} \quad (\%\ \underline{O})_{sat.} = 0.23$$

Therefore: $$(\%\ \underline{O}) = 0.14$$

PROBLEM NO. 20

SILICON AND SULPHUR DISTRIBUTION BETWEEN STEEL AND SLAG

Carbon steel is melted in a dicalcium silicate crucible at 1600°C under a $CaO-SiO_2$ slag and a $CO-CO_2$ atmosphere in which $P_{CO} = 1$ atm.

(a) Show that, if the carbon content of the steel is specified, the state of the system is completely defined.

(b) Find the composition of the slag in equilibrium with the steel and the crucible and the activities of CaO and SiO_2, using the relationships given in Fig. 88.

(c) Calculate, for a steel containing 1% C:

(i) The percentage of silicon and oxygen in the steel.

(ii) The mole fraction of FeO in the slag, assuming $\gamma_{Fe} = 1$ and taking the value for γ_{FeO} from Fig. 92.

(iii) The composition of the gas phase in equilibrium with this steel.

(d) If the steel contains a small amount of sulphur, calculate the partition coefficient for sulphur between the slag and steel in C, using the value of $\gamma_{CaS} = 18$ obtained from Sharma and Richardson[17] for 1500°C and assuming that it is unchanged at 1600°C.

Data: Free energies of reaction, melting, and solution:

$$(Si) + [O_2] \rightleftarrows \langle SiO_2 \rangle \qquad \Delta G_1^o = -222{,}800 + 47.6\ T \text{ cal}$$

$$2\langle C \rangle + [O_2] \rightleftarrows 2[CO] \qquad \Delta G_2^o = -53{,}400 - 42.0\ T \text{ cal}$$

$$\langle C \rangle + [O_2] \rightleftarrows [CO_2] \qquad \Delta G_3^o = -94{,}200 - 0.2\ T \text{ cal}$$

$$2(Fe) + [O_2] \rightleftarrows 2(FeO) \qquad \Delta G_4^o = -111{,}250 + 21.7\ T \text{ cal}$$

$$(SiO_2) + 2(CaO) \rightleftarrows \langle Ca_2SiO_4 \rangle \qquad \Delta G^o_5 = -\ 24{,}400 - 5.8\ T\ \text{cal}$$

$$2\langle Ca \rangle + [O_2] \rightleftarrows 2\langle CaO \rangle \qquad \Delta G^o_6 = -\ 302{,}650 + 47.3\ T\ \text{cal}$$

$$2\langle Ca \rangle + [S_2] \rightleftarrows 2\langle CaS \rangle \qquad \Delta G^o_7 = -\ 258{,}870 + 45.6\ T\ \text{cal}$$

$$\langle C \rangle \rightleftarrows (C)_{1\%} \qquad \Delta G^o_8 = 5{,}400 - 10.1\ T\ \text{cal}$$

$$(Si) \rightleftarrows (\underline{Si})_{1\%} \qquad \Delta G^o_9 = -\ 28{,}000 + 5.5\ T\ \text{cal}$$

$$[O_2] \rightleftarrows 2(\underline{O})_{1\%} \qquad \Delta G^o_{10} = -\ 56{,}460 - 1.14\ T\ \text{cal}$$

$$[S_2] \rightleftarrows 2(\underline{S})_{1\%} \qquad \Delta G^o_{11} = -\ 63{,}040 + 10.54\ T\ \text{cal}$$

It is assumed that all solutes in the steel follow Henry's Law and that the free energies of melting of CaS and CaO are the same.

SOLUTION

(a) This system contains five components (Fe, Si, C, O and Ca) and four phases (metal, crucible, slag and atmosphere) in equilibrium. The variance of the system is then three. If the temperature and partial pressure of CO are specified, then there must be a unique value for the third variable, the concentration of carbon in the steel.

(b) As shown by the isothermal section of the $CaO\text{-}SiO_2\text{-}Al_2O_3$ phase diagram given in Fig. 88, the liquid $CaO\text{-}SiO_2$ slag in equilibrium with the compound $2CaO \cdot SiO_2$ contains 42 mol. % SiO_2 and 58 mol. % CaO. The activity of SiO_2, referred to the pure solid, is 2×10^{-2}. The activity of CaO can be calculated from the

$$(SiO_2) + 2(CaO) = \langle 2CaO \cdot SiO_2 \rangle$$

equilibrium. The equilibrium constant of this reaction, calculated from the data, is 1.29×10^4 at 1600°C. As the activities of $2CaO \cdot SiO_2$ and SiO_2 are equal to 1 and 0.02 respectively, the activity of CaO will then be:

$$a_{CaO} = \left[\frac{1}{0.02 \times 1.29 \times 10^4} \right]^{\frac{1}{2}} = 0.062$$

The activity coefficient, γ_{CaO}, is then equal to:

$$\gamma_{CaO} = \frac{a_{CaO}}{N_{CaO}} = \frac{0.062}{0.58} = 0.107$$

(c) (i) Two equations are required in order to determine the percentage of silicon and oxygen present in a 1% C steel. Two successive equilibria can be considered, for which the free energy changes can be calculated from the data:

$$(\underline{C})_{1\%} + (\underline{O})_{1\%} = [CO]: \qquad \Delta G^o = -\ 3870 - 10.33\ T = -\ RT \ln K_{(C-O)}$$

$$\langle SiO_2\rangle + 2(\underline{C})_{1\%} = 2[CO] + (\underline{Si})_{1\%}: \qquad \Delta G^o = 130{,}600 - 63.8\ \ T = -\ RT \ln K_{(Si-C)}$$

At 1600°C, $K_{(C-O)} = 492$ and $K_{(Si-C)} = 0.054$. Since the activities of C, O, and Si, are equal to their weights % (Henry's Law), that of SiO_2 in the slag is 0.02 and $P_{CO} = 1$ atm, then:

$$(\underline{\%\ O}) = \frac{P_{CO}}{K_{(C-O)} \times \%\ C} = \frac{1}{492} = 2.03 \times 10^{-3}$$

and
$$(\underline{\%\ Si}) = \frac{K_{(Si-C)} \times a_{SiO_2} \times (\%\ C)^2}{P^2_{CO}} = \frac{0.054 \times 0.02}{1} = 1.08 \times 10^{-3}$$

(ii) The mole fraction of FeO in the slag can be calculated from the free energy of the reaction:

$$(Fe) + (\underline{O})_{1\%} \rightleftarrows (\underline{FeO})_s: \qquad \Delta G^o = -\ 27{,}400 + 11.42\ T$$

The oxygen activity is equal to the wt.% (2.03×10^{-3}), $a_{Fe} = N_{Fe} \simeq 1$ and $\gamma_{FeO} = 4$ from Fig. 92, curve (d) at 1600°C. The equilibrium constant $K_{(Fe-O)}$ is 5, and consequently:

$$N_{FeO} = \frac{K_{(Fe-O)} \times N_{Fe} \times (\%\ O)}{\gamma_{FeO}} = \frac{5 \times 1 \times 2.03 \times 10^{-3}}{4}$$

$$= 2.55 \times 10^{-3} = 0.255 \text{ mol. }\%$$

(iii) The composition of the atmosphere in equilibrium with the melt can be calculated from the free energy of:

$$[CO] + (\underline{O}) \rightleftarrows [CO_2] \qquad \Delta G^o = -\ 39{,}280 + 21.37\ T \text{ cal mole}^{-1}$$

The equilibrium constant is 0.82 at 1600°C, and since $P_{CO} = 1$ and $a_{(O)} = (\%\ O) = 2.03 \times 10^{-3}$, $P_{CO_2} = K \times P_{CO} \times (\%\ O) = 1.67 \times 10^{-3}$ atm.

The balance is CO, i.e. the atmosphere is about 99.83% CO.

(d) The sulphur equilibrium between metal and slag can be written:

$$(CaO)_s + (\underline{S})_m = (CaS)_s + (\underline{O})_m \qquad \text{(i)}$$

Since the free energy of melting of CaS is assumed to be the same as that for CaO, the standard free energy and the equilibrium constant for (i) are equal to those for the reaction involving solid CaO and CaS:

$$\Delta G^{o} = 25{,}180 - 6.69\ T\ \text{cal mole}^{-1} \quad \text{and} \quad K_{(i)} = \frac{(\%\ O)_m \times \gamma_{CaS} \times N_{CaS}}{(\%\ S)_m \times a_{CaO}}$$

At 1600°C, $K_{(i)} = 3.3 \times 10^{-2}$. The partition coefficient for sulphur K_{ps}, can be defined by $K_{ps} = (\%\ S)_s/(\%\ S)_m$.

With $$(\%\ S)_s = \frac{m_s \times 100}{m_{CaO} + m_{SiO_2}} = \frac{N_{(CaS)} \times 3200}{N_{(CaO)} \times 56 + N_{(SiO_2)} \times 60} = 55.5\ N_{(CaS)}$$

then $$K_{ps} = \frac{55.5\ N_{(CaS)}}{(\%\ S)_m}$$

Substituting $N_{(CaS)_s}/(\%\ S)_m$ by its value as a function of $K_{(i)}$, we obtain:

$$K_{ps} = \frac{55.5 \times K_{(i)} \times a_{(CaO)_s}}{(\%\ O)_m \times \gamma_{(CaS)}} = 3.1$$

This value agrees rather well with that of Fig. 126 (in which $(\%\ S)_s/(\%\ S)_m$ is given as a function of $(\%\ FeO)_s$ for a low basicity slag $(i \sim 1.5)$ and $N_{FeO} \sim 0.25$ (mol. %).

The low value of K_{ps} follows to some extent from the low value of $a_{(CaO)_s}$ in a slag of this basicity, but it is principally due to the high value of $\gamma_{(CaS)_s}$: the CaS is only slightly soluble in the slag, and if the sulphur content of the system is high, CaS separates as a solid phase ($a_{CaS} \simeq 1$). With slags containing up to 2 ∿ 3% S, the CaS is in solution; above this level it is present as a separate solid phase.

PROBLEM NO. 21

APPLICATION OF MOLECULAR THEORY TO THE DEPHOSPHORISATION OF STEEL[8]

During refining, steel in an open hearth furnace is in equilibrium with a slag of the following composition:

Component:	CaO	SiO_2	FeO	Fe_2O_3	MgO	P_2O_5	MnO
Weight %:	20.76	20.50	38.86	4.98	10.51	1.67	2.51
Molar fraction:	0.232	0.215	0.339	0.0195	0.165	0.0075	0.022

(a) Applying the molecular theory, and assuming that the species that actually exist in the slag are: free CaO, $Ca_4P_2O_9$, $CaFe_2O_4$, $Ca_2Si_2O_6$, $Ca_4Si_2O_8$, FeO, MgO, and MnO, calculate the molar fraction of each one of these components.

The reactions of formation of $4CaO \cdot P_2O_5$ and $CaO \cdot Fe_2O_3$ are assumed complete. The following equilibrium will be assumed to exist for the silicates:

$$(Ca_4Si_2O_8) \rightleftarrows (Ca_2Si_2O_6) + 2(CaO)_{free}: \; K_1 = 1 \times 10^{-2}$$

The equilibrium constant is independent of the temperature. MgO and MnO are included with CaO.

(b) Using the following data and the previous results, calculate the oxygen and phosphorus contents of the steel in equilibrium with the slag at 1600°C.

Data:

$$\log(\% \underline{O})_{sat.} = -\frac{6320}{T} + 2.734$$

$$4(CaO)_s + 2(\underline{P})_{1\%} + 5(\underline{O})_{1\%} \rightleftarrows (Ca_4P_2O_9)_s: \; \log K_2 = \frac{71{,}667}{T} - 28.73$$

SOLUTION

(a) If MgO and MnO are considered together with CaO, the molar proportions of six compounds remain unknown: free CaO, $Ca_4P_2O_9$, $CaFe_2O_4$, Ca_2SiO_6, $Ca_4Si_2O_8$, and FeO. Of the equations which make it possible to elucidate the system, one is given by the equilibrium constant K_1, while the others come from the mass balance. For one mole of slag, the mass balance is:

For FeO: $$n_{FeO} = 0.339 \qquad (a)$$

For Fe_2O_3: $$n_{CaFe_2O_4} = 0.0195 \qquad (b)$$

For P_2O_5: $$n_{Ca_4P_2O_9} = 0.0075 \qquad (c)$$

For CaO(MgO + MnO): $$n_{CaO\,(free)} + 2n_{Ca_2Si_2O_6} + 4n_{Ca_4Si_2O_8}$$

$$= 0.232 + 0.165 + 0.022 - n_{CaFe_2O_4} - 4n_{Ca_4P_2O_9}$$

$$= 0.3695 \qquad (d)$$

For SiO_2: $$2n_{Ca_2Si_2O_6} + 2n_{Ca_4Si_2O_8} = 0.215 \qquad (e)$$

On the other hand, the equilibrium constant K_1 can be expressed in the form:

$$\left[\frac{n_{CaO_{free}}}{\Sigma n}\right]^2 \cdot \frac{n_{Ca_2Si_2O_6}}{n_{Ca_4Si_2O_8}} = 10^{-2} \quad \text{(f)}$$

Σn can be obtained by adding together these amounts:

$$\Sigma n = n_{CaO_{free}} + n_{FeO} + n_{Ca_2Si_2O_6} + n_{Ca_4Si_2O_8} + n_{CaFe_2O_4} + n_{Ca_4P_2O_9}$$

$$= n_{CaO_{free}} + 0.4735 \quad \text{(g)}$$

Finally, the molar fractions of the actual species are related to the number of moles by:

$$N_i = n_i/\Sigma n.$$

This solution makes it possible to calculate the number n_i of moles and the molar fractions of the different species that actually exist in the slag (Table 39).

TABLE 39

Species	Number of moles, n	Molar fraction, N
CaO (free)	0.0505	0.096
$Ca_2Si_2O_6$	0.056	0.107
$Ca_4Si_2O_8$	0.052	0.099
$Ca_4P_2O_9$	0.0075	0.014
$CaFe_2O_4$	0.0195	0.037
FeO	0.339	0.647
	Σn = 0.5245	1.000

(b) The amount of oxygen dissolved in iron in equilibrium with pure FeO can be calculated from:

$$\log(\%\ O)_{sat.} = -\frac{6320}{1873} + 2.734 = -0.640 \quad \text{and} \quad (\%\ O)_{sat.} = 0.23$$

When FeO is in solution in a slag ($a_{FeO} < 1$), then $(\%\ O)/(\%\ O)_{sat.} = a_{FeO}$.

In this example, $a_{FeO} = N_{FeO} = 0.647$ and:

$$(\%\ O) = 0.23 \times 0.647 = 0.149$$

At 1600°C, the equilibrium constant K_2 has a value of:

$$\log K_2 = \frac{71,667}{1873} - 28.73 = 9.55 \qquad \text{and} \qquad K_2 = 3.55 \times 10^9$$

With

$$K_2 = \frac{N_{Ca_4P_2O_9}}{(CaO(free))^4 \cdot (\%\ O)_m^5 \cdot (\%\ P)_m^2}$$

then P in the steel is 0.025%.

The importance of the dephosphorising function of this slag can be assessed from the following.

If there are 100 kg of slag per tonne of steel, the phosphorus content will be: % $P_2O_5 \times 62/142$ and the slag contains 0.73 kg $\underline{P}$, while each tonne of steel contains 0.25 kg $\underline{P}$. If this quantity of phosphorus had not been transferred to the slag, the amount in the steel would be equal to 0.1%, sufficient to render it useless for most purposes.

PROBLEM NO. 22

DEOXIDATION OF STEEL BY ALUMINIUM(9, 18)

A steel which contains 0.05 at. % of oxygen is deoxidised at 1600°C by addition of aluminium in the ladle. According to the amount of Al added, either alumina, hercynite ($Al_2O_3 \cdot FeO$), a mixture of both, or liquid FeO with Al_2O_3 in solution is formed.

(a) What can be deduced from the study of the variance?

(b) What phase(s) is (are) formed when Al is added to the steel?

(c) With the help of the following data, find the relationship between $N_{(\underline{O})}$ the fraction of oxygen in the steel and the fraction $N_{(\underline{Al})}$ of Al.

Draw the curve $N_{(\underline{O})} = fN_{(\underline{Al})}$ and comment on the shape.

(d) What amount of Al must be added to obtain the maximum deoxidation?

Data:

$$(Fe) + \tfrac{1}{2}[O_2] = (FeO) \qquad \Delta G^o_{(1)} = -\ 56,900 + 11.82\ T \qquad (1)$$

$$2(Al) + \tfrac{3}{2}[O_2] = \langle Al_2O_3\rangle \qquad \Delta G^o_{(2)} = -\ 401,500 + 76.91\ T \qquad (2)$$

$$\tfrac{1}{2}[O_2] = (\underline{O})^{\infty}_{Fe} \qquad \Delta G^o_{(3)} = -\ 28,000 + 6.0\ T \qquad (3)$$

$$(Al) = (\underline{Al})^{\infty}_{Fe} \qquad \Delta G^o_{(4)} = -\ 11,000 + 0.0\ T \qquad (4)$$

Reactions (3) and (4) correspond to a change of state (pure body → infinitely dilute solution with composition expressed in mole fraction).

$\underline{a_{(FeO)}}$ in Hercynite* = 3.0×10^{-1} (in equilibrium with $\langle Al_2O_3\rangle$) and (Fe,O)

$\underline{a_{\langle Al_2O_3\rangle}}$ in Hercynite = 1.

$$\varepsilon_{Al}^{O} = \varepsilon_{O}^{Al} = -105; \qquad \varepsilon_{Al}^{Al} = 6.7; \qquad \varepsilon_{O}^{O} = 0$$

SOLUTION

(a) Variance: When there are three phases in equilibrium (solid Hercynite, alumina, and liquid solution (Fe, O, Al), the variance is equal to 1. Thus, at a given T, the mole fractions of $(\underline{O})$ and $(\underline{Al})$ in the steel are fixed. When there are only two phases at equilibrium ($\langle Al_2O_3\rangle$ and (Fe, Al, O)), the variance is 2 and the fraction of $(\underline{O})$ is a function of that of $(\underline{Al})$.

(b) Consider the following equilibria (standard state for $(\underline{O})$ and $(\underline{Al})$ infinitely dilute):

$$(Fe) + (\underline{O}) = (FeO) \qquad \Delta G^{o}_{1600} = -\ 18{,}000 \qquad (a)$$

$$2(\underline{Al}) + 3(\underline{O}) = \langle Al_2O_3\rangle \qquad \Delta G^{o}_{1600} = -\ 185{,}162 \qquad (b)$$

Before adding Al in the ladle, we have a single phase (Fe, $\underline{O}$): If the concentration of $(\underline{O})$ is sufficient, a liquid slag (Al_2O_3, FeO) is formed when a small amount of Al is added and the activity of FeO in the slag corresponds to that in the steel. With the addition of more Al the solid phase, Hercynite, would appear and a_{FeO} and $a_{Al_2O_3}$ would be fixed as well as $a_{(O)}$ and $a_{(Al)}$ in the steel. First we would have the equilibrium between (Fe . O . Al) $\langle$Hercynite$\rangle$ and liquid (Al_2O_3 . FeO), then (Fe, O, Al), $\langle$Hercynite$\rangle$, $\langle Al_2O_3\rangle$ (Fig. 154). So long as there is Hercynite in the slag (with $\langle Al_2O_3\rangle$ or (Al_2O_3, FeO)), the percentage of the phases present in the slag will change without changing the composition of the steel. Starting from (a) and (b), and from data ($a_{FeO} = 3 \times 10^{-1}$, $a_{Al_2O_3} = 1$), it is possible to calculate successively $a_{(\underline{O})} \simeq N_{(\underline{O})}$ and $a_{(\underline{Al})} \simeq N_{(\underline{Al})}$ in equilibrium with Hercynite and $\langle Al_2O_3\rangle$.

$$N_{(\underline{O})} = 2.5 \times 10^{-3}$$

$$N_{(\underline{Al})} = 1.3 \times 10^{-7}$$

In the actual case $N_{(O)} = 5 \times 10^{-4}$, the value is too low for hercynite to be formed and the only oxide which appears is $\langle Al_2O_3\rangle$

*From equations (7.16) and (7.18) and $\langle FeO, Al_2O_3\rangle$ = (Fe) + (O)% + $\langle Al_2O_3\rangle$:

$$\Delta G^{o} = 43{,}810 - 18.21\ T\ \text{cal}^{(19)}.$$

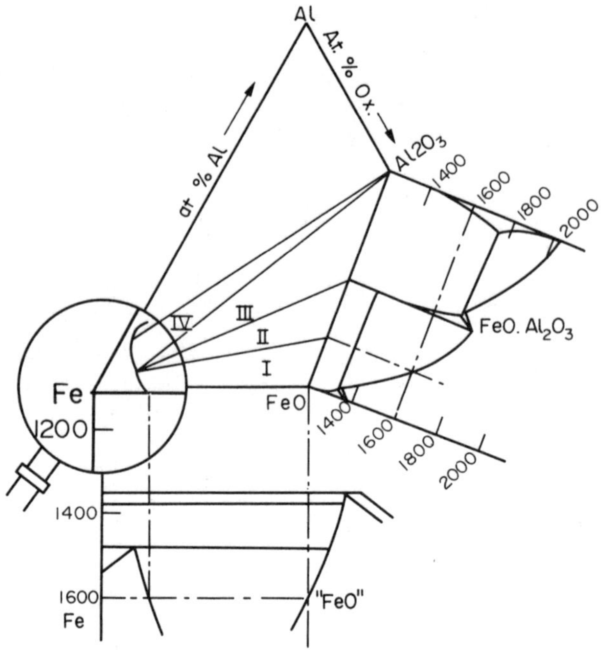

Fig. 154. Equilibria in the system Al-Fe-O
I. (Fe, O, Al) - (FeO, Al_2O_3);
II. (Fe, O, Al) - (FeO, Al_2O_3) - <FeO . Al_2O_3>
III. (Fe, O, Al) - <FeO . Al_2O_3> - <Al_2O_3>
IV. (Fe, O, Al) - <Al_2O_3>

find a relation between $a_{\underline{(Al)}}$ and $a_{\underline{(O)}}$:

$$\frac{\Delta G^o_{(b)}}{RT} = 3 \ln a_{\underline{(O)}} + 2 \ln a_{\underline{(Al)}} \tag{c}$$

with $\ln a_i = \ln N_i + \ln f_i$ and $\ln f_i = \varepsilon_i^j N_j + \varepsilon_i^i N_i$

Substituting the values of $\Delta G^o_{(b)}/RT$ at 1600°C, and ε_i^j in (c), we obtain the following relationship between $N_{(O)}$ and $N_{\underline{(Al)}}$:

$$3 \ln N_{\underline{(O)}} - 210\, N_{\underline{(O)}} = -2 \ln N_{\underline{(Al)}} + 301.6\, N_{\underline{(Al)}} - 49.43 \tag{d}$$

and can calculate

$$N_{\underline{(O)}} = f\left(N_{\underline{(Al)}}\right) \text{ (from } N_{\underline{(O)}} = 5 \times 10^{-4} \text{ and } N_{\underline{(Al)}} = 1.7 \times 10^{-6}$$

$N_{(\underline{Al})}$	1.7×10^{-6}	5×10^{-5}	10^{-3}	5×10^{-3}	10^{-2}	2×10^{-2}	5×10^{-2}
$N_{(\underline{O})}$	5×10^{-4}	5×10^{-5}	7.7×10^{-6}	3.9×10^{-6}	4.1×10^{-6}	7.1×10^{-6}	7.8×10^{-5}

It can be noted (Fig. 155) that the curve $N_{(\underline{O})} = f\left(N_{(\underline{Al})}\right)$ passes through a minimum obtained by differentiation of (d):

$$dN_{(\underline{O})}/dN_{(\underline{Al})} = \frac{-\left[2 - 301.6\,N_{(\underline{Al})}\right] \cdot N_{(\underline{O})}}{\left[3 - 210\,N_{(\underline{O})}\right] \cdot N_{(\underline{Al})}}$$

This minimum is at $N_{(\underline{Al})} = 2/301.6 = 6.62 \times 10^{-3}$ and $N_{(\underline{O})} = 3.8 \times 10^{-6}$. It is not possible to lower $N_{(\underline{O})}$ beyond this value with Al.* This minimum can be explained by the high value of ε_O^{Al}. The values of the activity coefficient $f_{(\underline{O})}$ decrease faster than those of $a_{(\underline{O})}$ and thus, $N_{(\underline{O})}$ will increase when further Al is added.

(c) The amount of Al added to obtain the minimum oxygen content can be calculated by adding Al used for deoxidation:

$$\frac{2}{3}\,(5 \times 10^{-4} - 3.8 \times 10^{-6}) = 3.4 \times 10^{-4}$$

to the Al dissolved in the steel (6.62×10^{-3}). It will be noted that it is necessary to add about 20 times the aluminium needed for deoxidation, i.e.:

$$7 \times 10^{-3} \text{ moles per mole of iron, or } 0.34\%$$

or else 3.4 kg of Al per tonne of iron.

But with only 2×10^{-3} moles of Al per mole of Fe, the oxygen content of the steel is little different from the minimum $(5.5 \times 10^{-6}$ instead of $3.8 \times 10^{-6})$.

*In practice, $(\%\ O)_m$ passes through a minimum and a maximum, before ultimately decreasing with a high Al content. This behaviour can be explained by using second order interaction coefficients.(18)

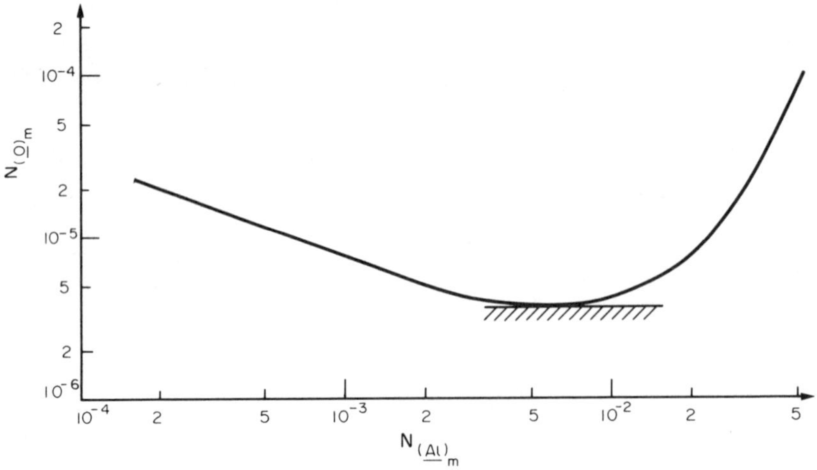

Fig. 155. Variation of $N_{(\underline{O})}$ with $N_{(\underline{Al})}$ in a steel at 1600°C

PROBLEM NO. 23

EFFECTS OF VARIATION IN FeO CONTENT AND CaO/SiO_2 RATIO IN THE SLAG ON THE REMOVAL OF PHOSPHORUS FROM STEEL

Liquid steel containing 0.155% P is blown with oxygen at 1600°C after the addition of 60 kg CaO per tonne of metal. Given the data below, calculate the partition coefficient of phosphorus between slag and metal (i.e. $(\% \, P_2O_5)_s/(\% \, P)_m$) for values of N_{FeO} from 0.1 to 0.5, and for basicity ratios, i, from 2 to 4. Plot $(\% \, P_2O_5)_s/(\% \, P)_m$ against N_{FeO} for each value of i. Comment on the form of the curves.

Data:

(a) $2(\underline{P})_{1\%} + 5(\underline{O})_{1\%} = (P_2O_5)$

$\Delta G^o = -168{,}600 + 133\,T$ cal.

The standard state for $(\underline{P})$ and $(\underline{O})$ is the infinitely dilute solution with the composition in weight %. It is assumed that $(\underline{P})$ and $(\underline{O})$ follow Henry's Law.

(b) $\log \gamma_{P_2O_5} = -1.12\,(22\,N_{CaO} + 12\,N_{FeO} - 2\,N_{SiO_2}) - \dfrac{42{,}000}{T} + 23.58$ (11)

(c) $\log (\% \, O)_{sat.} = -6320/T + 2.73$

(d) The activity coefficient of FeO in the slag is given by(12):

N_{FeO}	0.1	0.2	0.3	0.4	0.5
γ_{FeO}	1.7	1.5	1.3	1.2	1.1

SOLUTION

At 1600°C, log K for reaction (a) is equal to -9.4 from the data, i.e.:

$$-9.4 = \log \gamma_{(P_2O_5)_s} + \log \left[\frac{N_{(P_2O_5)_s}}{(\%\ P)_m^2} \right] - 5 \log (\%\ O)_m \qquad \text{(e)}$$

Each term of (e) can be expressed as a function of N_{FeO} and i. The principles of the calculation will be illustrated for i = 3 and each term of (e) will be expressed as a function of N_{FeO} only.

Expression of Log % O

$$(\%\ O)_m = a_{FeO} \times (\%\ O)_{sat.} = 0.23 \times \gamma_{FeO} \times N_{FeO}$$

because $(\%\ O)_{sat.} = 0.23$ at 1600° according to (c)

Then: $$\log (\%\ O)_m = -0.64 + \log \gamma_{FeO} + \log N_{FeO} \qquad \text{(f)}$$

Expression of Log $\gamma_{P_2O_5}$

$$\log \gamma_{P_2O_5} = -1.12 \left[22\, N_{CaO} + 12\, N_{FeO} - 2\, N_{SiO_2} \right] + 1.156 \qquad \text{(g)}$$

N_{SiO_2} and N_{CaO} can be eliminated from (g) because:

$$\frac{N_{CaO}}{N_{SiO_2}} = \frac{m_{CaO}}{m_{SiO_2}} \times \frac{60}{56} = 1.07i$$

and when i = 3, $N_{CaO}/N_{SiO_2} = 3.21$ or $N_{SiO_2} = 0.31\, N_{CaO}$.

$N_{P_2O_5}$ in the slag is so small that it can be neglected, and

$$N_{CaO} + N_{FeO} + N_{SiO_2} = 1$$

from which $N_{CaO} = \frac{1}{1.31} (1 - N_{FeO}) = 0.763\, (1 - N_{FeO})$

and $$\log \gamma_{P_2O_5} = -17.12 + 4.84\, N_{FeO} \qquad \text{(h)}$$

Expression of Log $\left[N_{(P_2O_5)_s}/(\%\ P)^2_m\right]$

The total quantity of phosphorus present in metal and slag is 50 moles per tonne of metal, i.e. $2n_{(P_2O_5)_s} + n_{(P)_m} = 50$. Because $n_{(P_2O_5)_s}$ and $n_{(P)_m}$ are small enough to be neglected relative to n_{CaO}, n_{FeO}, and n_{SiO_2} in the slag and n_{Fe} in the metal,

$$N_{(P_2O_5)_m} = n_{P_2O_5}/(n_{CaO} + n_{SiO_2} + n_{FeO})$$

and

$$(\%\ P)_m = (m_P \times 100)/m_{Fe}$$

Since the slag is based on 60 kg of CaO per tonne of metal, $n_{CaO} = 60{,}000/56 = 1071$; and for $i = 3$, $n_{SiO_2} = 20{,}000/60 = 333$. n_{FeO} and m_{Fe} depend on N_{FeO}:

$$N_{FeO} = n_{FeO}/\left(n_{CaO} + n_{FeO} + n_{SiO_2}\right)$$

or

$$n_{FeO} = \frac{N_{FeO}\left(n_{CaO} + n_{SiO_2}\right)}{1 - N_{FeO}} = \frac{1404\ N_{FeO}}{1 - N_{FeO}}$$

In the same way:

$$m_{Fe} = 10^6 - 56\ n_{FeO} = 10^6(1 - 1.078\ N_{FeO})/(1 - N_{FeO})$$

Consequently:

$$N_{(P_2O_5)_s} = \frac{n_{(P_2O_5)_s}(1 - N_{FeO})}{1404} = \frac{n_{(P)_s}(1 - N_{FeO})}{2 \times 1404}$$

and

$$(\%\ P)_m = \frac{31 \times 10^2\ (n_{(P)_m}(1 - N_{FeO})}{10^6(1 - 1.078\ N_{FeO})} \qquad \text{(i)}$$

Substituting for $\log \gamma_{P_2O_5}$, $\log (\%\ O)_m$, and $\log \left(N_{(P_2O_5)_s}/(\%\ P)^2_m\right)$ in equation (e), we obtain:

$$\log \left(\frac{n_{(P)_s}}{n^2_{(P)_m}}\right) = 2.96 - 4.84\ N_{FeO} + 5 \log \gamma_{FeO} - \log \frac{(1 - 1.078\ N_{FeO})^2}{(1 - N_{FeO}) \times N^5_{FeO}} \qquad \text{(j)}$$

For $i = 2$, (j) becomes:

$$\log \left(\frac{n_{(P)_s}}{n^2_{(P)_m}}\right) = 0.77 - 2.6\, N_{FeO} + 5 \log \gamma_{FeO} - \log \frac{(1 - 1.088\, N_{FeO})^2}{(1 - N_{FeO}) \times N^5_{FeO}} \qquad (j')$$

and for $i = 4$:

$$\log \left(\frac{n_{(P)_s}}{n^2_{(P)_m}}\right) = 4.21 - 6.1\, N_{FeO} + 5 \log \gamma_{FeO} - \log \frac{(1 - 1.074\, N_{FeO})^2}{(1 - N_{FeO}) \times N^5_{FeO}} \qquad (j'')$$

Since $n_{(P)_s} + n_{(P)_m} = 50$, the values of $n_{(P)_s}/n^2_{(P)_m} = k$, obtained from (j), (j') and (j") make it possible to calculate $n_{(P)_s}$ and $n_{(P)_m}$ for different values of N_{FeO} by solving the second order equation:

$$50 - n_{(P)_m} = k n^2_{(P)_m}, \text{ from which } n_{(P)_m} = \frac{-1 + \sqrt{1 + 200\,k}}{2\,k}$$

$(\%\ P_2O_5)_s$, $(\%\ P)_m$ and the partition coefficient $(\%\ P_2O_5)_s/(\%\ P)_m$ can be calculated from the relationships:

$$\%\ P_{(m)} = \frac{31\, n_{(P)_m} (1 - N_{FeO}) \times 100}{10^6 \left(1 - (1 + \varepsilon) N_{FeO}\right)}$$

(The values of ε are 0.078 for $i = 3$; 0.088 for $i = 2$; 0.074 for $i = 4$)

and $$(\%\ P_2O_5)_s = \frac{m_{P_2O_5} \times 100}{m_{CaO} + m_{FeO} + m_{SiO_2}} = \frac{\frac{142}{2} n_{(P)_s} \times 100}{m_{CaO} + m_{SiO_2} + 72\, n_{FeO}}$$

Using the values of n_{FeO} given above, we find:

for $i = 2$, $$(\%\ P_2O_5)_s = \frac{71\, n_{(P)_s} (1 - N_{FeO})}{900 + 231\, N_{FeO}}$$

For $i = 3$, $$(\%\ P_2O_5)_s = \frac{71\, n_{(P)_s} (1 - N_{FeO})}{800 + 211\, N_{FeO}}$$

For $i = 4$, $$(\%\ P_2O_5)_s = \frac{71\, n_{(P)_s} (1 - N_{FeO})}{750 + 201\, N_{FeO}}$$

The values obtained for $(\%\ P)_m$, $(\%\ P_2O_5)_s$ and $(\%\ P_2O_5)_s/(\%\ P)_m$ are given in Table 40, and $(\%\ P_2O_5)_s/(\%\ P)_m$ is plotted as a function of N_{FeO} for i = 2, 3, and 4 in Fig. 156 (compare with experimental results of Fig. 130[13]).

The maxima can be explained by the fact that FeO performs two functions: it is a basic and an oxidising component. $(O)_{Fe}$ increases with N_{FeO} in the slag, which favours reaction (a); but, as a basic oxide, FeO has an effect which is less than that of CaO. Further, the increase of N_{FeO} diminishes N_{CaO} and, consequently, the activity coefficient of P_2O_5 in the slag increases. These two effects of FeO are opposed. In the beginning, the oxidising effect of FeO is preponderant, but for higher contents of FeO, the diluting effect of CaO becomes more important.

A steel of good quality must contain less than 0.030 % P. This result can be obtained only with a slag strongly basic (i ≃ 4) and oxidising (N_{FeO} = 0.2).

TABLE 40

i	N_{FeO}	γ_{FeO}	$\log \frac{n_{(P)_s}}{n^2_{(P)_m}}$	$\frac{n_{(P)_s}}{n^2_{(P)_m}}$	$n_{(P)_m}$	$(\%P)_m$	$n_{(P)_s}$	$(\%\ P_2O_5)_s$	$\frac{(\%P_2O_5)_s}{(\%\ P)_m}$
2	0·1	1·7	-3·28	$5{\cdot}2 \times 10^{-4}$	48·8	0·153	1·2	0·083	0·54
	0·2	1·5	-2·25	$1{\cdot}1 \times 10^{-3}$	47·5	0·150	2·5	0·15	1·0
	0·3	1·3	-1·87	$1{\cdot}4 \times 10^{-2}$	34·1	0·110	15·9	0·82	7·4
	0·4	1·2	-1·58	$2{\cdot}6 \times 10^{-2}$	28·6	0·094	21·4	0·92	9·8
	0·5	1·1	-1·60	$2{\cdot}5 \times 10^{-2}$	29·0	0·099	21·0	0·73	7·4
3	0·1	1·7	-1·32	0·048	23·5	0·074	26·5	2·06	28
	0·2	1·5	-0·51	0·31	11·2	0·036	38·8	2·62	73
	0·3	1·3	-0·35	0·45	9·5	0·030	40·5	2·33	78
	0·4	1·2	-0·30	0·50	9·0	0·030	41·0	1·97	66
	0·5	1·1	-0·38	0·51	9·9	0·034	40·1	1·57	46
4	0·1	1·7	-0·195	0·64	8·1	0·025	41·9	3·48	139
	0·2	1·5	+0·488	3·07	3·9	0·012	46·1	3·31	276
	0·3	1·3	+0·517	3·28	3·7	0·011	46·3	2·84	258
	0·4	1·2	+0·446	2·79	4·1	0·013	45·9	2·35	181
	0·5	1·1	+0·230	1·70	5·1	0·017	44·9	1·87	110

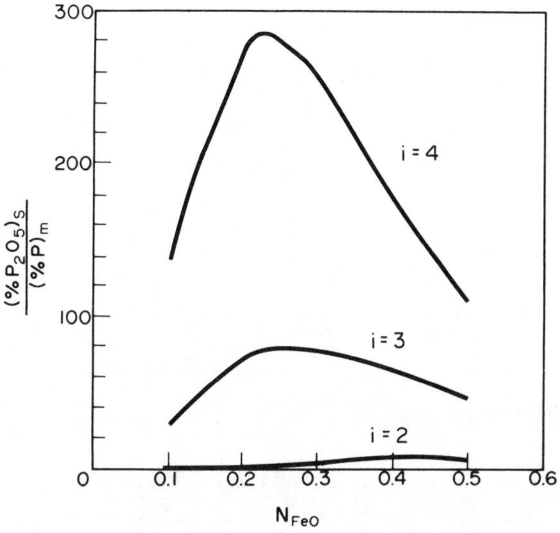

Fig. 156. Partition coefficient vs N_{FeO} for different basicities

THREE PROBLEMS CONCERNING COMPLEX EQUILIBRIA

(Problem Nos. 24, 25 and 26)(to be solved by computer)

Chemical equilibria at high temperatures are often complex, because the amount of each compound and the number of phases present depend on the conditions of temperature and pressure. To determine the amount of each compound formed, two methods of numerical resolution can be used:

(A) When the number and nature of the phases present are known, in order to determine the quantity of each of the n compounds formed, it is necessary to solve a non-linear calculation from n independent equations:

* For a system with c elements, it will be possible to write c equations of material balance.

* The $n - c$ other equations are supplied by the equilibrium constants of the reactions between the n compounds.

This method has been applied in the simple case of the cracking of methane (Problem No. 11). It will also be applied to the two following problems:

- Deoxidation of a steel by CH_4.

- The Ugine-Perrin process.

(B) When the phases which are formed are not known *a priori*, the method to be employed consists of minimizing the global free energy, G, of the system:

$$G = \sum_{i=1}^{c} n_i \mu_i$$

The chemical potential μ_i of the constituent i is equal to:

$$\mu_i^o + RT \ln a_i.$$

The activity a_i is equal to one for the condensed pure body, and to p_i (in atm) for the gases. The previous equation can be written:

$$\frac{G}{RT} = \sum_{i=1}^{c} n_i \left[\frac{\mu_i^o}{RT} + \phi_i \ln p_i \right]$$

The coefficient ϕ_i takes the value zero when i is a pure condensed body, and the value one for the gases. μ_i^o/RT is deduced from tables (JANAF Thermochemical Tables - see References of Chapter III):

$$\frac{\mu^o}{RT} = \frac{1}{R}\left[\frac{G^o - H^o_{298}}{T}\right] + \frac{H^o_{298}}{RT}$$

The conservation of the mass requires that the number of atoms of each element remains constant, i.e.:

$$\sum_{i=1}^{c} n_i \nu_{ji} = b_j = \text{constant}$$

ν_{ji} is the number of atoms j in the compound i, and b_j is the number of atoms j of the system.

This method will be applied to the "Production of TiC by reducing $TiCl_4$ by CH_4".

Note: In the three following problems, only the mode of solution and the numerical solutions are given. (For more copious detail, see references cited.)

PROBLEM NO. 24

DEOXIDATION OF LIQUID STEEL BY METHANE (14)

In order to deoxidise a steel which contains 0.1 weight % of oxygen in the presence of $P_{CO} = 1$ atm, natural gas CH_4 is blown through the porous bottom of a ladle which contains the steel at 1600°C. It will be assumed that CH_4 is completely cracked after crossing the metal bath, and that the latter is homogeneous and at equilibrium with the bubbles of gas which burst at the surface, at pressure P of 1 atm.

Write the equations which permit the determination of the contents of $(\underline{O})$, $(\underline{C})$, $(\underline{H})$ in the steel, and the partial pressures of CO, CO_2, H_2, H_2O, also the amount of exit gas as a function of the number of moles of CH_4 blown.

Solve these equations with a computer, and draw the curves $(\underline{O})$, $(\underline{C})$, $(\underline{H})$, P_{H_2},

P_{H_2O}, P_{CO}, and P_{CO_2} as a function of the amount of CH_4 blown per tonne of steel.

Data: Equilibrium constants of reactions at 1600°C.

The pressures are expressed in atm, and the composition of liquid phase in p.p.m.

$$(C) + (\underline{O}) = [CO] \qquad K_1 = 4.93 \times 10^{-6}$$

$$[H_2] = 2(\underline{H}) \qquad K_2 = 6.11 \times 10^{2}$$

$$[H_2] + (\underline{O}) = [H_2O] \qquad K_3 = 3.66 \times 10^{-4}$$

$$[CO] + (\underline{O}) = [CO_2] \qquad K_4 = 8.78 \times 10^{-5}$$

SOLUTION

The previous system consists of eight unknowns: $(\underline{H})$, $(\underline{C})$, $(\underline{O})$, P_{CO}, P_{H_2O}, P_{H_2} and Σn_i (partial pressures P_i are bound to n_i by the relations $P_i = n_i/\Sigma n_i$. The utilisation of P_i instead of n_i does not bring supplementary unknowns).

Four equations can be derived from the thermodynamic equilibrium:

$$P_{CO} = K_1 \cdot (\underline{O}) \cdot (\underline{C}) \tag{1}$$

$$P_{H_2} = \frac{1}{K_2} (\underline{H})^2 \tag{2}$$

$$P_{H_2O} = K_3 \cdot P_{H_2} \cdot (\underline{O}) = \frac{K_3}{K_2} (\underline{O}) \cdot (\underline{H})^2 \tag{3}$$

$$P_{CO_2} = K_1 \cdot K_4 \cdot (\underline{O})^2 \cdot (\underline{C}) \tag{4}$$

Moreover:
$$P = P_{CO} + P_{CO_2} + P_{H_2O} + P_{H_2} = 1 \tag{5}$$

Equations (6), (7) and (8) result from the mass balance. This balance must be established in differential form, because the composition of the bath and of the exit gas vary with the amount of CH_4 blown; C, H, or O, originating in the methane or the bath are present either in the bath or in the exit gas, i.e.:

For C:
$$dn_{CH_4} = dn_{(C)} + dn_{CO} + dn_{CO_2}$$

$$= \frac{d(\underline{C})}{12} + \frac{P_{CO} + P_{CO_2}}{P} d\Sigma n_i \tag{6}$$

For H: $$4dn_{CH_4} = d(\underline{H}) + \frac{2P_{H_2} + 2P_{H_2O}}{P} d\Sigma n_i \qquad (7)$$

For O: $$\frac{d(\underline{O})}{16} + \frac{P_{CO} + P_{H_2O} + 2P_{CO_2}}{P} d\Sigma n_i = 0 \qquad (8)$$

This system can be simplified by replacing the partial pressures of (6), (7) and (8) by their values from (1), (2), (3) and (4); then the system (6), (7), (8) is solved, giving n_{CH_4} increasing values, starting from zero up to the amount at which $(\underline{O})$ is sufficiently low, by small intervals dn_{CH_4}. The results make it possible to draw the curves of Fig. 157(a) and (b).

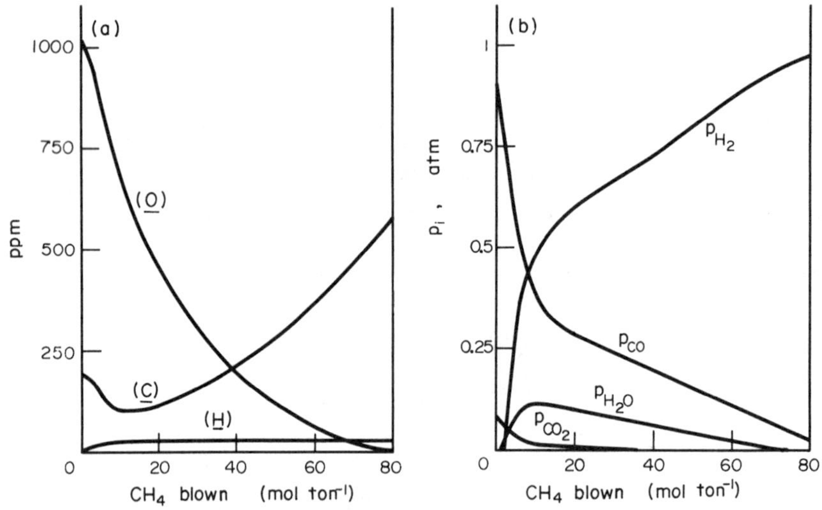

Fig. 157. Evolution of: (a) the composition of the steel[14]; (b) the partial pressures in the exit gas as a function of CH_4 blown

The decrease of $(\underline{C})$ at the beginning of the blowing can be explained qualitatively in the following way: The cracking of CH_4 produces 2 moles of H_2 and 1 mole of C, which reacts with $(\underline{O})$ to form CO. If the oxygen content was sufficient, the partial pressures would stabilise at 1/3 CO and 2/3 H_2. When the blowing starts, the pressure of H_2 increases; that of CO decreases rapidly as well as the product $(\underline{C}) \cdot (\underline{O})$ $\left[(\underline{C}) \cdot (\underline{O}) = P_{CO}/K_1 \right]$, more rapidly than the supply of C. Consequently, $(\underline{C})$ decreases before increasing again.

PROBLEM NO. 25

PRODUCTION OF HIGH GRADE Fe-Cr BY THE UGINE-PERRIN PROCESS[15]

Production of high grade Fe-Cr by the Ugine-Perrin process is carried out in two steps, at 2000 K:

Chromite, melted with lime (used in order to slag silica) in a slag arc furnace (30% Cr_2O_3) is cast into a ladle at the same time as intermediate Fe-Si-Cr.

Equilibrium is reached quickly, thanks to intimate mixing of slag and metal. Low C Fe-Cr and an intermediate slag which still contains Cr are obtained.

This slag is poured into a second ladle with Fe-Si-Cr from an arc reduction furnace (Fe: 15%, Cr: 40%, Si: 45%). The slag obtained, poor in Cr (Cr_2O_3 < 0.1%), is discarded. Intermediate Fe-Si-Cr is used to reduce the slag from the slag furnace.

Starting from the following data, calculate the optimum quantity m of Fe-Si-Cr to add in the intermediate ladle (the second) in order to obtain:

High grade Fe-Cr with a maximum of Cr and a minimum of Si.

A final slag with a minimum of Cr (take as a basis of calculation: 1 tonne of slag at 30% Cr_2O_3).

Data:

$$(Fe) + \frac{1}{2}[O_2] = (FeO) \qquad \Delta G^o = -56{,}900 + 11.82\ T$$

$$2\langle Cr\rangle + \frac{3}{2}[O_2] = \langle Cr_2O_3\rangle \qquad \Delta G^o = -270{,}550 + 61.35\ T$$

$$\langle Cr\rangle = (Cr) \qquad \Delta G_f = 5{,}000 - 2.3\ T$$

$$\langle Cr_2O_3\rangle = (Cr_2O_3) \qquad \Delta G_f = 6{,}200 - 2.3\ T$$

$$(Si) + [O_2] = (SiO_2) \qquad \Delta G^o = -223{,}800 + 46.1\ T$$

It will be assumed that the ratios:

$$\frac{\gamma_{SiO_2}}{\gamma^2_{FeO}} \quad \text{and} \quad \frac{\gamma_{SiO_2}}{\gamma^2_{Cr_2O_3}}$$

are near to unity.

For $N_{Si} < 0.5$, empirical formulae:

$$\ln \gamma_{(Si)Cr,Fe} = -6.7\,(1 - N_{Si})^2 \quad \text{and} \quad \left.\begin{matrix}\ln \gamma_{Fe} \\ \ln \gamma_{Cr}\end{matrix}\right\} = -6.7\,N_{Si}^2$$

will be chosen.

The Fe-Cr solution is assumed ideal. The first formula corresponds approximately to the curves $\log \gamma_{Si} = f(N_{Si} + N_C)$ (from Rein and Chipman, *Trans. A.I.M.E.* 233, 417, 1965).

SOLUTION

The diagram of Fig. 158 gives the route followed by the ore and the reductant. This system includes twelve unknowns which are the quantities (in kg) of Fe, Si, Cr, and their oxides contained in each ladle when the equilibrium is reached, and a parameter m which represents the amount of reductant to add. To solve this system, it is necessary to write twelve equations: Four are given by oxidation/reduction equilibria (two each ladle), and eight by the material balance of Cr, Fe, Si, O, in each ladle.

Equilibrium constants are the same in each ladle:

$$2FeO + Si = 2Fe + SiO_2 \qquad \ln K_1^{2000K} = 16.4$$

$$2Cr_2O_3 + 3Si = 4Cr + 3SiO_2 \qquad \ln K_2^{2000K} = 25.5$$

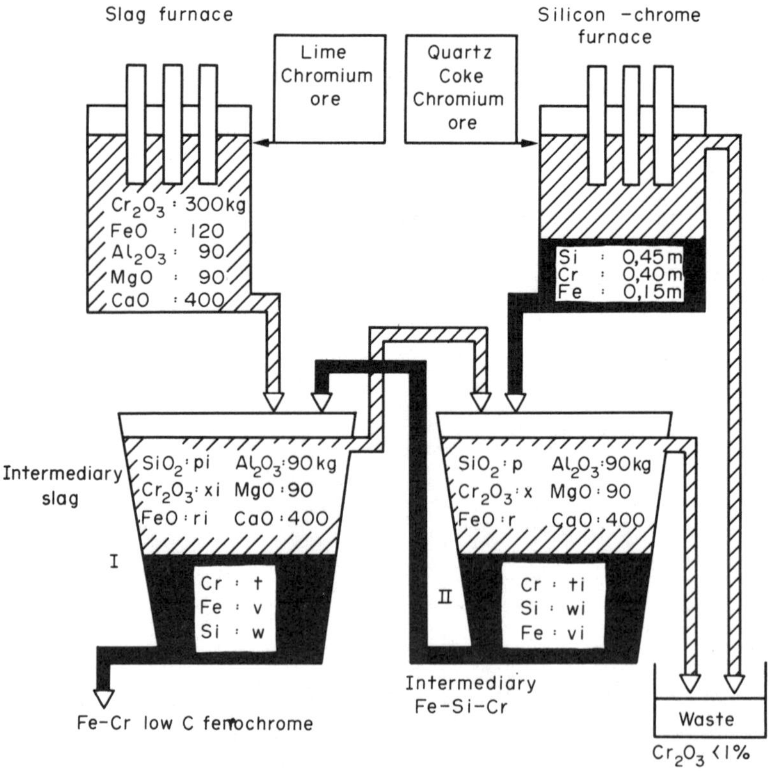

Fig. 158. Production of high grade Fe-Cr by the Ugine-Perrin process

With $$\ln K_1 = \ln \frac{\gamma_{Fe}^2 \cdot N_{Fe}^2 \cdot \gamma_{SiO_2} \cdot N_{SiO_2}}{\gamma_{Si} \cdot N_{Si} \cdot \gamma_{FeO}^2 \cdot N_{FeO}^2} \left\{ \begin{array}{l} \text{final ladle} \\ \text{intermediate ladle} \end{array} \right.$$

and $$\ln K_2 = \ln \frac{\gamma_{Cr}^4 \cdot N_{Cr}^4 \cdot \gamma_{SiO_2}^3 \cdot N_{SiO_2}^3}{\gamma_{Si}^3 \cdot N_{Si}^3 \cdot \gamma_{Cr_2O_3}^2 \cdot N_{Cr_2O_3}^2} \left\{ \begin{array}{l} \text{final ladle} \\ \text{intermediate ladle} \end{array} \right.$$

$\ln \gamma_i$ are given by data. The mole fraction of a metal or an oxide is:

$$N_i = \frac{m_i/M_i}{\Sigma m_i/M_i}$$

m_i is the mass of the body i (i.e. t, v, w; p_i, x_i, r_i; p, x, r; t_i, v_i, w_i);
M_i is the corresponding atomic mass.

Equations of material balance (see Fig. 158):

The quantity of an element entering a ladle is equal to the quantity which goes out. For example, the balance of Si in the left ladle is:

$$p_i \times \frac{28}{60} + w = w_i$$

or else: $$p_i = \frac{60}{28} (w_i - w)$$

In the same form, the balance of Si in the other ladle, and that of Fe, Cr, O, are:

$$x_i = 300 - \frac{152}{104} (t - t_i)$$

$$r_i = 120 - \frac{72}{56} (v - v_i)$$

$$t_i = 0.4 \text{ m} + \frac{104}{152} (x_i - x)$$

$$w_i = 0.45 \text{ m} + \frac{28}{60} (p_i - p)$$

$$v_i = 0.15 \text{ m} + \frac{56}{72} (r_i - r)$$

oxygen

$$p_i = \frac{30}{72} (120 - r_i) + \frac{90}{152} (300 - x_i)$$

$$p - p_i = \frac{30}{72} (r_i - r) + \frac{90}{152} (x_i - x)$$

With these last eight equations, it is possible to eliminate eight unknowns, keeping the four which are in the four equations (two for each ladle) of $\ln K_1$ and $\ln K_2$.

The system is solved, giving to the parameter m increasing values from 50 to 300 kg of Fe-Si-Cr. The result is shown in Fig. 159. We can see that the best results are obtained with m = 240 kg Fe-Si-Cr per tonne of chromite containing 30% of Cr_2O_3.

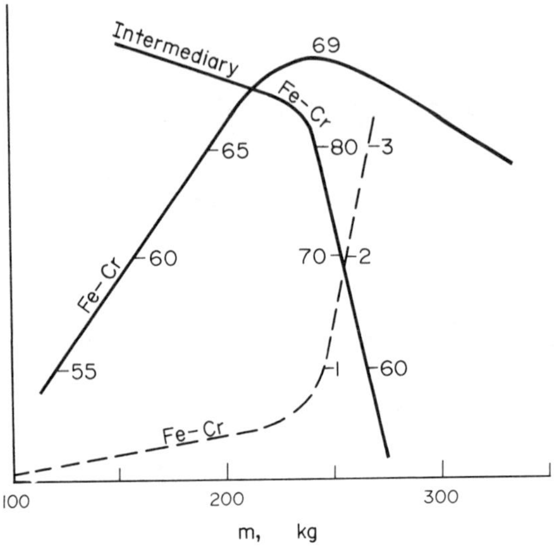

Fig. 159. Weight percent of Cr ——— and of Si – – – in the intermediate and the final ladle as a function of the amount of Fe-Si-Cr added

PROBLEM NO. 26

PRODUCTION OF TiC BY REDUCING $TiCl_4$ BY CH_4 (16)

In order to produce titanium carbide, TiC, by reducing $TiCl_4$ by CH_4, these two gases, carried by H_2, are heated in a reactor at high temperature (between 1000 and 2000 K). The main reaction:

$$[TiCl_4] + [CH_4] = <TiC> + 4[HCl]$$

is incomplete and, at equilibrium and according to the conditions, the following compounds can be obtained: unreacted $[TiCl_4]$, $[CH_4]$, $[H_2]$, and <C>, <Ti>, <TiC>, $[TiCl_3]$, $[TiCl_2]$, [TiCl], [HCl], $[Cl_2]$, i.e. eleven compounds.

(1) Knowing that it is not possible to produce <Ti> and <C> at the same time (see diagram Ti - C), what is the variance of the system according to the number of condensed phases? In each case, what parameters can be fixed?

(2) Since the number of condensed phases is not known, the only available method consists of minimising the overall free energy of the system, taking into account

the mass balance. Indicate the data for supply to the computer to solve this system when the pressures of the gases which enter the reactor are:

$$P^*_{TiCl_4} = 10^{-4}$$

$$P^*_{CH_4} \quad \text{from } 10^{-6} \text{ to } 1$$

$$P^*_{H_2} = P - P^*_{TiCl_4} - P^*_{CH_4}$$

Calculate the number of moles of the different compounds formed, and draw the curves of efficiency for TiC and Ti at 1800 K as a function of the pressure of methane when $P = 1$ atm.

SOLUTION

(1) Variance: The independent constituents are four: Ti, Cl, H and C, and the number of phases varies from one to three: 1 gaseous phase; 0, 1, or 2 condensed phases (but never 3); then the variance is 3, 4, or 5. When there is only a gas phase, besides T and P, three of the six following ratios can be fixed by fixing the initial conditions:

$$Ti/Cl = 1/4 \qquad Cl/H = \frac{2\,P^*_{TiCl_4}}{1 - P^*_{TiCl_4} + P^*_{CH_4}} = 4\ Ti/H$$

$$Cl/C = 4\ Ti/C = \frac{4\,P^*_{TiCl_4}}{P^*_{CH_4}} \qquad C/H = \frac{P^*_{CH_4}}{2(1 - P^*_{TiCl_4} + P^*_{CH_4})}$$

The asterisk refers to the entry gas. The ratio Ti/Cl is constant, and consequently is fixed. When one or two phases condense, it is not possible to fix the partial pressure of a gas which contains one of the condensed elements. Thus, if TiC and C (or TiC and Ti) are condensed, the only ratio to be fixed from initial conditions is Cl/H. The variables that initial conditions make it possible to fix are indicated in Table 41.

(2) The data to be supplied to the computer for the calculation of the minimum of the free energy at 1800 K:

$$G/RT = \sum_{i=1}^{t} n_i \left[\mu_i^o/RT + \phi_i \ln P_i \right]$$

are given by Table 42. From these values, the standard potential μ_i^o of each constituent can be calculated:

$$\mu_i^o = T\left(\frac{G_T^o - H_{298}^o}{T}\right) + H_{298}^o$$

TABLE 41

Number of Solid Phases	Nature	V	Independent Variables
0		5	P, T, Ti/Cl, Cl/H, C/H
			P, T, Ti/Cl, Cl/H, Cl/C
			P, T, Ti/Cl, Cl/C, C/H
1	TiC	4	P, T, Cl/H, (Ti - C) / H
	C		P, T, Cl/H, Ti/Cl
	Ti		P, T, Cl/H, C/H
2	TiC + Ti	3	P, T, Cl/H
	TiC + C		P, T, Cl/H

ϕ_i is equal to one for gases; zero for the condensed phases.

For 1 mole of $TiCl_4$, CH_4 and H_2 entering the reactor, the mass balance of one element (Ti, for example) is:

$$n_{TiCl_4} + n_{TiCl_3} + n_{TiCl_2} + n_{TiCl} + n_{<Ti>} + n_{<TiC>} = 10^{-4}$$

The results of the calculation for 1800 K are summarised in Table 43, and the efficiency, η, for Ti and TiC are plotted in Fig. 160 as a function of $P^*_{CH_4}$. For a small initial pressure of CH_4, Ti and TiC condense ($P^*_{CH_4} < 2.5 \times 10^{-5}$); then only TiC ($2.5 \times 10^{-5} < P^*_{CH_4} < 2 \times 10^{-3}$); and finally TiC and C($P^*_{CH_4} > 2 \times 10^{-3}$). For higher pressures of $TiCl_4$, a thermodynamic study indicates that the range of <TiC> increases to the detriment of those of <TiC> + <Ti> and <TiC> + <C>.

TABLE 42

Species	ϕ_i	$(G_T^o - H_{298}^o)/T$ cal mol^{-1} deg^{-1} at 1800 K	ΔH_{298}^o cal mol^{-1}
[$TiCl_4$]	1	- 108.540	- 182,400
[CH_4]	1	- 56.794	- 17,895
[H_2]	1	- 38.022	- 0
[$TiCl_3$]	1	- 94.451	- 128,900
[$TiCl_2$]	1	- 80.448	- 67,500
[TiCl]	1	- 68.995	+ 36,900
[Cl_2]	1	- 61.633	0
[HCl]	1	- 51.581	- 22,063
<C>	0	- 5.055	0
<Ti>	0	- 14.200	0
<TiC>	0	- 16.253	- 44,000

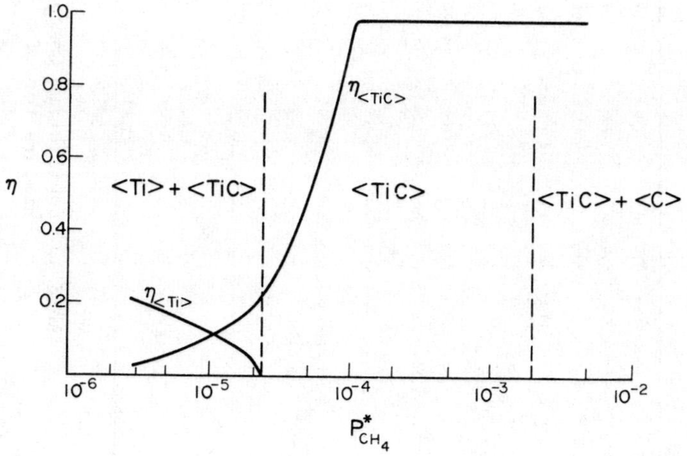

Fig. 160. Efficiency in Ti (<TiC> + <Ti>) as a function of the initial P_{CH_4}.

(Note: The efficiency in TiC is about 100% when $P_{CH_4} \geq P_{TiCl_4}$)

	<TiC> + <Ti>			<TiC>					<TiC> + <C>		
$TiCl_4^*$	1.0×10^{-4}	1.0×10^{-4}	1.0×10^{-4}	1.0×10^{-4}	1.0×10^{-4}	1.0×10^{-4}	1.0×10^{-4}	1.0×10^{-4}	1.0×10^{-4}	1.0×10^{-4}	1.0×10^{-4}
CH_4^*	3.0×10^{-6}	1.5×10^{-5}	2.0×10^{-5}	2.5×10^{-5}	3.0×10^{-5}	1.0×10^{-4}	2.0×10^{-4}	1.0×10^{-3}	2.0×10^{-3}	1.0×10^{-1}	5.0×10^{-1}
$[H_2]$	0.999 674 87	0.999 674 87	0.999 674 87	0.999 673 91	0.999 669 06	0.999 600 08	0.999 503 44	0.999 884 435	0.998 675 01	0.998 710 53	0.998 807 39
$[HCl]$	$2.491\ 1 \times 10^{-4}$	$2.491\ 3 \times 10^{-4}$	$2.491\ 2 \times 10^{-4}$	2.510×10^{-4}	2.609×10^{-4}	3.965×10^{-4}	$3.997\ 2 \times 10^{-4}$	$3.997\ 8 \times 10^{-4}$	$3.994\ 3 \times 10^{-4}$	3.638×10^{-4}	$2.668\ 3 \times 10^{-4}$
$[TiCl_4]$					1.0×10^{-8}	1.0×10^{-8}					
$[TiCl_3]$	9.0×10^{-8}	1.0×10^{-7}	9.0×10^{-8}	9.0×10^{-5}	8.0×10^{-8}		3.0×10^{-8}				
$TiCl_2$	7.464×10^{-5}	7.463×10^{-5}	7.463×10^{-5}	7.368×10^{-5}	6.881×10^{-5}		1.0×10^{-8}				
$[TiCl]$	1.25×10^{-8}	1.24×10^{-6}	1.24×10^{-6}	1.22×10^{-6}	1.08×10^{-6}	2.0×10^{-8}	3.0×10^{-8}				
$[CH_4]$	1.0×10^{-8}	1.0×10^{-8}	1.0×10^{-8}	1.0×10^{-8}	1.0×10^{-8}	1.70×10^{-6}	9.341×10^{-5}	$6.121\ 3 \times 10^{-4}$	7.242×10^{-4}	$7.243\ 1 \times 10^{-4}$	$7.244\ 5 \times 10^{-4}$
$[CH_2H_2]$							3.33×10^{-6}	1.437×10^{-4}	2.012×10^{-4}	$2.012\ 8 \times 10^{-4}$	$2.013\ 0 \times 10^{-4}$
<Ti>	2.100×10^{-5}	9.031×10^{-6}	4.029×10^{-6}								
<TiC>	2.986×10^{-6}	1.498×10^{-5}	1.998×10^{-5}	2.498×10^{-5}	2.998×10^{-5}	9.829×10^{-5}	9.990×10^{-5}	9.998×10^{-5}	9.998×10^{-5}	$9.999\ 5 \times 10^{-5}$	9.999×10^{-5}
<C>									$7.716\ 4 \times 10^{-4}$	$9.866\ 121 \times 10^{-2}$	0.498 210 51
$n_{Ti} = \frac{<T_i>}{[TiCl_4]^*}$	0.21	0.09	0.04	0							
$n_{TiC} \frac{<TiC>}{[TiCl_4]^*}$	0.03	0.15	0.2	25	30	98	100	100	100	100	100

TABLE 43 Equilibrium Composition as a Function of the Initial Mixture for T = 1800 K and P = 1 atm (accuracy 10^{-8}) (16)

REFERENCES

1. E. Desre, *Elements de Thermodynamique Metallurgique*, No. 3. ENSEEG-INP, Grenoble (1962).

2. L.S. Darken and R. Gurry, *Physical Chemistry of Metals*, McGraw-Hill Co., N.Y. (1953).

3. J. Mackowiak, *Physical Chemistry for Metallurgists*, Allen and Unwin, London (1965).

4. Ibid.

5. E. Desre, op. cit.

6. A. Rist and G. Bonnivard, *Rev. Metallurgie*, 60, No. 1 (Jan. 1963).

7. L. Coudurier, I. Wilkomirsky and G. Morizot, *Trans. Inst. Mining and Metallurgy*, Sec. C, 79 (1970).

8. S. Mudd Series, *Basic Open Hearth Steelmaking*, A.I.M.E. Soc., N.Y. (1964).

9. F. Vachet, Thesis for D. Eng., ENSEEG-INP, Grenoble (1965).

10. E.M. Levin, C.R. Robbins and H.F. Murdie, *Phase Diagrams for Ceramists*, The American Ceramic Society, Columbus, Ohio (1964).

11. E.T. Turkdogan and J. Pearson, *J.I.S.I.* 175, 398-401 (1953).

12. C. Fetters and J. Chipman, *Trans. A.I.M.E.* 145, 95-112 (1941).

13. K. Balajaiva et al, *J.I.S.I.* 153, 115-145 (1946).

14. N. Meysson and A. Rist, *Revue de Metallurgie*, No. 2 (Feb. 1965).

15. J.P. Riegert, Thesis for D. Eng., ENSEEG-INP, Grenoble (1973).

16. C. Bernard, Y. Deniel, A. Jacquot, P. Vay, M. Ducarroir and M. Jaymes, *J. of Less Common Metals*, 40, 165-171 and 173-183 (1975).

17. R.A. Sharma and F.D. Richardson, *Trans. A.I.M.E.* 233, 1586-92 (1965).

18. G.R. St. Pierre, *Met. Trans., B*, 8B, 215 (1977).

19. C.K. Kim and A. McLean, *Metal-Slag-Gas Reactions and Processes*, p. 284, Ed. Foroulis and Smeltzer, The Electrochemical Society Inc., New Jersey (1975).

20. O. Kubaschewski and C.B. Alcock, *Metallurgical Thermochemistry*, 5th Ed., Pergamon Press, Oxford (1979).

SYMBOLS

A	Constituents of a solution; affinity of a reaction.
a	Activity.
C, C^*, C_o	Concentration, concentration at equilibrium, initial concentration.
C_p, C_v	Specific heat at constant pressure; constant volume.
D	Diffusion coefficient.
e	Thickness.
e_i^j	Interaction parameter of j upon i.
f	Activity coefficient (Henry's Law); relative penetration.
F	Helmholtz free energy; fraction of transformed solid.
$\mathcal{F}$	Faraday constant.
G, G^o	Gibbs free energy; standard Gibbs free energy.
$\bar{G}_A, \bar{G}_A^M$	Partial molar free energy of A; partial molar free energy of mixing of A.
G^M	Integral free energy of mixing.
$\bar{G}_A^{ex.}, G^{ex.}$	Excess partial free energy of A; excess integral free energy.
$\bar{G}^{id.}, G^{id.}$	Ideal partial free energy of A; ideal integral free energy.
H	Height; enthalpy (the same notation as for the free energy).

i	Constituent of a solution; index of basicity.
j	Constituent of a solution.
k, k'	Specific constant of velocity.
K, $K_{eq.}$	Equilibrium constant.
M	Molar weight; constituent of a solution (metal).
m	Mass.
n	Number of moles.
N	Molar fraction.
p, p^o, P, p^*	Partial pressure; vapour pressure of a body at saturation total pressure; equilibrium pressure.
q	Number of moles per unit volume.
Q	Heat quantity
r	Radius; velocity of a reaction.
R	Radius; gas constant.
S	Surface; entropy (the same notation as for G).
t	Time.
T	Temperature.
U	Internal energy.
V	Volume.
X	Distance; thermodynamic quantity.
z	Height; valency
α, β	Phases.
γ	Activity coefficient (Raoult's Law).
δ	Thickness.
ε	Porosity.
ε_i^j	Interaction parameter of j upon i.
ϕ	Flow or flux.
η	Viscosity; efficiency; coefficient of utilisation.

μ, μ^o	Chemical potential; standard chemical potential.
ν	Stoichiometric coefficient of a reaction.
ρ	Number of atoms per particle.
τ	Constant with dimension of time

[A]	Gas.
(A)	Liquid.
<A>	Solid.
$\underline{A}$	In solution.
$(\underline{A})_m$	In solution in metal phase (liquid).
$(\underline{A})_s$	In solution in slag (liquid).

INDEX